Animal Architecture

Oxford Animal Biology Series

Titles

Animal Eyes
M. F. Land, D-E. Nilsson

Animal Locomotion
Andrew A. Biewener

Animal Architecture
Mike Hansell

The Oxford Animal Biology Series publishes attractive supplementary text-books in comparative animal biology for students and professional researchers in the biologiacal sciences, adopting a lively, integrated approach. The series has two distinguishing features: first, book topics address common themes that transcend taxonomy, and are illustrated with examples from throughout the animal kingdom; secondly, chapter contents are chosen to match existing and proposed courses and syllabuses, carefully taking into account the depth of coverage required. Further reading sections, consisting mainly of review articles and books, guide the reader into the more detailed research literature. The Series is international in scope, both in terms of the species used as examples and in the references to scientific work.

Animal Architecture

Mike Hansell
Institute of Biomedical and Life Sciences,
University of Glasgow

OXFORD
UNIVERSITY PRESS

Great Clarendon Street, Oxford OX2 6DP

Oxford University Press is a department of the University of Oxford.
It furthers the University's objective of excellence in research, scholarship, and education by publishing worldwide in

Oxford New York

Auckland Bangkok Buenos Aires Cape Town Chennai
Dar es Salaam Delhi Hong Kong Istanbul Karachi Kolkata
Kuala Lumpur Madrid Melbourne Mexico City Mumbai Nairobi
São Paulo Shanghai Taipei Tokyo Toronto

Oxford is a registered trade mark of Oxford University Press
in the UK and in certain other countries

Published in the United States
by Oxford University Press Inc., New York

First published 2005

A Catalogue record for this title is available from the British Library

British Library Cataloguing in Publication Data

Library of Congress Cataloging in Publication Data

(Data available)

ISBN 0–19–850751–8 (hbk.)

ISBN 0–19–850752–6 (pbk.)

1 3 5 7 9 10 8 6 4 2

Typeset by Newgen Imaging Systems (P) Ltd., Chennai, India
Printed in Great Britain
on acid-free paper by
Biddles, King's Lynn

To my colleagues, current and former.

Acknowledgements

I have always tried to relate the research I have conducted on any species of builder to building biology as a whole. In this book I try to create a synthesis of that biology. That is a rather daunting task, and one that I would have found very difficult to undertake without the encouragement and wisdom of others. I am fortunate enough to work in an environment where there is a strong curiosity about the natural world and a readiness to share ideas.

In preparing this book I have been greatly helped by discussions with colleagues concerning general biological principles and matters of detail. This has had important beneficial effects on the clarity of what I have said and in the way I have said it. I am especially indebted to Graeme Ruxton for giving me the benefit of his breadth of knowledge and understanding, and encouraging me simply by sharing my delight in the biology of animal builders. There is hardly a part of this book that he has not read in draft and which has not benefited from his constructive critical comment.

Several other colleagues have been kind enough to read parts of the book in draft and offer helpful advice on improvement. In this respect, I am very grateful to Stuart Humphries, Malcolm Kennedy, Sarah Brown, and Martin Burns. Friends in other institutions have also been kind enough to read over and comment upon draft text. They are Bob Jeanne, Holly Downing, and Bill McGrew; I would like to thank them for their helpful suggestions and encouragement. For discussions on particular matters and for translations of passages of published research I am indebted to my colleagues Reudi Nager, Neil Metcalfe, Geoff Hancock and Felicty Huntingford.

This book is reliant upon good support of the text by the figures. I could not have had a better person working on the graphics and their presentation than Liz Denton. She has an eye for clear presentation and insists, with firm diplomacy, line drawing, lettering, layout or other detail is presented in the best possible way. This has greatly helped the presentation of my ideas.

The figures have been obtained from a variety of sources. Those from published sources are acknowledged in the relevant figure captions, however, I would like to personally thank some people for their assistance over certain figures. Firstly I am most grateful to Margaret Mullen of the University of Glasgow Electron Microscope Unit for her professionalism in the preparation of the scanning electron micrographs in Fig. 2.4 and 5.6. I am grateful also to the University of Glasgow Media Services (Photographic Unit) for a photograph used in Fig. 5.6, and those in Fig. 5.8. I also wish to thank Bob Jeanne for allowing me to reproduce the photograph of the nest of the wasp *Leipomeles* in Fig. 1.5, Gary Taylor who kindly sent to me the lerp photographs used in Fig. 2.5, and Guy Theraulaz who supplied me with the simulated wasp nests of Fig. 4.10. I am most grateful to Karin Schuett for the spider silk thread attachment micrographs in Fig. 5.7, to Samuel Zschokke for the micrograph of spider web radius in Fig. 5.9, and to Leticia Avilés for the photograph of *Tapinillus* spiders shown in Fig. 8.8.

I also gratefully acknowledge permission to use from the following sources:

Fig. 2.6 from Mattson, S. and Cedhagen, T. (1989). Aspects of the behaviour and ecology of *Dyopedos monacanthus* (Metzger) and *D. porrectus* Bate, with comparative notes on *Dulichia tuberculata* Boeck (Crustacea: Amphipoda: Podoceridae). *Journal of Experimental Marine Biology and Ecology*, **127**, 253–272. Copyright 1989 with permission of Elsevier.

Fig. 2.8 from Flood, P.R. (1991). Architecture of, and water circulation and flow rate in the house of the planktonic tunicate *Oikopleura labradoriensis*. *Marine Biology*, **111**, 95–111. With permission of Springer-Verlag GmbH.

Fig 3.5 from Fitzgerald, T.D. and Clark, K.L. (1994). Analysis of leaf rolling behaviour of *Caloptilia serotinella*. *Journal of Insect Behaviour*, **7**, 859–872 by permission of Kluwer Academic/Plenum Publishers.

Fig. 3.10 from Evans, H.E. and West Eberhard, M.J. (1970). *The wasps*. The University of Michigan Press, Ann Arbor ©. Fig. 5.2 (a) and (b) from Wiggins (1977) by kind permission of the Royal Ontario Museum ©.

Fig. 4.11 from Hoftmann, A.A. (1994). *Behaviour genetics and evolution*. In *Behaviour and Evolution* (eds. P.J.B. Slater and T.R. Halliday) pp. 7–42 by permission of Cambridge University Press.

Fig. 5.5 from Jeanne, R.L. (1975). Adaptiveness of social wasp nest architectue. *Quarterly Review of Biology*, **50**, 267–287. With permission of The University of Chicago Press.

Fig. 7.8 from Büttner, H. (1996). Rubble mounds of the sand tilefish *Malacanthus plumieri* (Bloch, 1787) and associated fishes in Colombia. *Bulletin* of *Marine Sciences*, **58**, 248–260 by permission of *Bulletin of Marine Science*.

Fig. 8.1 and 8.2 from - Stuart, A.E. and Hunter, F.F. (1998). End products of behaviour versus behavioural characters: a phylogenetic investigation of pupal cocoon construction and form in some North American blackflies (Diptera: Simuliidae). *Systematic Entomology*, **23**, 387–398 ©.

Preface

My aim in this book is to investigate and celebrate the biology of animal architecture. I believe that by writing this comprehensive overview, it can be seen that this is a coherent biological topic which gives us important insights. I last did this 20 years ago (Hansell 1984), so it is interesting for me not only to see how much the subject has developed during that time, but also how my views have changed too.

Animal builders are patchily distributed through the animal kingdom, but research effort is also unevenly distributed within that. Spiders in particular have received a lot of research attention, from the level of their building material to the functional design of webs and their foraging ecology. Bird nests still remain rather under-researched, but there is a flurry of exciting research on bowerbird displays. The book reveals a need for more information in a number of areas, for example, on the composition and properties of self-secreted building materials other than silk, and the mechanical properties of nearly all structures other than spider webs. On the other hand we now have a much better understanding of how simple organisms can build large complex structures, and there have been developments in the ecological and evolutionary concepts of niche construction and ecological inheritance to which studies of animal builders have contributed.

This book recognises three broad categories of structure: homes, traps and displays. Chapter 1 looks at the functional role of these: homes to protect builders from the hostile forces of the physical and biological world, foraging and feeding assisted by burrowing or by the use of nets or webs, and structures for intraspecific communication, in particular the displays of bowerbirds.

Chapter 2 tests predictions relating to building materials: that self-secreted materials will tend to be more standardised and more complex than collected materials and that, because of this, they will tend to be more characteristic of dynamic structures like traps than of static ones, exemplified by houses. In fact, collected materials prove to be quite highly standardised, while the synthesis of self-secreted materials does show some flexibility.

The process of building is examined in Chapters 3 and 4. Building anatomy is shown to be generally unspecialised for delicate manipulative skills but modified for power in many burrowing species. Building behaviour is found to be simple and repetitive. These findings support predictions I have previously made (Hansell 1984, 2000). The creation of very large and complex structures is shown to be possible largely through a dialogue between the builder and the developing structure in which building actions in response to local stimuli change the stimulus situation; complex architecture is an emergent property of self-organising processes. These principles apply equally well to building by large workforces of social insects as to single individuals. Animal tools are considered in the light of these findings, because they are generally regarded as important in the context of human evolution, in spite of being small and often of simple construction. Some tool makers are found to show evidence of advanced learning and cognition, but assessment of these attributes in builders generally suffers from lack of evidence.

Mechanics, growth, and design are the subjects of Chapter 5. Animal homes show how building rules can be conserved while the structures grow with the size of the individual or colony occupying them. Spiders' webs provide models for the study of engineering in tension, while display structures, in particular those of bowerbirds, provide possible models for the investigation of the evolution of an aesthetic sense. In Chapter 6, the cost of home building and its trade-offs with other life history traits is examined using examples of birds' nests and caddis cases; on trap building costs and their consequences, spider webs again supply the majority of the evidence.

Buildings change the world both for builders and organisms that associate with them. These are the themes of Chapters 7 and 8. Predictions (Hansell 1987*a*, 1993) that builders, as ecosystem engineers, will tend to stabilise habitats, resist extinction, and promote biodiversity are examined. The last of these is clearly supported, although this is found to be largely through facultative associations by organisms with constructed habitat niches. The limitations of animal built structures as evidence of phylogeny is discussed, and the concept of a key adaptation examined with the conclusion that arthropod silk has the strongest claim to this title. Evidence that building has contributed to social evolution (Hansell 1987*a*) is found to be inconsistent. Finally, builders are seen to alter the course of their own evolution through ecological inheritance, the passing on to their descendents of habitats that they have altered.

Contents

1 Functions

1.1 Introduction and who builds

The functions of animal built structures are essentially only three: to create a protected home, to trap prey, and for intraspecific communication. To this could be added the incidental creation of burrows by obligate subterranean foragers, and the manufacture or at least use of tools by some species. The overwhelming number and diversity of builder species fall into the category of homemakers, although this includes homes built for personal use, for the family, or for the offspring alone (Table 1.1).

The most common functions of these homes are protection against extremes of temperature and the threat of predation. However, their protective walls may generate secondary problems, gas exchange, for example, which must be solved by further adaptations, some architectural, so adding to the complexity of the structure. In addition, these living spaces may, through evolution, incorporate new functions, such as food storage or fungus cultivation. Homes, therefore, vary greatly from a simple protective wall to a complex differentiated residence. Species that build prey capture devices, are very limited in their taxonomic distribution but ecologically widespread and important. Communication structures are simply rare, tool construction rarer still.

The common attribute of all these structures is that they extend the control of the builder over some aspect of the environment. This chapter considers first the ability of homes to regulate the interior physical and also the biological environment, before looking at the nature and mode of operation of prey capture and finally communication devices.

Table 1.1. The functions of animal-built structures

	Homes	*Traps*	*Foraging burrows*	*Communication structures*	*Tools (use of)*
Human	****	****		****	****
Other Mammals	**		*		*
Birds	***			**	*
Reptiles & amphibians	*			*	
Fish	*				
Urochordates (Appendicularia)	*	*			
Molluscs	*	*			
Insects	***	**	*	*	*
Spiders & allies (Arachnids)	**	***		*	
Crustaceans	*		*	*	
Other Arthropods	*				
Annelids	*	*	*		
Other invertebrate phyla (including sponges, platyhelminths & echinoderms)	Little of note in any category				

Note: The table shows the broad taxonomic distribution of their occurrence, and the extent of their prevalence and complexity indicated by a 'star rating', where **** = complex, and almost universal in that group, and * = infrequent and/or modest.

1.2 Places to live: control of the physical environment

1.2.1 Temperature

Regardless of whether an organism is endothermic or exothermic, temperature extremes are stressful. Both endotherms and ectotherms have temperature optima. For the former, environmental temperature matching, which is maintained by the animal itself, minimises the metabolic costs of thermoregulation; for an ectotherm it means above all maximising developmental rate. The two most common sources of heat for organisms are the sun and their own metabolism. Architecture may provide ways of trapping or conserving these to help maintain environmental optima, more exceptionally it may provide a method of dissipating them if temperatures become high.

Heat moves from one location to another through convection, conduction, and radiation. Placing a wall around the body will limit the movement of heat by all three of these mechanisms. Exposed to the sun, the wall shades the resident, but can also act as a heat sink to spread the delivery of the heat energy over time. Choice of site, architecture, and its orientation will influence exposure to the sun's radiation and to convective cooling by wind. The material of the wall will determine rates of conductive heat transfer, whether that of the sun from the outside, or of metabolic heat generated within by the builder.

Architecture oriented with respect to solar radiation is seen in webs of the forest-dwelling, orb web spider *Micrathena gracilis*. These show a predominantly north–south orientation in shaded sites but an east–west orientation in open sites. Measurement of body temperature of the spiders shows that it is significantly influenced by the orientation of the web (Biere and Uetz 1981). Grigg (1973), by physically rotating the mound to alter the north-south mound axis of the termite *Amitermes meridionalis* to east–west, demonstrated that the result was a less uniform mound temperature through the day and a higher midday temperature. Surveys of the orientation of *Amitermes* mounds throughout their range, however, reveal slight but significant regional variations with, for example, the orientation of most coastal A. *meridionalis* significantly to the west of more inland mounds. This is interpreted by Jacklyn (1991, 1992) as an adjustment to compensate for the convection cooling effects of coastal winds by exposing the mounds to increased solar radiation.

Insect cocoons are demonstrably varied in function, providing both examples of anti-predator adaptations (Section 1.3.1) as well as protection against the physical environment. Mature, overwintering larvae of *Leguminivora glycinivorella* (Tortricidae) spin silk cocoons 1–3 cm below the soil surface. The cocoons have no effect in insulating larvae against freezing temperatures. However, whereas larvae unprotected by the cocoon will freeze at –3 to –4°C, inside the protection of the cocoon, larvae will supercool to about –22°C before freezing (Sakagami *et al.* 1985). The pupal cocoon, of the moth *Gynaephora groenlandica* is a double-layered structure apparently designed to accelerate pupal development in the short summers of the arctic. The outer layer is of pale translucent silk, the inner much darker. Its effect is to shorten the pupal period by at least 10 days. As the next generation must reach the second larval instar in order to overwinter, this could be an important gain (Lyon and Cartar 1996). The temperature inside the cells of the open combs of *Polistes* wasps remains a little above ambient. Experiments with paper models show that the addition of a dark oral secretion to the comb seen in some species, further assists in raising brood cell temperature (Hozumi and Yamane 2001*a,b*).

Ishay and Barenholz-Paniry (1995) and Ishay *et al.* (2002) attribute a much more complex role to the silk caps covering the pupal cells of the Oriental hornet *Vespa orientalis*. They claim that it acts first as an insulator between the pupa and the air outside the cell, and second that it acts as a thermostatic regulator by virtue of its thermoelectric properties. As the nest heats up during the day, some of the heat energy is converted and stored as electric charge, while excess heat is removed by water evaporating from the pupal cell wall. As the ambient temperature cools, the silk absorbs water and by virtue of its microcapillary properties, conveys it around the cocoon; meanwhile, the electric charge in the cocoon is released as current, which in the process generates heat. This effect, they believe, maintains the pupal cell temperature close to the optimum of 28–31°C.

Caterpillars of the small eggar (*Eriogaster lanestris*) build communal tents in early spring. These are effective in maintaining an internal temperature above ambient even without the effects of solar radiation (Ruf and Fiedler 2000). This is because inside the tent the caterpillars can more effectively conserve metabolic heat, as multiple layers of silk reduce heat loss by convection and reflect back radiated heat. These authors interpret the function of the tent as a method of shortening larval development time by raising body metabolism. Tents of the eastern tent caterpillar (*Malacosoma americanum*) are sited on the south-eastern aspect of trees. In this position they catch the morning sun and, when exposed to solar radiation, maintain an internal temperature at least 4°C above the external. This could benefit the larvae by enabling them to exploit the high quality leaf diet available in early spring (Fitzgerald and Willer 1983).

1.2.2 Oxygen and respiration

Protection against temperature extremes and against predation risk, the two functional premiums in house building, both necessitate the creation of a physical barrier between the builder and the outside world. Critical among secondary outcomes of this is limitation of gas exchange, leading to reduced levels of oxygen (*hypoxia*), and enhanced levels of carbon dioxide (*hypercapnia*). Roper *et al.* (2001), reviewing studies covering 15 species of burrowing mammals, confirm the general occurrence of depressed O_2 and elevated CO_2 levels in burrows, with extreme reported values of 14.5% O_2 compared with 20.8% ambient, and 9% CO_2 compared with 0.1% ambient. However, while stressing the need for further studies to explain observed variations, they conclude that respiratory gas levels in such burrows may often be at levels close to ambient.

There are two possible adaptive responses of home dwellers to the problems of hypoxia and hypercapnia, one behavioural and one physiological. The behavioural solution is ventilation. The energy required to achieve this is in some cases provided by the active pumping behaviour of the occupant; in others the energy provided by the natural movement of the fluid (air or water) outside the dwelling is used by the architecture to induce a ventilating current through it. Physiological responses are varied and, although a feature of the organism alone, show how through evolution both organism and architecture have adjusted to one another.

Withers (1978) developed models to predict the effectiveness of diffusion alone in allowing gas exchange sufficient for the survival of burrowing subterranean and subaquatic organisms. These models incorporate key variables such as substrate porosity, burrow dimensions, number of burrow occupants, and respiration rates. They predict, for example, that only very small subaquatic burrowers will be able to rely on passive diffusion alone. Withers (1978) also predicts that, in subterranean burrows, ectotherms will not require mass flow of the medium unless, like termites, they live in very large colonies.

Fossorial endotherms may encounter excessive O_2 or CO_2 gradients, especially under conditions of crowding or low soil porosity due to rain. A prevalence of these conditions could lead to selection for physiological adaptations and/or mass flow systems. Models developed by Wilson and Kilgore (1978) and by Gupta and Deheri (1990) identified diffusion though the soil as a strong influence on the rate of gas exchange in the burrow, with burrow architecture and air temperature as subsidiary influences. The former predicted that normathermic eutherian mammals of greater than 0.5 kg would be largely precluded from a fossorial lifestyle.

Physiological adaptations in mammals to burrow dwelling are broadly correlated with the extent to which they spend their lives underground. The most widespread adaptation is an elevated O_2-affinity of haemoglobin (lowered P_{50}) (Boggs *et al.* 1984), while other blood adaptations include a reduced Bohr effect (displacement of the dissociation curve to the right with lowered pH or raised CO_2) and raised haematocrit (ratio of red cell volume to blood volume).

Burrow architecture alone has allowed a wide variety of organisms to make use of the energy in the fluid medium passing over the burrow exits to power a ventilation system. Burrows of black-tailed prairie dog (*Cynomys ludovicianus*) have exits of two different kinds, lower (104 mm) rounded mounds and steep-walled, taller (186 mm) craters. When the wind flows over them, regardless of its direction, air is drawn in through mound apertures and out through the craters (Vogel *et al.* 1973). This can be explained as a consequence of the Bernoulli principle. According to this, the pressure of a steadily moving fluid must decrease whenever its velocity increases so that its total energy remains the same (Vogel 1978). Whenever horizontally moving air meets a mound it will speed up so, as an air stream passing over an aerofoil, creates reduced pressure over the upper compared with the lower aperture. However, even when a fluid passes over a flat surface, the boundary layer will create a velocity gradient. Fluid will be induced to pass up a tube suspended vertically in this gradient due to the pressure difference resulting from the greater sheer stress being exerted upon the fluid in the tube at the top entrance compared to the bottom. This effect is termed *viscous entrainment*. Concerns about the relative contributions of these two influences can be ignored if they are referred to collectively as *induced flow* mechanisms.

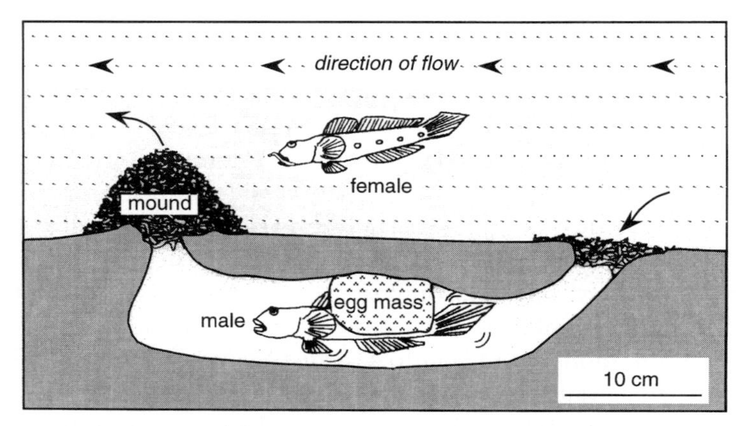

Fig. 1.1. Longitudinal section of the spawning burrow of the goby *V. longipinnis*, indicating the direction of water flow through the burrow, induced by the passage of a current over the mound. (Adapted from Takegaki and Nakazono 2000.)

Equivalent induced flow burrow architecture is also built in the marine environment, for example, by the goby *Valenciennea longipinnis*. While males are fanning the eggs inside the burrow, the female builds a large mound over one entrance (Fig. 1.1), continuing to maintain it throughout the incubation period (Takegaki and Nakazono 2000). In the absence of the male, the water continues to flow through the tunnel, venting at the mound end, its speed positively correlated with the current outside.

Wither's (1978) prediction that only very large colonies of subterranean ectotherms will require a ventilation system is supported by the construction of specialised apertures in the nest mounds of some ants and some advanced termites (*Macrotermes*). Leaf-cutter ants, *Atta vollenweideri*, form underground colonies, among the largest known for any animal. The evidence of these colonies above ground is the presence of multiple entrances scattered over a shallow dome of disturbed soil. Kleineidam *et al.* (2001) found that air emerged from openings at the centre of the dome and entered through those at its edge. Convection caused by heat within the colony is insufficient to explain this air movement, while the positive correlation between wind speed and ventilation airflow supports an induced flow interpretation.

A similar design is found in the nest mounds of the termite *Macrotermes subhyalinus* (Darlington 1984), but mounds of *Macrotermes jeanneli* illustrate an alternative open ventilation design. The mound is topped by a single chimney, which towers several meters above it. All exhaust gasses leave through the chimney, with air apparently entering at the mound base through the soil or many small foraging apertures. Darlington *et al.* (1997) calculated that large mounds may have an outflow of 100,000–400,000 ld^{-1}, which includes 800–1,500 ld^{-1} of CO_2 and 0.5–1.3 ld^{-1} of CH_4. The net water

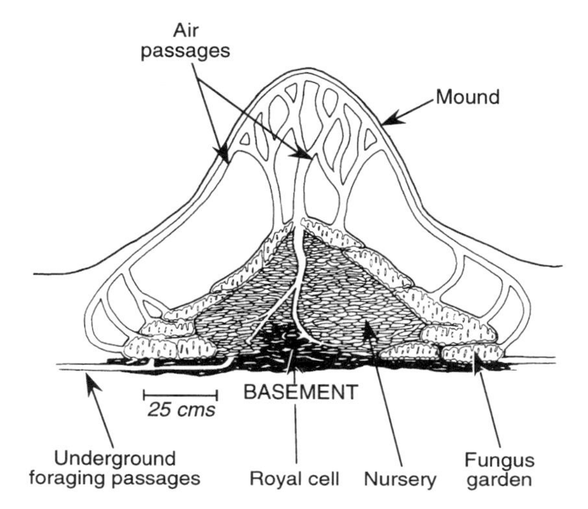

Fig. 1.2. Vertical section through a mature mound of the termite *M. michaelseni*, showing air passages connecting the interior of the nest to spaces lying just below the outer mound wall. (Adapted from Darlington 1985.)

loss was 5 ld^{-1}. The relative merits of the low mound and high chimney designs are unstudied, however, *M. subhyalinus*, a species characterised by variable nest architecture, exhibits both (Darlington 1984). Mounds of the subterranean termite *Odontotermes transvaalensis* also have a single chimney rising to as much as 2 m. Scott Turner (1994) showed experimentally that airflow was generally out of the chimney and resulted from induced flow, with convection only of secondary importance, and forced flow down the chimney possible only in gusty conditions. Here, ventilation was found to have little effect on controlling nest temperature.

Macrotermes species have also evolved a ventilation system powered in a completely different way. It is an enclosed system of channels where air movement results from temperature differences within the mound. The mound of *Macrotermes michaelseni* exemplifies this in a simple form. A system of passages originates within the hive area at the base of the mound, then travels up under the outer wall of the mound linking with one or more rising up the mound core (Fig. 1.2) (Darlington 1985).

Luscher (1955), using the mounds of *Macrotermes natalensis (bellicosus)* as a model, was the first to produce an explanation of how closed termite ventilation systems worked. He proposed that the heat at the core of the mound, generated by termites and their associated fungus gardens, rose up through the colony into an open space in the top of the mound. From there it was forced into narrow channels carrying the air near to the mound surface to allow gas exchange by diffusion, and down into a basement below the hive before it was drawn back into the living space above.

Korb and Linsenmair (2000) tested the Luscher (1955) hypothesis on *M. bellicosus* mounds in two different Ivory Coast habitats, one savannah the

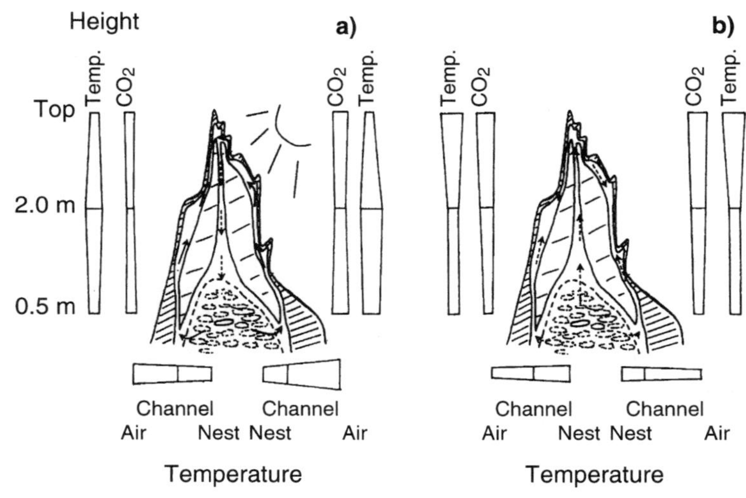

Fig. 1.3. Suggested ventilation mechanisms in *M. bellicosus* mounds located in savannah. (a) Daytime: externally driven ventilation, (b) Night-time: internally driven ventilation. Bars show the relative temperature and CO_2 levels at different heights (0.5 m, 2.0 m, and top), and different locations. (Air = air temperature, Channel = temperature inside the peripheral air channels, Nest = nest temperature.) Arrows inside the mounds show registered ventilation currents (solid lines), supposed currents (dashed lines). (Adapted from Korb and Linsenmair 2000.)

other galley forest. The architecture of the two is different. Korb and Linsenmair (2000) found that during the day air did circulate within the savannah mounds, but in the direction opposite to that predicted by Luscher (1955). This occurred because, under the influence of the sun, external temperatures became hotter than the mound core, so the air rose in the peripheral channels and apparently descended down the mound centre (Fig. 1.3). During the night ambient temperatures fell below that of the core, allowing air to rise up the mound core in a manner similar to that predicted by Luscher (1955), although evidence of actual circulation was weak. In gallery forest mounds, ambient temperature was greater than that of the mound core during the day but less at night. However in both, core temperature was greater than that of the air channels allowing air circulation by convection. CO_2 concentrations at the top of the mound proved to be less during the day than the night, but this may be in part due to the greater diffusion rate of CO_2 at higher temperatures. In fact the nature of gas movement across the mound wall and indeed the nature of the porous wall itself remains to be studied.

Korb and Linsenmair (1998, 1999) suggest that differences in mound design between the two habitats demonstrate that there is a trade-off between thermoregulation and gas exchange. Below optimal ambient temperature, surface area of the mound should decline exponentially to reduce loss of heat to the

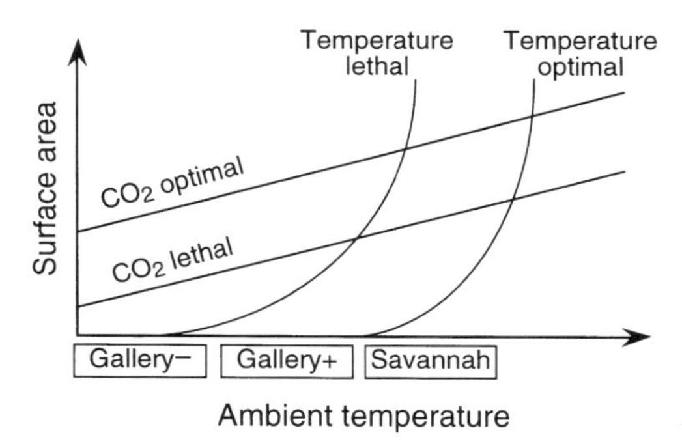

Fig. 1.4. Suggested model for the interaction of thermoregulation and gas exchange in determining surface area (*y* axis) of *M. bellicosus* mounds in relation to ambient temperature (*x* axis). Ambient temperature influences the surface area of mounds via thermoregulation and gas exchange. The two pairs of curves show the mound surface areas that yield optimal and lethal nest temperatures, and optimal and lethal CO_2 concentrations. Below the optimum temperature, heat loss from the mound can be counteracted by a reduction in its surface area, provided this does not lead to dangerous levels of CO_2. Along the *x*-axis the habitats are arranged according to ambient temperature: Gallery– = dense gallery forest regions without mounds; Gallery+ = open stands in the gallery forest with dome-shaped mounds; Savannah = shrub savannah with cathedral-like mounds. (Adapted from Korb and Linsenmair 1999)

environment and to yield a constant internal temperature. But, mound surface area should increase linearly with increasing ambient temperature to guarantee gas exchange to yield constant CO_2 concentration. This model predicts that, in savannah, the mounds do not need to be insulated against heat loss and mound surface is determined mainly by gas exchange. In open gallery forest, termites should build compact nests to reduce loss of internal heat, but too much gas exchange is needed to have optimal area to minimise heat loss. As a consequence, gallery forest mound design is a compromise between thermoregulation and gas exchange (Fig. 1.4).

1.2.3 Humidity control and water management

Habitats vary along a continuum for almost completely dry to entirely watery. Architecture may be used to deal with the control of water availability at any point along this. At the extreme dry end the priority is protection from desiccation, at intermediate points rainfall has various damaging effects, while at the other extreme, water can itself be used as a protective barrier.

Water conservation has been established as one of the functions of insect larval cases and pupal cocoons. Larvae of the moth *Tinea pellionella* require an environment of 95% relative humidity to prevent desiccation in a habitat of between 55% and 84%. This they achieve by permanently carrying around a larval case (Chauvin *et al.* 1979). Pupae of the hymenopteran parasitoid

Cotesia glomerata (Braconidae) normally loose more than half their body weight during the pupal period in spite of an investment of 35% of their body weight to the creation of a silken cocoon. Removal of the cocoon results in only 32% survival at 50% relative humidity compared with 95% with the cocoon present (Tagawa 1996). The silk cocoon of the leek moth (*Acrolepiopsis assectella*) is capable of absorbing up to two-thirds of its weight in water in a saturated atmosphere reducing the rate of water loss of pupae during development and affecting subsequent survival and fertility (Nowbahari and Thibout 1990).

When water vapour turns to rain, animal architecture must address a new range of problems. Protection against structural damage to mud and paper architecture is shown in the presence of features encouraging water run-off. The termites *Cubitermes*, for example, build mushroom-shaped, mud mounds with small downward projections from the edge of the roof, and nests of *Procubitermes niapuensis* are apparently protected by chevron ridges across tree trunks above them (Emerson 1938). No experimental evidence supports these interpretations, although both occur in areas of high tropical rainfall. The nests of stenogastrine wasps, which are made of a fragile paper (Hansell 1987*a*), are usually protected from rain by choice of a nest site, however, the stick-like nests of *Metischnogater* species have one or more cone-shaped 'hats' located on the nest suspension above them, which may function to deflect water from the fragile nest (Turillazzi 1991).

Some species live at the interface between aquatic and terrestrial environments, using their architecture to allow them to breathe air and take advantage of the protection provided by water. Beaver lodges sited in the middle of an artificial lagoon are the most obvious example of this. The mudskipper *Periophthalmodon schlosseri* lives in the upper intertidal zone in burrows which remain permanently full of water. The water in the burrows is severely hypoxic, however, at the base of the burrows are one or more side tunnels that lead to chambers partly filled with air carried down in gulps by the fish themselves (Ishimatsu *et al.* 1998). The roofs of these chambers are used for spawning and it appears that the air may provide oxygen for embryonic development. The fiddler crab *Uca uruguayensis* maintains an air-filled burrow at all times plugging the entrance before the tide comes in (Iglesia *et al.* 1994).

The subsocial, intertidal staphylinid beetle *Bledius spectabilis* spends all stages of its life inside burrows in salt marsh mud that are submerged at high tide. The burrow chamber of an adult beetle is about 60 mm deep and about 5–6 mm wide, but there is an entrance tube about 14 mm long, which has a diameter of only about 2 mm. This is just narrow enough that a meniscus is formed as the incoming tide covers it, giving the burrow resident enough time to plug the entrance with mud before increasing water pressure causes burrow flooding. Oxygen consumption estimates indicate that a sealed air-filled burrow can provide oxygen to supply the beetles for about 4 h whereas if it becomes flooded with seawater, the oxygen only lasts about 8 min (Wyatt 1986). The water spider *Argyroneta aquatica* constructs an underwater silk tent, which it fills with air to form a protective refuge. Here the surface area of the bubble acts as a gill; as the spider consumes oxygen from the air, its partial pressure within the bubble falls below that in the surrounding water, ensuring its constant replacement (Scott Turner 2000).

1.3 Places to live: control of the biological environment

1.3.1 Defence against predators and parasites

To a predator, homes contain potential prey in the form of the builders, frequently their offspring and sometimes also stored or cultivated food (Hansell 1993). Protection of homes by means of their architecture essentially takes two forms, avoidance of detection and prevention of invasion after detection has occurred. These may be used in combination so that, once detected, the structure may still offer resistance. However, a general prediction is that the larger the structure, the less likely that concealment will be effective. More specifically the prediction is that smaller structures will be selected for smallness, while larger structures may benefit from the addition of further protective material, and so be relatively large. This should be most evident in protection against visually hunting predators.

Principles of camouflage have been developed through the study of animal colouration (Cott 1940); these may be summarised as *concealment, crypsis*, and *masquerade*. Concealment hides an object from view, with crypsis the object resembles its background, while in masquerade it resembles a discrete non-edible object (Endler 1981). Animal architecture provides examples of all these, although not always with evidence of their effectiveness.

The lichen covering the exterior of the nests of some small bird species have been interpreted as causing them to resemble part of a branch (Collias and Collias 1984), however, Hansell (1996*a*) testing the predictions of a 'branch matching' hypothesis against a 'cryptic' hypothesis of matching the background beyond the nest site, found more support for the latter. Nests like those of paradise flycatcher (*Terpsiphone*) and the pied monarch (*Arses (telescopthalmus) kaupi*) have just a few pale lichen flakes scattered over the exterior of the nest. I have hypothesised that this creates the impression of light passing through an insubstantial object unworthy of a predator's attention Hansell (1996*a*, 2000), a concealment device termed *disruptive camouflage* (Cott 1940).

The nests of a variety of ground nesting birds are composed of materials that match their background and, where the incubating bird remains visible, its colouration is cryptic as well (Martin 1993). The eggs of cryptic ground nests tend also to be cryptic, while conspicuous ground nests tend to have non-cryptic eggs (Götmark 1993). Solís and Lope (1995) observed that the eggs and nest material of the stone curlew *Burhinus oedicnemus* matched the ground surrounding the nests and that removal of nest materials induced replacement with the same type of materials. Poorly ground-matching eggs were more heavily predated than well-matched ones. Apparent use of masquerade is seen in the design of nests like that of the red-faced spinetail *Cranioleuca erythrops*. These have hanging streamers of loose vegetation material above and below the nest chamber causing them to resemble structureless debris (Hansell 2000). All these bird nest examples are from among the smallest species.

Avoidance of detection in the design of wasp nests is also characteristic of the smallest species. Nests of the polistine wasp *Parachartegus colobopterus* attached to the trunk of a tree exhibit crypsis since the carton envelope covering the brood combs closely matches the colour of the trunk (Schremmer 1978). Other wasp species exhibit masquerade in their nest building. Colonies of Stenogastrinae (Vespidae) rarely exceed 10 individuals so the nests are small and capable of disguise. The designs vary greatly between species (Turillazzi 1991), however, some are elongate and stick-like, attached to hanging strands of vegetation. An extreme example of this is that of *Metischnogaster drewseni* where each elongate cylindrical cell is attached directly to the cell below it with only a very small side entrance to each. As specialised in terms of nest construction is the leaf-mimicking envelope of the polistine wasp *Leipomeles*. The brood combs, which are attached to the underside of a leaf are concealed by a smooth paper envelope into which is built a pattern of darker lines resembling the veins of the leaf itself (Fig. 1.5) (Wenzel 1998). The

Fig. 1.5. *Leipomeles* wasp species that uses building materials of different darkness to create an impression of leaf veins in the nest envelope. (Photo R. L. Jeanne.)

lepidopteran larva *Aidos amanda* apparently mimics being parasitised, by incorporating into the silken wall of the pupal cocoon, one or two holes that resemble the exit holes of parasitic hymenoptera (Epstein 1995).

Dilution of risk may be the benefit enjoyed by bird species that construct empty nests close to one that that contains brood. Brown cachalotes (*Pseudoseisura lophotes*) build a succession of nests through the year; one consequence of this is that during the breeding season one territory will contain a number of abandoned nests (Nores and Nores 1994). Breeding Pink-legged Graveteiros (*Acrobatornis fonsecai*) build several nests together in the same tree, some of which are apparently dummies (Whitney *et al.* 1996), although the effectiveness of these against nest predation is yet to be shown experimentally.

The social Vespidae illustrate how differences in ability of colony members to actively protect the nest are linked to the method of protection provided by the nest architecture. Two distinct tactics have evolved, both related to heavy predation of colonies by ants in the tropics (Jeanne 1991*a*). Nests of small colonies founded by a single independent queen make use of an 'ant guard' protecting access to an open comb; larger colonies, founded by a swarm, enclose the comb with an envelope. Single foundress colonies are ancestral to the swarm founded type (Carpenter 1991).

The diminutive nests of stenogastrine wasps are mostly open combs but many are suspended from fine strands of vegetation, restricting access by ants. In addition, some species add a sticky secretion from Dufour's gland around the suspension (Hansell 1981; Pardi 1981 and Turillazzi). These do appear to deter ants although apparently through their stickiness rather than any repellent effect (Turillazzi 1991). About 500 species in five genera of polistine wasps, all with small colony sizes, build nests with open combs suspended below one or more fine petioles constructed out of nest material (Gadagkar 1991). In all these five genera are species that coat the nest petiole with an ant repellent secretion produced from a gland on the sixth abdominal sternite: *Mischocyttarus* (Jeanne 1970), *Parapolybia* (Kojima 1992), *Ropalidia* (Kojima 1983), *Polistes*, and *Belonogaster* (Gadagkar 1991).

The loss of the chemical defence in the Polistinae seems to be linked more to swarm founding than to the appearance of the envelope. The New World genera *Apoica* and *Agelaia* are swarm founders, yet neither builds a nest envelope while both lack the sternal gland for producing ant guard repellent (London and Jeanne 2000). Interestingly, *Nectarinella* and *Leipomeles* are both swarm founders yet surround the nest envelope with sticky hairs (Schremmer 1977; Wenzel 1991). Jeanne (1991*b*) suggests that this specialised defence results from them being very small wasps, where active nest defence against ants is unlikely to be effective.

The pervasive nature of ant predation in the tropics has apparently led to the evolution of other purely mechanical constructed defences by a variety of insects. The larvae of a number of tortoise beetles (Chrysomelidae) construct protective portable cases or shields (Hansell 1984). The larva of *Hemisphaerota cyanea* emits faecal pellets in strands, which it glues to abdominal spines. This creates a thatch, which is retained and enlarged at each moult to create a defensive shield against insect predators (Eisner and Eisner 2000). Nymphs of the West African assassin bugs *Paredocla* and *Acanthaspis* disguise themselves by placing a cover of dust and soil particles, plus larger items including the corpses of the ants on which they prey to create a 'backpack'. In this case, experiments show that, although the dust may help disguise against detection by ants, the main advantage of the materials is protection against predators such as spiders and geckos (Brandt and Mahsberg 2002).

Another important cause of mortality among terrestrial arthropods is attack from insect parasitoids. This is reflected in the defensive architecture of spider egg cocoons, some examples of which are complex and incorporate more than one kind of silk. The suspension of egg cocoons of *Mecynogea lemniscata* and *Argiope aurantia* on silken threads appears to protect them from generalist predators such as ants, but in both spider species, the eggs are also protected by a layer of flocculent silk, which is then enclosed in a thin but dense outer layer. This silk cocoon appears to provide a specialist defence against parasitoids. Damage to the outer silk layer of *M. lemniscata* cocoons greatly enhances parasitism by a species of mantispid, while the flocculent silk layer protects against ichneumonid parasitism, being thicker than the ovipositor of this insect is long (Hieber 1992).

Some design features of the homes of birds and mammals have also been interpreted as mechanical protection, although they await empirical confirmation. Stone ramparts built by female blackstarts (*Cercomela melanura*) to constrict the entrance to the nest cavity are dislodged in over one-third of nest predation attempts. Leader and Yom-Tov (1998) hypothesise that the effort involved in the removal of this obstruction gives the nesting female time to escape. Woodrats *Neotoma lepida* construct dens composed of heaps of desert vegetation. The dens vary in composition, however, where spiny segments of *Opuntia* cactus are available, they tend to be laid on the top of the pile. Woodrats in dens composed largely of *Opuntia* survive longer than those in dens composed of yucca leaves (Smith 1995). Temperature fluctuations inside the surface lodges of the Karoo bush rat (*Otomys unisculatus*) are markedly greater than within the subterranean burrow systems of the largely sympatric, Brants' whistling rat (*Parotomys brantsii*). However, differences in the habitat ranges of the two species support the view that protection from predation offered by the surface lodges allows the bush rat to colonise areas where the hard ground is unsuited to burrowing (Jackson *et al.* 2002).

Infestations of nests by parasites can seldom be eliminated, so animals generally deal with them in one of two ways, nest abandonment or parasite control measures. Female cliff swallows (*Hirundo pyrrhonota*) may respond to high infestations of swallow bug (*Oeciacus vicarious*) in their own nests by dumping eggs in nests of close neighbours with lower infestations than their own (Brown and Brown 1991); high levels of ectoparasites in the colony may cause mass abandonment of clutches with some birds re-nesting elsewhere (Brown and Brown 1986). The incorporation of green plant materials in their nests by a number of bird species has been variously interpreted, one hypothesis being that it serves to control populations of ectoparasites through fumigation by plant volatiles. This *nest protection* hypothesis makes a number of testable predictions. These are discussed in *Section 6.2.1.2*.

The sets of European badgers (*Meles meles*) contain many bedchambers and the residents change their sleeping place every 2 or 3 days. Badgers treated to remove fleas, ticks, and lice change sleeping places less frequently, leading to the conclusion that the function of the behaviour is to control ectoparasite populations (Butler and Roper 1996). A similar change is shown by Brants' whistling rat (*Parotomys brantsii*) after treatment to remove fleas (Roper *et al.* 2002). This is a species in which an individual occupies a burrow system that may extend over about 70 m^2, and contain up to six nest chambers.

1.3.2 Food storage

The storage of food inside the home reduces fluctuations in food availability over time. With a food store, survival is ensured and development of young may even be continued uninterrupted when foraging is impossible or suspended due to its high cost. However, food stores themselves have costs in terms of construction and maintenance, in particular protection against theft and decay, factors that should select for storage of foods of high resource value. Where reproduction and growth are being supported, high protein content should be favoured, sugars, starch, or oils where energy is a priority.

The decomposition problem of meat storage is typically overcome by the immobilisation of live prey. The mole *Talpa europea* decapitates worms before storing them in the wall of its burrows (Funmilayo 1979). Wasps of the families Pompilidae and Sphecidae, and sub-family Eumeninae (Vespidae) store live arthropod prey with the aid of a paralysing injection. The storage of sugars from nectar is shown by so-called honey pot ants, but they avoid the construction of containers by the use of special workers (*repletes*) that can store large amounts of fluid in their distended crops. This specialisation is expressed most fully in some desert-dwelling species, for example, *Melophorous* species in Australian, and *Myrmecocystus* species in North America (Hölldobler and Wilson 1990).

The superfamily Apoidea evolved from sphecid-like ancestors to become fully vegetarian, substituting pollen for arthropod prey. The Apidae, a family which includes the most eusocial bees, bumblebees (Bombini), honeybees (Apini), and stingless bees (Meliponini), store both nectar and pollen in containers made wholly or largely out of wax. Honey, regurgitated nectar with the sugars partially digested, provides an energy source for the colony, while pollen provides the protein for larval growth (Michener 1974). In bumblebees, where colonies are generally in the region of 50–400 adult workers, the storage of pollen and nectar allows the colony to survive interruptions in foraging over several days of inclement weather. Honeybee colonies are much larger; over 20,000 adult workers of *Apis mellifera* and *Apis dorsata* (Roubik 1989). The large amount of honey stored by *A. mellifera* allows a colony of workers to survive winter food shortage or a season of drought, and found colonies by swarming. This has surely helped the species to extend its range over habitats from temperate to tropical (Michener 1974).

Colonies of stingless bees (Meliponini) range in size from a few thousand individuals to a few tens of thousands. Their nests are characterised by distinctive storage pots for pollen and for nectar that are located apart from brood comb (Fig. 1.6) (Michener 1974). These bees generally mix the wax with plant resin, which can become dissolved in the honey possibly enhancing its storage properties. This may in part explain the generally lower sugar content in honey of this subfamily (70% sugar), compared with that of the honeybees (Roubik 1989).

Although relatively low in nutrient content, plant leaves and stems are stored by some species. Beavers (*Castor*) immerse branches of felled trees in lagoons to give access to winter food under the ice (Richard 1961). Hay is

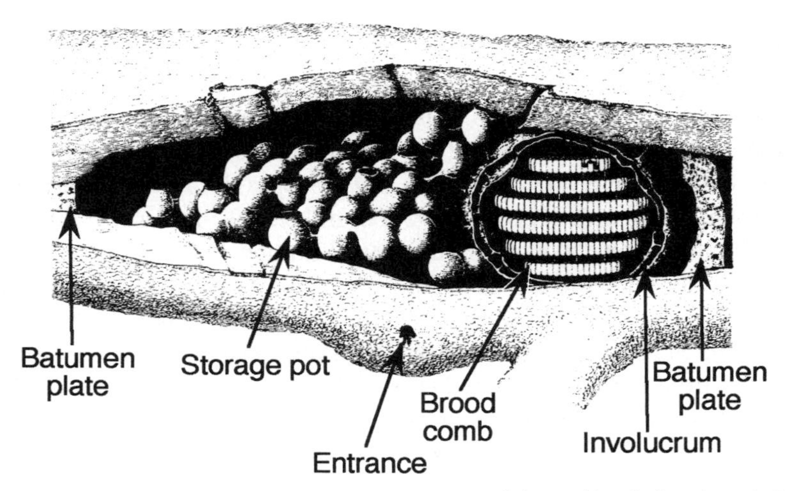

Fig. 1.6. Nest of the stingless bee *Melipona interrupta grandis* located in a hollow branch. The distinctive globular storage pots are of different design and located separately from cells of the brood comb. (Adapted from Michener 1974 after Camargo 1970.)

gathered by certain termites and stored in the nest notably by *Nasutitermes, Drepanotermes, Amitermes,* and *Trinervitermes* (Lee and Wood 1971). *Amitermes meridionalis* and *Amitermes laurensis,* which live in a habitat of seasonal rainfall, gather large quantities of plant fragments to store in their 2–3 m high mounds, apparently to sustain them over the dry season (Jacklyn 1991).

The most widely used storage food is seeds, for which two reasons are particularly evident: First seeds are nutrient rich, encapsulating protein for embryo growth and a concentrated energy source sometimes in the form of oils (which yield 38.9 kJg^{-1} compared with 17.1 for carbohydrates); second, in semi-arid habitats (the very habitats where plant food abundance is highly seasonal), seeds are designed to survive for long dry periods without decomposition. Predictably therefore, seed storage has evolved in a number of species that live in desert environments, notably ants and small mammals. Evidence in support of this *'survival between seasons of plenty'* explanation for granaries is, however, better supported by evidence from rodents than from ants.

Among the ants, seed storing specialists are found chiefly among the family Myrmecinae (Hölldobler and Wilson 1990). Typical of these are species of *Monomorium (Chelander), Pheidole,* and *Messor*. Seed storage is also widespread in small borrowing rodents (Vander Wall 1990). The most highly developed seed storage behaviour is shown by the kangaroo rats (*Dipodomys*) and this does give some support to the hypothesis of survival between seasons of plenty. The size of *Dipodomys* granaries can become substantial with an average granary size of 55 l for *Dipodomys spectabilis* at the autumn peak. During the winter these stores do become depleted (Monson 1943).

Further evidence suggests that the granaries for kangaroo rats may store water in addition to food. Reichman and Rebar (1985) found that *Dipodomys spectabilis* preferred eating slightly mouldy, over dry, or very mouldy seeds. They interpreted this as the rodents taking advantage of some nutritional benefits of fungal metabolism while avoiding high levels of mycotoxins. Reichman *et al.* (1986) showed that *D. spectabilis* exhibited management of the granaries, apparently to achieve a preferred level of mouldiness. This *nutritional* hypothesis of granary management is, however, challenged by Frank (1988*a,b*). He found that *Dipodomys merriami* prefer moist over dry seeds, and that *D. spectabilis* presented with seeds of the same moisture content but differing mouldiness, prefer the non-mouldy seeds over slightly or very mouldy (Fig. 1.7). As kangaroo rats are entirely dependent upon food as their water source, he concludes that slightly mouldy seeds are tolerated for their generally higher moisture content than non-mouldy ones.

Four hypotheses for the function of seed storage by ants were tested by Mackay and Mackay (1984) in a study on three *Pogonomyrmex* species. No evidence was found to support the 'survival between seasons of plenty' hypothesis. During the non-productive season (winter), the ants became inactive and

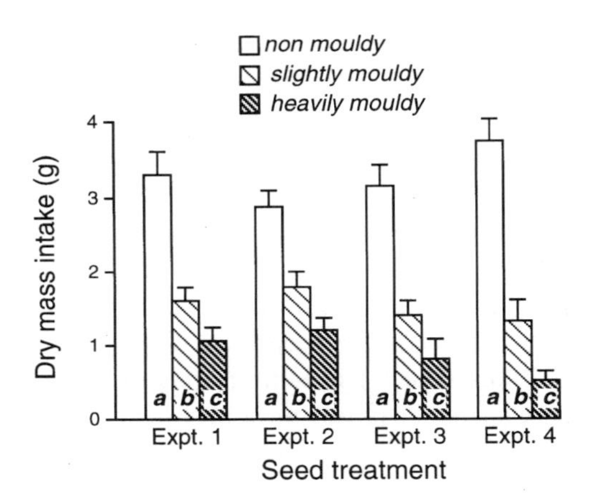

Fig. 1.7. Kangaroo rats *D. spectabilis* presented with seeds of the same moisture content, prefer non-mouldy over slightly mouldy or highly mouldy seeds. This is confirmed in four replications of the experiment, where bars within each experiment that bear a different lower case letter are significantly different. (Adapted from Frank 1988*b*.)

stores were not depleted. No evidence was found of storage to permit sudden production of new larvae, nor was there support for a hypothesis of insurance against an unpredictable climate, although survival of colonies of the desert ant *Messor (Vermessor) pergandei* for 12 years of drought has been attributed to seed storage (Tevis 1958). Mackay and Mackay (1984) did, however, find evidence supporting the fourth hypothesis, risk of predation. When predation on foraging *Pogonomyrmex* became too great, colonies stopped foraging completely, forcing invertebrate predators such as spiders to disperse in search of more active colonies (Mackay 1982).

1.3.3 Cultivation

Species that have evolved this practice are overwhelmingly cultivators of fungi on composted plant material. This form of cultivation appears to have evolved to deal with one of two problems associated with the digestion of plants: the breakdown of cellulose cell walls, and the detoxification of plant secondary compounds. A consequence of overcoming either of these within the digestive system will be an enlarged gut mass and associated metabolic costs. Locating this part of the digestive process outside the body may save some of these costs. Structures built for cultivation are nonetheless rare, possibly because symbioses between cultivator and culture are difficult to evolve (Hölldobler and Wilson 1990).

The breakdown of organic marine sediments, or rather the cultivation of microbes upon them, is a possible function of the walls of some marine crustacean burrows. The decapod crustacea *Upogebia littoralis* (Ott *et al.* 1976) and *Calocaris macandreae* (Nash *et al.* 1984) both bring organic material into the burrow. This practice could provide a substrate for bacterial growth and enhance meiofauna production in the burrow walls. The related *Jaxea nocturna* excavates a simple system of branching Y-shaped burrows, which can reach a depth of 90 cm. Behavioural and anatomical evidence indicate it is a deposit feeder and omnivorous scavenger, but this flexible feeding style could benefit from gardening (Nickell and Atkinson 1995; Nickell *et al.* 1998).

Fungus cultivation occurs in three insect orders, beetles, termites, and ants. Two beetle families are fungus cultivators: Scolytidae and Platypodidae. The majority of the former and all the latter are xylomycetopagous and are referred to as ambrosia beetles after the fungus upon which the larvae feed. They burrow into dead or dying wood to create a gallery where the fungal inoculum will grow (Beaver 1989).

Only a minority of species of termites and ants exhibit fungus cultivation and the two groups of insects show a complete geographical separation, with the fungus cultivating termites found through the tropics of Africa and Asia, while such ants are confined to the New World, again predominantly in the tropics. Six families of termites show no fungus cultivation but depend for the digestion of complex polysaccharides such as cellulose and lignin upon symbiotic intestinal protozoa. These are referred to as the 'lower termites'. The 'higher termites' are represented by a single family (Termitidae), and it is in one of its four subfamilies (Macrotermitinae) that the fungus cultivators occur. The fungus is grown in special subterranean chambers (Fig. 1.2), upon fresh faecal material produced in part from the ingestion of dead plant material, which is then acted upon by termite digestive enzymes and symbiotic gut bacteria, but not protozoa. The faecal material produced allows rapid invasion by the mycelium of the fungus gardens which is then also consumed by the termites. The fruiting bodies of the cultivated fungus appear periodically on the exterior of mounds confirming that one genus alone has evolved this symbiosis, the basidiomycete *Termitomyces*. The influence of fungus-growing termites on ecosystems can be considerable. Studies in West Africa have shown that over 25% of all plant decomposition is accomplished by Macrotermitinae compared with 7.5% for other termites (Wood and Sands 1978).

The key reason for the success of fungus growing in ants may be different from that in termites. They use fungi to process fresh plant material; this assists them in overcoming some of the problems of plant toxins faced by tropical defoliators, so expanding the range of food plants available to them (Hölldobler and Wilson 1990). Fungus-growing ants belong to the tribe Attini of the Myrmecinae. Their colony sizes vary as does their level of fungus cultivation. Species of *Cyphomyrmex* and *Mycocephurus*, for example, live in small colonies and cultivate fungus on insect faeces and corpses, or on fruit. Uniquely in the animal kingdom, however, 15 species of *Atta* and 24 species of *Acromyrmex* grow fungi on plant material collected live. These are the leaf cutter ants. In colonies of *Atta*, adults feed mainly on nectar, and as scavengers or predators, while the fungus is harvested for larval development (Hölldobler and Wilson 1990).

Wilson (1980; 1983 *a,b*) argues that a rare combination of circumstances was needed for leaf cutter ants to evolve. This was: an ant with a large-bodied morph for cutting live plant tissue with smaller worker morphs for plant processing tasks. Large, medium, and small morphs are found in *Atta* species, exhibiting this range of specialised tasks (Bass and Cherrett 1994). As with the Macrotermitinae, the influence of leaf cutter ants over their environments may be considerable. The sizes of their colonies dwarf human comparison as they exceed that of almost any human city and even the populations of some smaller countries. Nests of *Atta volleneweideri* reach 3–6 m below ground (Jonkman 1980). It is estimated that leaf cutter ants (*Atta* and *Acromyrmex*) may harvest between 12–17% of tropical forest leaf production (Hölldobler and Wilson 1990).

1.4 Feeding: burrowers and net builders

There are two quite different construction methods used to obtain food; one is to dig tunnels, the other is to fabricate traps. Exemplars of the former are the fossorial mammals, adapted as they are to a completely subterranean life. They dig extensive burrow systems in search of either invertebrate prey, as in the case of the hairy-tailed mole (*Parascalops breweri*) (Fig. 1.8) (Hickman 1983), or of roots and tubers as do the fossorial rodents. This feeding method is not considered further here, but the specialised anatomy associated with it is examined in Section 3.8, while its economics is discussed in Section 6.3.1, alongside that of trap building species.

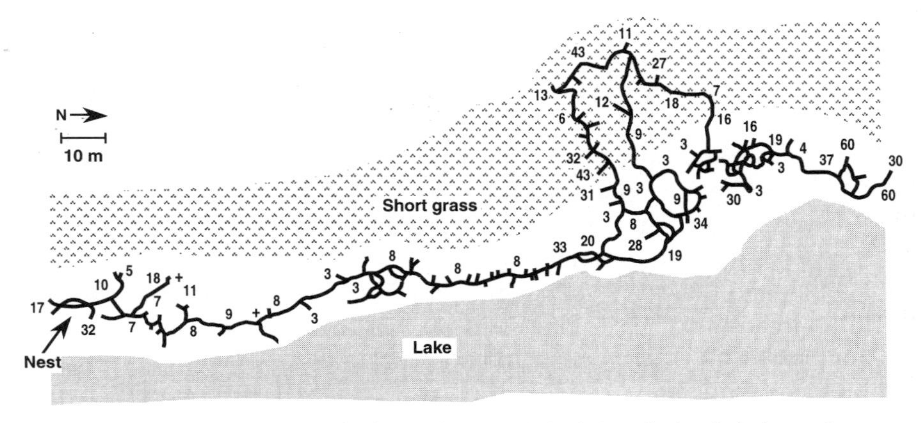

Fig. 1.8. A 550 m burrow system of a fossorial mammal, the hairy-tailed mole (*P. breweri*). Numbers indicate the depth in centimetres, and + the position of trees. (Adapted from Hickman 1983.)

Prey capture architecture is a method of increasing the prey encounter rate. This may be achieved by extending the reach or capture space surrounding the predator, or by ensuring that it to occupies space where the flux of prey is high. The taxonomic distribution of constructed capture devices (Table 1.1) is interesting when compared with that of builders. It is much more limited, and is confined to the invertebrates. Among the vertebrates, trap builders were apparently absent until the recent history of man. The reasons for this needs investigation, but whereas a house can just be a barrier between the builder and the outside world, a trap has a dynamic relationship between itself and the prey; the prey needs to approach the trap in a particular orientation to it, and then needs to be restrained by it. Traps are therefore more complex than homes and need to be more precisely engineered.

Humans use learning, innovation, and manipulative skill to create traps. No other vertebrate apparently has this level of development of brain and behaviour. Table 1.1 shows that invertebrate trap builders have solved the problem, but in a quite different way, having the sophistication in the material of the trap (Section 2.4), not in the construction behaviour. Virtually all non-human trap builders use self-secreted materials, and the capture principle they adopt is the net. The exceptions are simple in design and operation, as well as rare; examples are the conical pit traps of ant lions (Myrmeleontidae), and 'worm lions', larval diptera (Rhagionidae) (Tusculescu *et al.* 1975; Lucas 1982).

For net building suspension feeders, prey flux may be enhanced by placing the net across a moving fluid or, where flow is low or absent, by using the body as a pump. The marine polychaete worm *Chaetopterus*, which lives in a U-shaped burrow in marine mud, constructs a net to filter food from water driven through by means of fan-like notopodia. The mucus bag has a mesh averaging 0.76×0.46 μm^2 composed of coarser mucus fibre bundles of 0.1–0.2 μm diameter crossed by regularly spaced filaments of only 5–15 nm diameter (Flood and Fiala-Médioni 1982).

Gleba cordata, a marine, midwater, planktonic gastropod mollusc, uses slowly flowing water currents, but more particularly gravity to power the flow of particles into its net. It feeds on tiny planktonic organisms that adhere to its mucus web. The rate of decent of the particles must be slow, so the size of the net is correspondingly large. The organism is around 55 mm in length and the web about 350 mm across (Gilmer 1972). These fragile capture devices are only possible in the placid midwaters of the ocean, but similar foraging devices are produced by planktonic gastropods belonging to other families. *Cavolinia uncinata* creates a heart-shaped balloon of mucus to which prey items adhere. The web is then drawn into the mouth and, together with adhering particles, consumed (Gilmer and Harbison 1986).

Net-spinning caddis larvae exploit the power provided by flowing freshwater to maximise food capture. Capture nets are produced by larvae of three caddis families Hydropsychidae, Polycentropodidae, and Philopotamidae (McGavin 2001). The net of *Parapsyche cardis*, has a coarse-mesh net averaging $237 \times 382\ \mu m^2$ for fifth instar larvae, which is located in rapid water currents (Malas and Wallace 1977). Finer mesh nets are located in slower currents. That of *Macronema transvesum*, with a mesh of $25 \times 3\ \mu m^2$, is stretched across the chamber of a sand grain case, which has an inhalant current directed into the slowly flowing water (Wallace and Sherberger 1975).

The simplest spider prey capture architecture serves to extend the prey detection range of the predator. *Liphistus desultor* has radiating silk threads from the tunnel entrance inside which the spider sits concealed by a trap door. *Atypoides vivesi* attaches radiating leaves to the tunnel entrance with silk to extend the detection range of the resident spider (Coyle 1986). In the desert spider *Ariadna* (Segestriidae) prey detection is extended through a circle of six or seven very similar quartz pebbles attached to the burrow lip with silk (Henschel 1995). From such simple communication devices evolved three-dimensional arrays of threads that delayed the escape of prey, for example, the webs of Theridiidae and Linypiidae, and the two-dimensional adhesive webs of Araneoidea. Spider webs, and particularly those of araneoid spiders, have been a focus for research on all aspects of animal building biology, as will be seen throughout this book.

Spiders that spin two-dimensional webs depend largely upon the powered flight of insects to come to them. One consequence of this is the potential for catching very large prey, larger even than the predator itself. This has led to the evolution webs that combine the specialised features of absorbing the energy of impact of large, fast moving prey and of then preventing their escape. The second problem has been made easier by prey wrapping behaviour and a venomous bite inherited from more primitive spiders. Never the less, orb web spinning spiders have evolved two quite different prey capture principles, the ultra-fine silk of cribellate spiders and the viscous droplets of ecribellate spiders (Section 2.4.3).

Another problem associated with active potential prey is their ability to detect and avoid the web. The orb webs of *Mangora pia* have silk threads designed for high impact and have a high fibre density. Webs of *Theridiosoma globosum* by contrast have a low fibre density and are designed to resist only low impact. Analysis of the video of flight paths of fruit flies (*Drosophila melanogaster*) under different lighting conditions shows that, even under high light intensity, the silk threads of *T. globosum* are hardly visible to the flies. *Drosophila*, however, do detect light reflected from the web of *M. pia* at about 7.0 cm even though this is at a much greater distance than they can distinguish single threads (Craig 1986). This shows that selection is likely to have occurred on any aspect of the webs that would make them less visible. One of these is likely to have been the nature of the silk itself.

Craig and Bernard (1990) found that the silk of primitive, mygalomorph spiders exhibited greater reflectivity of UV wavelengths than light of longer wavelengths. By contrast, none of the web silks spun by ecribellate orb web builders (Araneoidea) have a peak of reflectance in the UV, and species that site their webs in brightly lit habitats, have a reduced UV signature to their silk (Craig 2003)

The most visible aspect of ecribellate orb webs proves to be the light scattered by the sticky droplets of the capture spiral. In a study of 22 species, there is seen to be a positive correlation between web visibility and droplet size, and this in turn is associated with foraging ecology. In general the species that produce the brighter webs are nocturnal foragers, or diurnal foragers in shaded locations (Craig 2003).

Mangora pia, Micrathena schreibersi, and *Argiope argentata* are three neotropical orb web spinners; webs of the first are typically found in sites of constant dim illumination, the second in conditions of fluctuating light intensity, the third of constantly well-illuminated sites. When one-half of the web is dusted with corn starch to make it more visible, insect impacts are less on the treated than the untreated side of the webs of all three species, demonstrating the web visibility does affect their capture success. The light reflecting properties of the viscid spiral silk of the three species suggests that spiders locating webs in well-lit sights need to use less visible silk. The capture threads of the webs of *A. argentata* are found to have a strongly directional component to their reflections and so become visible over a rather narrow angle, giving them a low detectability in their typically well-lit websites. The capture threads of the other two species scatter light much more broadly, making their webs conspicuous whenever a shaft of light strikes them in their typically dimly lit sites (Craig 2003).

The orb webs of *Nephila* often have a distinctive yellow colour the function of which has been hypothesised to be either for concealment or advertisement. Osaki (1989) recorded seasonal changes in *Nephila* web colour, with webs appearing more yellow with the onset of autumn, a change attributed to maintaining low web detectability. Craig *et al.* (1996) found that in *Nephila clavipes* web colour was dependent upon web location which, in this species, varies widely; individuals spinning yellow silk webs in well-lit sites, would spin webs of white silk if relocated to dimly lit locations. These colour variations broadly correspond to the spectral properties of the website, and so seem to be an adaptation to reduce web visibility (Craig 2003).

The varied evidence of selection for reduced web visibility highlights the problem of explaining the highly visible patches or patterns found on the orb webs of certain species. These have been variously called *stabilimenta, devices*, and *decorations* (Fig. 1.9). They vary in their composition, although the ones made entirely of silk are the most studied (Herberstein *et al.* 2000).

Fig. 1.9. A species of *Argiope* resting at the hub of its orb web, on a cross-shaped stabilimentum. (Photo M. H.Hansell.)

Several hypotheses have been advanced to explain the function of these prominent features. The term *stabilimentum* arose from the proposal that their role was structural, however, the presence of these silk devices in the webs of some diurnal but no nocturnal species has led to the conclusion that they have some visual function. Three such hypotheses are: prevention of web damage by birds (Eisner and Nowicki 1983), concealment of the spiders from predators, and the attraction of insect prey (Herberstein *et al.* 2000).

The first two of these hypotheses predict that the increased conspicuousness of the web due to the stabilimenta will reduce prey capture due to web avoidance; the prey attraction hypothesis claims that it is the conspicuousness of the web that attracts prey to it. Evidence in support of these predictions is contradictory, although mostly supportive. Webs of *Cyclosa cornica* containing stabilimenta, attract more insect prey than those without (Tso 1998). *Argiope aetherea* webs in poorly lit sites have a greater incidence of stabilimenta compared with well-lit locations (Elgar *et al.* 1996). Craig and Bernard (1990) found that, while capture threads of the webs of *A. argentata* have low UV reflectance, the silk devices in the same webs had high UV reflectance. Looking at the pattern of prey capture of webs in the field they found that decorated webs intercepted more prey than undecorated ones and that webs with only one of the possible four silk bands, received more insect impacts on the half of the web with the decoration than the undecorated half. Very similar results on the UV reflectance of the silk and on prey capture have been demonstrated for the uloborid (Cribellate) spider *Octonoba sybotides* (Watanabe 1999*a*).

Craig and Bernard (1990) propose that the UV reflection of web decorations can exploit two rather different responses of insects. The first is a more general 'open-space' response shown by many insects, where attraction to UV produced by the sky and sun allows escape from more enclosed spaces. The second is a more specific attraction of insect pollinators to the UV advertisements of flowers, and that the pattern of decorations in the webs of *Argiope*, with the spider located at hub, might mimic the UV guides of some flowers that attract pollinators such as bees.

Individual *A. argentata*, rather than regularly spinning the full X-shaped stabilimentum, which should be most attractive to insects, frequently use less than the four silk bands, and in different spatial arrangements. Craig (2003) points out that this is consistent with an attempt to disrupt avoidance learning by insects such as bees that are likely to return regularly to the same foraging patch. This prey attraction hypothesis was, however, not supported by observations of Blackledge and Wenzel (2000). They observed that, while it is true that the stabilimenta of *A. aurantia* webs do reflect more UV than other silk in the web, they also reflect strongly in the blue and green part of the spectrum. This, they demonstrate by an ingenious bee learning experiment, has the effect in the natural habitat of making the stabilimenta relatively cryptic to insects, while their relative conspicuousness to vertebrates lends support to the predator defence hypothesis.

Visually conspicuous webs risk attracting predators as well as prey. Individuals of *A. argentata* that show high rates of web decorating with stabilimenta have lower survivorship in the field (Craig 2003). Seah and Li (2001) also showed that the jumping spider *Portia labiata*, a specialist predator of web spiders, was more attracted to a web of *Argiope versicolor* with a stabilimentum than one without. Its learning skills were such that it could even be trained to go to a web with one stabilimentum configuration rather than another (Fig. 1.10).

Prey attraction is a feature of the extremely reduced web of *Mastophara* species, but in this case the attractant is olfactory. Late instar and adult female spiders swing at passing moths with a single large sticky droplet at the end of one thread (Yeargan and Quate 1997). The prey are males of certain species of nocuid moth attracted by a female pheromone mimic produced by the spider. This 'bolas' capture technique has now been reported from six genera of the Mastophorine located in both the Old and New World (Stowe 1986). Quite independently dipteran larvae of the genus *Arachnocampa* (Mycetophilidae) use luminescent organs on their abdomens to attract insect prey to their webs of

Web A Web B	Presence/absence of web builder *A. versicolor*	Predator spider choices	
		Web A first	Web B first
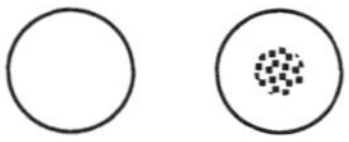	present	13	**27**
	absent	12	**28**
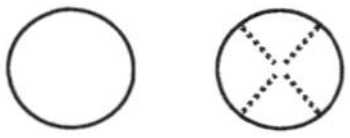	present	9	**31**
	absent	7	**33**
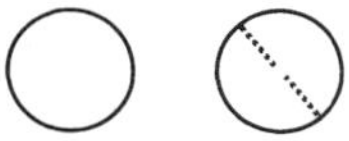	present	8	**32**
	absent	10	**30**
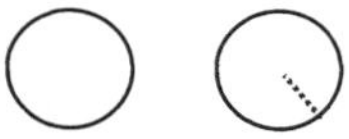	present	13	**27**
	absent	12	**28**

Fig. 1.10. Web stabilimenta of *A. versicolor* can attract predators. In all four experiments where one web with a stabilimentum and one without were presented to the predatory spider *P. labiata*, it was able to learn to associate the presence of a stabilimentum with prey (bold number in any row indicates a significant difference). (Adapted from Seah and Li 2001.)

vertical hanging threads set with sticky droplets (Broadley and Stringer 2001). Prey attraction can therefore be seen as an additional accompaniment to trap building, enhancing prey encounter rates.

Differences in web design between species are generally associated with differences in preferred web or prey specialisation. The Australian orb weaving spiders *Nephila plumipes* and *Eriophora transmarina*, for example, are similar in size and catch similar sized prey (Herberstein and Elgar 1994). The former, however, is predominantly diurnal in its foraging while the latter is nocturnal, exposing them to different potential prey, diptera and hymenoptera predominating in the day but lepidoptera at night. The web area of the nocturnal species is somewhat larger and its viscid thread has more adhesiveness than diurnal *Nephila*, nevertheless the total weight of the web of *Nephila* is greater than that of an *Eriophora* of comparable size (Fig. 1.11). There is a marked difference in the capture success of the two species, with webs of *Eriophora* being twice as successful as those of *Nephila*. This is also associated with a marked difference in web construction frequency, with *Eriophora* building a new web each night, while *Nephila* webs last several days.

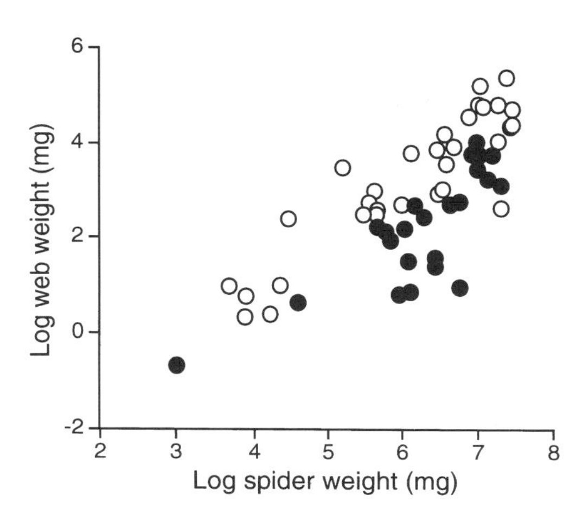

Fig. 1.11. The total web weight is significantly correlated with spider weight for both the diurnal spider *N. plumipes* (o circles), and the nocturnal *E. transmarina* (• circles). The slopes of the regression curves do not differ between the two species, but the elevation (web weight) of *N. plumipes* is higher than that of *E. transmarina*. (Adapted from Herberstine and Elgar 1994.)

Nocturnal species specialising in the capture of lepidoptera exhibit two types of orb web modification, one an elongation of the orb web, the other the concentration of the adhesive material to a very limited number of large sticky droplets. Both of these are designed to overwhelm the detachable scales that moths use to escape from sticky webs. The web shape of the diurnal *N. plumipes* is almost circular, while that of the nocturnal *Eriophora* is somewhat elongated in the vertical axis (Herberstein and Elgar 1994). The web of the nocturnal *Scoloderus cordatus* has the top sector of the orb greatly elongated to create a ladder-like capture surface (Stow 1978); in a New Guinea species it is the sector below the hub that forms the ladder (Robinson and Robinson 1972). Stowe (1978) found streaks of discarded lepidopteran scales adhering to the web surface above the point of prey capture, confirming the hypothesis of Eberhard (1975) that the method of capture in these webs is to allow a struggling moth to roll down the web surface until all its protective scales are lost, and it is firmly stuck.

1.5 Intraspecific communication

Even the slightest social interaction requires communication. Animals have repeatedly evolved simple mechanisms for communication over distance using visual, auditory, and olfactory signals, yet very seldom does animal architecture serve a communication function (Table 1.1), their additional costs apparently adding little to already effective and economical mechanisms. Nevertheless, communication aided by architecture has evolved in the context of both territoriality and courtship. Here they follow a pattern predicted for animal communication generally, that the former, involving mainly male–male rivalry, will be simple, while the latter, largely concerned with female choice, may exhibit great complexity.

The spherical balls of mud placed round the burrow entrance by both males and females of the fiddler crab species *Uca tangeri* seem to perform a dual function. Removal of the balls increases level of male–male agonistic behaviour, while females are more likely to approach a dummy male with mud balls than one without (Olivera *et al.* 1998).

Males of the fiddler crab *Uca musica* build semicircular hoods over their burrow entrances. By manipulating the presence of these hoods at the entrance of male burrows, Christy *et al.* (2002) demonstrated that a female was more likely to approach a courting male when a hood was also present. When the hood was displaced 3 cm away from the burrow opening, females often approached the hood rather than the displaying male. A possible explanation for this is that males benefit because the hoods or pillars stimulate female approach in some other context, for example, seeking shelter, a *sensory trap hypothesis*. This predicts that hoods will attract females when presented with a model predator, and that such structures will not be species-specific signals in the courtship context. These predictions are supported. Female *U. musica, U. beebei* (where males construct pillars), and even *Uca stenodactylus*, a species in which males build no structures, approach either hoods or pillars when presented with a model predator (Christy *et al.* 2002, 2003*a,b*).

The design of architecture to enhance the propagation of auditory signals is best understood from the auditory communication signals of certain Orthoptera, in particular male mole crickets. The burrow of males of the mole cricket *Gryllotalpa australis* consists of a bulb connected through a constriction to a flared horn (Fig. 1.12). The male sings by rubbing his wings together while positioned at the constriction, facing into the bulb. The burrow dimensions are such as to resonate with a second modal frequency of 3.0 kHz, forming a standing wave that has a pressure minimum at the constriction. The effective length of the horn is close to 1/2 the wavelength, and that of the bulb 1/4 the wavelength of 3.0 kHz. The bulb seems therefore to act as a tuned cavity, effectively extending the length of the horn by reflecting the wave so that it is in phase with the sound pressure in the horn (Daws *et al.* 1996).

Bailey and Roberts (1981) found a correlation between the resonance frequency of the burrow of species of the frog *Heleioporus* (Leptodactylidae) and the peak frequency of the sound they produce. The frogs sit in a chamber near to the entrance of the burrow. This position corresponds closely to a position which maximises the output of the dominant song frequency. The burrows of male *Eupsophus emeliopugini* (Leptodactylidae) appear to have an architecture to assist the receiver rather than the broadcaster of the call. The fundamental resonance frequency of burrows (814 Hz) does not correspond to the fundamental resonance frequency of the calls emitted by the frogs (1062 Hz), but rather to the dominant frequency of calls reaching the burrows after the

a)

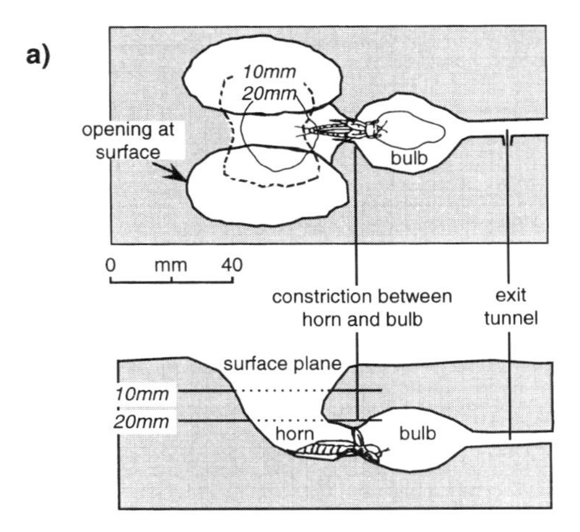

b)

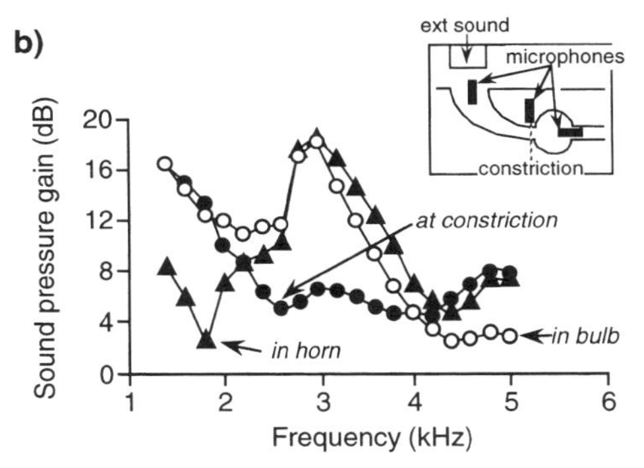

Fig. 1.12. The singing burrow of the male mole cricket *G. australis.* (Adapted from Daws *et al.* 1996.) (a) Plan and vertical section of the burrow to show the position of the male when singing. (b) Sound pressures, expressed as gain in decibels, against driving frequency in kilohertz, measured in a model burrow when driven by an external sound source. The position of the sound source and three recording microphones is shown in the inset.

higher frequencies have been selectively attenuated by passage through a humid atmosphere (Penna and Solis 1999).

In male three-spined sticklebacks nest quality measured as 'compactness' and 'neatness' has been found to be correlated with male physiological condition. Neater nests are built by males with larger kidneys, which secrete the glue-like protein used in nest building. This, in turn is positively correlated with androgen levels. Nest compactness is negatively correlated with spleen enlargement, an indicator of immune stress (Barber *et al.* 2001). Here, the nest may provide information about itself, but clearly it could also provide information to females about male quality. In the monogamous black wheatear (*Oenanthe leucura*) mated pairs carry large numbers of stones to place in nest cavities. An average of 1–2k of stones is carried largely by the 35–40 g male. There is no evidence here that the stones actually contribute to nesting success, however, stone-carrying effort appears to be used by females to assess male quality. Females start reproduction earlier, lay larger clutches, and rear more young for males that carry many stones.

Females of the rufous bush robin (*Cercotrichas galactotes*) lay significantly more eggs in nests of greater mass (Palomino *et al.* 1998). Both sexes build the nest, however, the result seems to indicate that females make a post-copulatory decision on the level of investment based on nest quality. Attributes of male nest building are also used by females to adjust parental investment in another monogamous species, the magpie (*Pica pica*). Although both sexes build the substantial domed twig nest, males make more trips to collect nesting materials. Experimental manipulation of nest size shows that females produce a larger clutch for a larger nest (Soler *et al.* 2001). Magpies breed only once a year but, if the clutch is destroyed, will nest again, often building a new nest. De Neve and Soler (2002) showed that female investment in the replacement clutch was in this case not based on the size of the original or replacement nest, but upon the speed of replacement nest construction.

In *Lek polygyny*, males display collectively in adjacent territories which contain no apparent resources. This system allows females to make comparisons between males. In lekking species where males build display objects, notably the bowerbirds (Ptilonorhynchidae), questions can therefore be asked about what information the structures provide to females. Structures that perform a courtship display role are now known for a few birds other than bowerbirds, and also some cichlid fish. The so-called 'bowers' built by some mouth-brooding cichlids are mating platforms evolved from nests (Tweddle *et al.* 1998).

Males of the cichlid *Cyathopharynx furcifer* construct mounds, which are generally somewhat clustered. Females visit several of these before spawning to one male. Males without bowers are not offered eggs (Karino 1996). In the *Capadichromis eucinostomus* species complex, males defending a sand 'bower' attract more females than those simply defending a rocky ledge. Their reproductive success is also negatively correlated with the number of parasitic cestodes found in the liver (Taylor *et al.* 1998).

Bowerbird males build true bowers that are distinct from and unrelated to the nests that are built by females. In some of these species, bowers display a level of complexity in structure and ornamentation that raises questions about male self-awareness and the experience of females in viewing them; this is discussed in Section 5.5. In this chapter the challenge is to explain why a female might benefit by selecting a male with a particular sort of bower. Broadly this might be in one of two ways: indirectly by choosing a male whose qualities will be expressed in her offspring, or choosing some quality of the male display from which she directly benefited herself.

Of the indirect benefit hypotheses, the *runaway* hypothesis (Fisher 1930) proposes that a quite arbitrary male character, if preferred by some females, could become exaggerated over generations until checked by some counter selection. The *good genes* hypothesis asserts that a female's choice ensures that her offspring of whichever sex are better able to survive and therefore reproduce.

Of the direct benefit hypotheses, the *proximate benefit* hypothesis proposes some immediate benefit to a female (Ryan 1997). More than one kind of proximate benefit has been advanced to account for some aspects of bower design. Humphries and Ruxton (1999) argue that the relative simplicity of bowers like that of the spotted bower bird (*Chlamydera maculata*), where males are widely spaced, compared with the more complex bowers of the closely spaced satin bowerbirds (*Ptilonorhynchus violaceus*), has the proximate benefit for females, because it enables them to remember bower details more readily, and make a comparison more easily. The best evidence of proximate benefit is, however, *threat reduction*. This hypothesis proposes that bowers have been selected to provide protection to females from the aggression of resident males.

The bower of the spotted bowerbird is a broad avenue of thin grasses in which the female stands while observing through one wall the courting male vigorously running at her from 3-4 m away. This design can be explained as offering threat reduction since a female may watch the male while partially screened by the avenue wall (Borgia 1995*b*). Females prefer the most vigorously displaying males but, if one of the bower walls is experimentally removed, the female positions herself so that the remaining bower wall separates her from the displaying male (Borgia and Presgraves 1998). Males of McGregor's bowerbird (*Amblyornis macgregoriae*) place themselves behind the maypole of the bower when the female first arrives (Borgia 1997).

Bowerbird species that do not build a threat reduction structure help to explain how this factor could have influenced bower evolution. Males of the tooth-billed bowerbird (*Scenopoetes dentirostris*), which represent the ancestral state (plesiomorphic) for the maypole builders (Kusmierski *et al.* 1997), display on a bare court on the forest floor decorated with leaves laid pale-side-up. At the arrival of a female, the male hides behind a nearby tree and displays through vocalisation before later emerging (Frith and Frith 1994). At display courts of *Archboldia papuensis*, the male initially responds to the arrival of the female by prostrating itself motionless (Frith *et al.* 1996).

The good genes hypothesis also receives support from the bower displays of several species. Video monitoring of visits by marked female satin bowerbirds (*P. violaceus*) to male bowers confirm that each female may visit several bowers, that some males obtain more copulations than others, and that female preference is positively correlated with certain measures of bower structure and ornament (Borgia 1985*a*). Similar findings have been obtained from studies of the spotted bowerbird (Borgia 1995*a*; Madden 2003*a,b*). Ornament stealing and damage to bowers by neighbours is known in the satin bowerbird (Borgia 1985*b*; Borgia and Gore 1986), the spotted bowerbird (Borgia and Mueller 1992), and vogelkop bowerbird (*Amblyornis inornatus*) (Pruett-Jones and Pruett-Jones 1982).

This competition between males gives a female a visible indication of male quality, provided that this competition is a significant cost to the males. In the spotted bowerbird this appears to be the case. Males of this species preferentially display green *Solanum* berries, and the numbers of which are a good predictor of mating success (Madden 2003*a*). However, males that have the numbers of their berries artificially enhanced, attract more damage to their bowers from rival males and actually remove berries to restore their total to the previous level. It appears that, faced with the costs incurred from bower damage from rivals, males adjust their displays to reflect their local status (Madden 2002).

2 Building materials: nature, origins, and processing

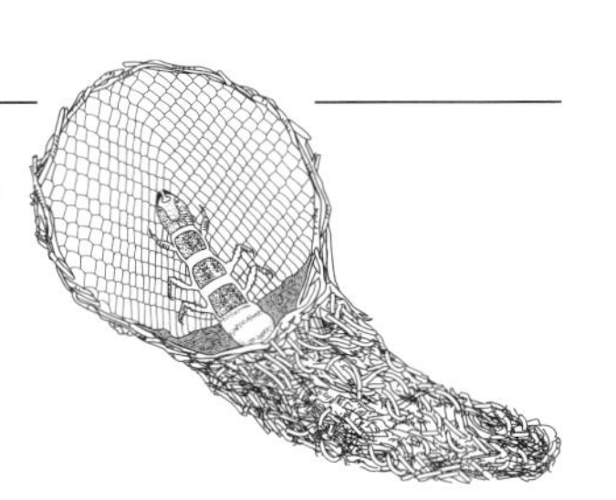

2.1 Introduction

Architecture results from the application of behaviour to materials. Animals are surrounded by potential materials, mineral and organic, where *mineral* is represented by mud and stones, and *organic* includes those of animal, plant, and fungal origin (Hansell 2000). These materials must not simply be effective after construction, they must be capable of efficient manipulation during construction. Constructional considerations may therefore impose constraints upon what materials are usable, and so in turn on what structures are possible. This chapter deals largely with structural materials, although materials with other functions are incidentally mentioned. It is not a comprehensive list of the materials used by animals but concentrates on key materials and examines their role in the creation of built structures.

The chapter distinguishes two categories of materials in particular, those that are collected and those that are secreted, predicting certain differences between them. It also predicts the characteristics of materials that will be required in the construction of homes as opposed to prey capture devices, and whether these are more likely to be fulfilled by collected or secreted materials.

2.1.1 Complexity of materials: some predictions

There are two fundamental differences between collected and self-secreted materials. The first is that collecting is a behavioural process, whereas secreting a material is a physiological one. The second is that collected materials are not primarily created for the building purpose, whereas self-secreted building materials are. A stick used by a bird for nest building is chosen from those

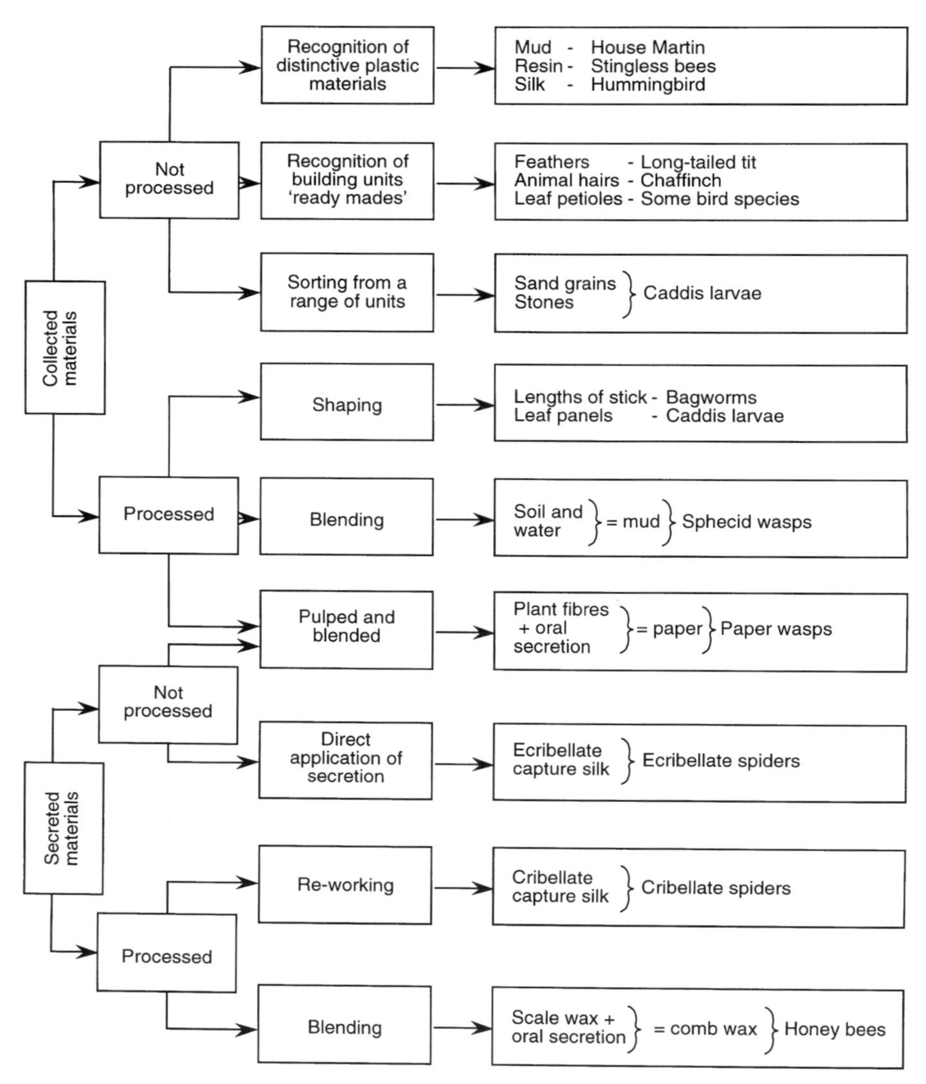

Fig. 2.1. The origins of available materials and the pathways they take prior to incorporation in the built structure, including processing and blending with other materials.

available on or under local trees. Natural selection will refine the choices made by the birds, but the sticks themselves will continue to owe their properties to selection operating upon trees.

Materials collected from the environment may be applied to the building without modification, or they may be processed in some way to make them suitable for building, for example, by cutting or shaping, or by mixing with another material as when soil particles are mixed with water to make mud. For self-secreted materials, although some may be secreted in a directly usable form, others need some processing before construction can begin, as in the blending of scale wax with oral secretion in the creation of comb wax by honeybees. Processing may also bring together different material components to form a composite material, as with wasp nest papers, many of which combine collected plant fragments with a salivary secretion. Figure 2.1 summarises the possible origins of materials and pathways they may take to inclusion in a building.

The standardisation of collected building units may be achieved in one of two ways: sorting or shaping. Mineral particles are hard to modify, so the builder must employ a selection mechanism to ensure that only suitable particles are used for building. For certain plant materials, on the other hand, the builder may be able to fashion building units to a particular size or shape. Some plants, or animals, even the builder itself, may grow structures (such as body hairs) that have uniform dimensions and are available in sufficient numbers to allow construction with them.

Self-secreted building materials are selected to enhance the fitness of the secretor. Not all such materials have evolved primarily for building; the down feathers of a female eider duck (*Somateria mollissima*), which line her nest as insulation, being such an example. However, for materials that are secreted as building materials, like spider web silks, natural selection will act through their effectiveness in that role, upon the physiological processes by which they are produced. This should lead us to expect extreme standardisation in secreted materials, unless some competing selection pressures favour flexibility of synthesis. There are two possible circumstances where selection could favour flexibility. The first is that the secretion has a generalised function, so favouring flexibility in its composition. The second is a constraint in the availability of raw materials needed to produce the secretion, that is, the same constraint at the physiological level that may be faced by collectors of building materials at the behavioural level. These differences in the origins of collected and secreted materials do allow us to test the validity of a prediction of product specificity from self-secreted materials compared with collected materials. In the event of failure of this prediction, we should look for evidence in secreted materials of either functional flexibility or of dietary constraint.

Two further predictions can also be tested regarding the likely differences between materials used in home building and in the making of prey capture devices. The reasoning behind these predictions is that traps are dynamic in their operation while homes are static, as a consequence the precision with which the former are made at whatever level, from the molecular up to that of the whole structure, needs to be much greater.

The three predictions to be examined in this chapter are therefore:

(1) that the glandular origin of self-secreted materials will exercise a quality control over them that ensures uniformity in their composition and performance, that will not be matched by collected materials;
(2) that self-secreted materials will be a more prominent feature of traps than of houses;
(3) that self-secreted materials used in traps will exhibit greater complexity than those used in making houses.

2.1.2 Specialisation in the selection of materials

Although time spent engaged in nest material collection and processing is a cost, there are reasons why animals could gain from obtaining or making materials of uniform consistency. Standardised building units could be an advantage at the level of selection, construction, and in the completed structure. Collecting and processing of materials will be simplified by building material uniformity, as will construction behaviour. Standardisation of materials will produce a more regular structure, which will probably enhance its durability.

Not all materials are structural. Structural materials may dominate in the building of traps, but homes fulfil a variety of functions, each of which may benefit from a material of specialised properties. More than one material is required to construct the nest of the long-tailed tit (*Aegithalos caudatus*), with separate and distinctive materials used to camouflage the outside, give structural integrity to the nest wall, and insulate the interior (Hansell 2000). However, there are potential disadvantages in the collection of very specific building materials that could, in some circumstances, lead to the acceptance of more varied materials. Specificity restricts possible sources. This could increase building costs in terms of the time and energy needed for locating materials compared with more relaxed selection criteria. At the functional level, the creation of a regular structure by the use of standardised building units, could make it more easily recognised by visual predators, where relaxed selection criteria might produce a more motley and therefore less obvious structure; such an anti-predator hypothesis for irregular construction is untested.

2.1.3 Mechanical properties of materials

Silk fibres bear loads in tension. The *stress* (σ) placed upon a thread $\sigma = F/A$ or force per unit cross-sectional area. As a result of that stress, the thread initially becomes elastically deformed. The amount that the thread extends under stress is referred to as the *strain* (ε), where $\varepsilon = \Delta L/ L_0$, where L_0 is the initial length of the fibre, and ΔL is the change in its length (Gosline *et al.* 1999).

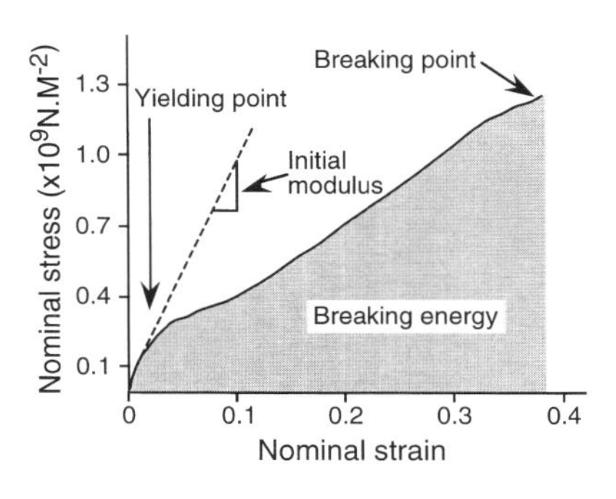

Fig. 2.2. The stress to strain response of a silk fibre under tension to breaking point. Young's modulus is the ratio of stress to strain. The area under the curve (shaded) represents the energy required to break the thread. (Adapted from Vollrath and Knight 2001.)

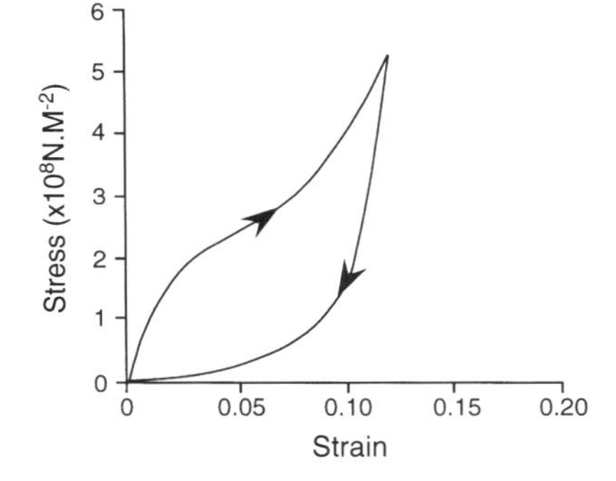

Fig. 2.3. Demonstration of the hysteresis of the frame thread of a spider web. The area between the curves represents the amount of energy lost as heat in the extension–relaxation cycle. (Adapted from Gosline *et al.* 1986.)

It conveniently turns out that for many materials the relationship between stress and strain is a straight line: stress/strain = constant E. This is referred to as the *modulus of elasticity* or *Young's modulus*. Young's modulus, therefore, can be said to express the *stiffness* of a material, and will have a large value (steep slope of stress/strain curve) for a material of high stiffness, and vice versa (Gordon 1976) (Fig. 2.2).

Strength is the maximum amount of *stress* that a material can bear before failure (σ_{max}). The value of ε at the point of failure (ε_{max}) is the thread's extensibility. The amount of energy expended to cause failure of the material is given by the area under the curve up to the point of failure (Fig. 2.2). This is sometimes referred to as *work of fracture* or *toughness* (Gordon 1976).

The energy required to extend an elastic material may not all be stored in it, some may be dissipated as heat. This can be seen in the discrepancy in the stress/strain curve in the relaxation compared with the extension phase (Fig. 2.3). The ratio of the energy dissipated, compared with the energy absorbed is the measure of *hysteresis* of the material.

The mechanical properties of a material may differ from point to point, or depending upon the axis of application of the stress. Such materials are called *anisotropic*, as opposed to *isotropic* materials, where these properties are all the same regardless of direction. Wasp nest paper is normally added by being spread along the growing edge of a brood cell or the envelope, load by load. This method of application of the material may cause plant fibres to be oriented in a predominant direction or give rise to slight weaknesses along the junction between neighbouring loads (Fig. 2.4). Both of these can give rise to anisotropy, with, for example, the envelope paper of *Dolichovespula* species showing greater strength in tension when tested in the direction of load application, than at right angles to it (Cole *et al.* 2001).

The failure of a material in tension is brought about by the propagation of a crack through it. Cracks concentrate stress at their apex, potentially causing the run-away effect seen, for example, in glass, which gives it low toughness. Much greater toughness is shown by the composite material of glass fibres embedded in resin (fibreglass). This difference comes about because in the composite material, as each fibre is broken, the crack then tends to spread along the axes of other fibres ahead of the crack, so dispersing rather than concentrating the stress and enhancing the energy required to break the material. Composite materials may also be able to combine the virtues of strength in compression as well as tension; concrete reinforced with steel rods is such a material, as is mud in which are embedded plant fibres, a material used by some birds to make their nests (Hansell 2000).

Although these various terms are illustrated largely with respect to materials under tension, they can be equally applied to materials in compression. In practice, it is easier to study threads in tension than more massive pieces of material in compression. In the case of animal building materials, it is the self-secreted ones characteristic of trap construction that occur in the form of threads in tension, and this is certainly one reason why they have received the most attention as building materials.

2.2 Materials obtained by selecting and sorting

2.2.1 Animal materials

Birds provide the majority of examples of collectors of building materials of animal origin. Fur and feathers are used by birds to provide thermal insulation of the nest, mostly against heat loss to the colder exterior. In a few of these species, adults when breeding remove some of their own feather insulation to line the nest they have created for their offspring. The female common eider duck, for example, makes a nest that is almost entirely of her own down feathers. This is using self-secreted rather than collected materials, however, a larger number of species of birds insulate the nest cup with feathers of other bird species or the hair of mammals, that is, with collected materials of animal origin (Hansell 2000).

Only one structural building material of animal origin features as a collected material of any significance. It is arthropod silk, and the builders that make use of it are small birds. It is such an important material in bird nest construction that it is found in the nests of at least some species of 56% of passerine families (Hansell 2000). Its merits as a building material are that it is very strong for its weight and that, used as a 'looped' material of a *Velcro* fabric, it makes construction easy (Section 3.3.1).

2.2.2 Plants, lichens, and fungi

A flowering plant is a differentiated structure; one species may produce potential building materials of several shapes and with varied properties: leaves, leaf petioles, stems, flowers, seeds, and roots. Considering all flowering plant species, this amounts to a very wide range of possible building materials of different sizes and shapes; lower plants such as ferns and mosses, offer additional specialist structures, while including lichens and fungi for convenience in this section, illustrates further possibilities.

The stems of many plants act as vertical pillars that support the weight of flowers or leaves without buckling or without snapping when exposed to lateral forces of wind. Branches spreading from a trunk are cantilevered beams. These components of plants offer rigidity. Leaves on the other hand are characterised by their flexibility but, since they have to survive shaking by the wind, they may combine flexibility with strength in tension. This is particularly evident in plants such as the grasses (Graminiae) and palms (Palmae) where leaves are elongate and strengthened by long parallel veins. Roots are not required to be rigid as they are generally supported in the ground but may have strength in tension to save the plant from toppling over. A common feature therefore of stems, roots, and many leaves is that they offer elongate building units, some flexible, some rigid.

Woody sticks, generally dead, are the typical nest material of many larger bird species (Hansell 2000). The variable lengths and widths of twigs available to species that select rather than shape them probably results in either the rejection of a significant proportion of them or a tolerance of some variation in them, although there is very little systematic documentation of this. Some plants do, however, offer standardised stick-like building units in the form of leaf petioles. These become available when the leaves of certain trees fall and the leaf blades separate from the petiole.

Elongate, flexible plant materials are highly suited to the building style of birds (Section 3.3.4). Among the flexible materials recognised in my classification of bird nest building materials are grass leaf, palm fronds, palm fibres, and horsehair fungus (*Marasmius*, Basidiomycetes). All these are characterised by strength in tension, and all are used to make complete hanging basket nests (Hansell 2000). The leaves of dicot plants, which have branching veins and are not characteristically elongate, provide different potential as building material. Their leaves are typically broad to intercept the maximum solar radiation, and provide builders with an opportunity to use leaves to cover surfaces. Perhaps the simplest way to do this is to roll up a living leaf to create a shelter (Section 3.6), another is to cut panels from them which are stuck together to form a shelter, as do some caddis larvae (Section 2.3.2).

2.2.3 Mud, soil, and stones

Mud is a plastic material that can be moulded when wet into structures of quite delicate detail which, on drying, hardens into a durable shape. Mud is used by a varied range of taxa, although not a large number of species. It has been assessed as a vital material in the nests of about 5% of bird species (Rowley 1970). In these it is not usually the sole material, but mixed with a fibrous material to give the toughness of a composite material, able to withstand both compression and tension stresses effectively.

Mud is a material that most obviously can bear loads in compression, however, nests of the house martin (*Delichon urbica*) and the cliff swallow (*Hirundo pyrrhonota*) are both built on vertical walls or rock overhangs, where their weight is not supported from below. The addition of grass or horsehair into the mud of both these nests probably provides the strength for parts of the nest that are necessarily in tension (Hansell 2000). The most massive of rock-attached mud nests is that built by the white-necked rockfowl (*Picathartes gymnocephalus*) a bird of about 200 g that builds a nest of 2 kg.

Mud may of course vary in its water content, mineral, or organic composition, and so the mud might be selected by a builder only when of a consistency appropriate to the type of structure being built. This is suggested by the observation that, where barn swallows and cliff swallows are nesting in the same location, the shallow bracket-shaped nests of the barn swallow are built of mud with less sand and more silt than are the retort-shaped nests of the cliff swallows (Kilgore and Knudsen 1977). The authors speculate that this allows cliff swallows an improved ability to mould a more complex shape, albeit at some cost to the nest strength. These results need confirmation and the predictions testing, as this would provide an important example of a trade-off between constructional and structural considerations.

An appropriate distinction between mud and stones as building materials is not their absolute size, but whether the mineral particles are gathered singly or as a load. This allows even very small particles of less than 1 mm across, collected by a polychaete worm *Sabella* to create its tube, to be considered as stones alongside 6.7 g stones collected by a male black wheatear (*Oenanthe leucura*) in its display to females (Moreno *et al*. 1994).

The filter feeding polychaete *Sabella pavonina* uses what is essentially a mechanical sorting device. It is dependent upon the size of the groove in its gill filaments, to determine which particles are passed by ciliary currents towards the mouth (Nicol 1931). Periodically this stream is not ingested but is combined with mucus and applied to the anterior rim of the worm's dwelling tube by the ventral collar fold of the worm as it rotates upon its axis within the tube. The sediment burrowing polychaete *Pectinaria toreni*, having settled out of the plankton, initially builds a tube entirely from its own secretion, but soon begins to form a tube of sand grains held together with the secretion. As the worm grows, so the diameter of the tube grows correspondingly, but not the size of the particles themselves (Vovelle 1973). The increase in the size of the sand grains instead corresponds to that of the building organ, a selection device associated with the buccal apparatus.

An obviously more behavioural and less mechanical selection mechanism is employed by caddis larvae in selection of mineral particles for case building. The selection of sand grains for the roof of the tube of *Silo pallipes* is a three-stage process, including manipulation at the site where a new particle will be attached (Hansell 1968*a*). Selection for size but not shape was demonstrated here, however, larvae of the caddis *Chaetoteryx villosa* have been shown to select flattened particles in preference to more spherical ones, and to alter their selection criteria on the basis of recent experience (Nepomnyashchikh 1993).

2.3 Materials created by processing

2.3.1 Processing mud

A variety of invertebrates, notably insects, make structures out of mud using the same technique as birds. But, while for birds the water content of mud is a matter of selection, choosing mud of appropriate stiffness from the edges of ponds or streams, for a eumenine wasp, such as *Zeta abdominale*, it is a matter of processing. The wasp first brings the water in the crop to the location of the soil to create the desired consistency of mud, which is then carried to the site of nest construction (Taffe 1983).

The mud-like nest materials of mound-building termites also show evidence of selection and processing. Garnier-Sillam and Harry (1995) found that, in the nest wall of the termites of *Noditermes lamanianus* and *Cubitermes fungifaber*, the materials had a high clay content compared with the control soil profile, indicating soil particle selection. Mound wall materials of *Thoracotermes macrothorax* and *Crenetermes albotasalis*, however, showed no evidence of raw material selection. In addition, nest materials of *N. lamanianus*, *T. macrothorax*, and *C. fungifaber* were enriched with organic matter and exchangeable cations, in particular Fe, Al, Ca, which were judged to enhance structural stability of the mound wall. In *C. albotasalis*, where the material is not enriched with organic matter or cations, the mound walls did not differ from control soils in their stability, notably resistance to water erosion. The source of the organic component of the mound material was evidently substantially from the guts of the termites themselves. *T. macrothorax* and other soil feeding termites form complex organomineral associations in their guts. These on excretion are combined with the raw soil material.

2.3.2 Processing: cut leaf and stick

Animals with strong jaws can shape tough, rigid plant stems by gnawing through them. Beavers use their massive incisors for cutting through the trunks of small trees, larvae of some bagworm species (Psychidae, Lepidoptera) cut uniform lengths of straight woody stems, which are stuck together with silk to form a cylindrical protective case.

The flexibility of plant leaves is greatest when the material is fresh. Weaverbirds (Polceinae) exploit this by making nests out of fresh strips of grass or palm leaf. The village weaver (*Ploceus cucullatus*) is able to standardise its plant material building units to quite a high degree by biting across a few veins of elephant grass leaf and then flying off, tearing a long strip of fresh green material as it goes (Collias and Collias 1964*a*). The same technique is found among the New World weavers, the Icterinae (Fringillidae), for example, the Montezuma oropendola (*Gymnostinops montezuma*) (Skutch 1954).

Species that feed on leaves are capable of using their jaws to cut from the leaf, standard building units that are fitted together to cover some surface, typically the body surface of the builder in the form of a portable case. An example is the caddis larva, *Lepidostoma hirtum*, that cuts near rectangular pieces from dead leaves of deciduous trees that have fallen into the water. These panels are wider than long, convex at the anterior and sides, and concave at the back (Hansell 1972, 1973). Panel dimensions, however, are not inflexibly determined. When all four sides of the case are experimentally cut back to same level, so removing the half-panel phase difference in the joints between panels of neighbouring sides, larvae restore the phase difference by cutting a wider size range of panel length, while maintaining the same mean value (Hansell 1974).

Another caddis species, *Glyphotaelius pellucidus*, use the same materials to cut near circular building units. As the larva cuts the disc, the positions of its legs and of its body change indicating that cutting is not a simple stereotyped process like using a geometrical compass to define a circle. Neither immobilising the joint between the head and pronotum, nor amputating the distal segments of one prothoracic leg disrupts the cutting process, indicating possible redundancy in sensory feedback regulating the cut (Rowlands 1985).

Some leaf eating lepidoptera larvae also cover themselves in larval or pupal cases made from cut leaf panels. The portable larval case of *Vespina nielseni* (Icurvariidae) is a purse of two regular, oval-shaped pieces of leaf silked together (Okamoto and Hirowatari 2000).

2.3.3 Propolis

Propolis is a substance to which a variety of functions have been attributed. Hepburn and Kurstjens (1984), testing the justification of its title 'bee glue', found propolis to be comparable in strength to beeswax, but more tough and less stiff. It does have sticky properties but its more obvious mechanical role in nests of honeybees is filling cracks and repairing damage to comb (Johnson *et al.* 1994).

Propolis is not even partly secreted by the bees themselves, but consists of plant resins collected, in Europe, principally from species of poplar (Konig 1985). In the megachilid bees, plant resins are used as glue to aid in the construction of cell partitions (Frolich and Parker 1985) but in honeybees, evidence suggests that the resins are used for their antiviral, antibacterial, and anti fungal properties to promote nest hygiene (König and Dustmann 1989; Bankover *et al.* 1995). A wide range of organic molecules have been isolated from European honeybee propolis, but phenolic compounds such as flavonids, phenolic acids and their esters have been identified as significant in their antibacterial effects (Bankova *et al.* 1995). A similarly large range of organic compounds has also been found in honeybee propolis from Brazil. The main part of their antibacterial activity is due to phenolic compounds (Bankova *et al.* 1995; Marcucci *et al.* 1998).

2.3.4 Carton and paper

The industrial manufacture of paper by humans involves the processing of timber to create wood pulp. In this process the individual fibres become somewhat weaker but they also become softer and more flexible allowing greater contact between fibres to be made. It is the capillary action of the water drawing the fibres together to help the formation of hydrogen bonds between them, which is largely responsible for the strength of the paper. Wetting the paper reverses this process making the paper weak when wet (Biermann 1993).

The nests of arboreal ants and termites are composed of fragments of plant material held together by a glandular secretion of the insects themselves. These are only known at a broad descriptive level, but this at least confirms that these materials exhibit variation within both insect orders. In the ant genus *Polyrachis*, nests are made of fragments of dead vegetation held together with larval silk (Hölldobler and Wilson 1990). Nests of the neotropical ant

Hypoclinea sp. (Dolichoderinae) are located on trees of *Pirinari excelsa* and made of wool-like fabric composed of the convoluted hairs harvested from leaves of the tree. Prance (1992) speculates that this is a symbiotic relationship in which the ants provide the plant some protection from phytophagous insects; if so, this would provide an example in which a collected material could be exposed to forces of natural selection exerted by the collector. Carton from the terrestrial mounds of the termite *Coptotermes kishori* is composed of grasses, wood fragments, soil mineral particles, and excreta (Reddy 1983).

Among the social wasps of the family Vespidae, the material referred to as paper consists of fragments of plant material shaped by mandibulation, either adhering together or bound by some secretion of the wasp itself. Its virtue is that it can be made into moulded sheets of a light material, strong in tension. Paper is used to make nests by species from the three subfamilies of social Vespidae, the Stenogastrinae, Polistinae, and Vespinae. In addition materials made of plant fragments cemented together are used for nest construction by sphecid wasps of the genus *Microstigmus*.

The Sphecidae are a family of hymenoptera that are generally non-social. They exhibit a variety of nest building habits in particular digging burrows or constructing cells of mud (Matthews 1991). *Microstigmus* is an exception; females of this genus gather plant fibres which are then bound together by filaments of a secretion from glands at the tip of the abdomen (Matthews 1968). *Microstigmus* species are also unusual among Sphecidae in being eusocial, and the five or so females in a colony of *Microstigmus comes* have been observed to cooperate in the gathering of plant materials on the underside of the frond of the palm *Chrysophila guagara* to form a tiny bag no more than 20 mm across, which is then lowered from the palm frond on a fine petiole of the same material (Mattews and Starr 1984). These nest suspensions are characteristic of *Microstigmus* species although the plant materials vary and may not be so specific; those of *Microstigmus coupei*, for example, include fragments of bark, lichen, and sand grains (Melo and Matthews 1997).

The nest materials of the Stenogastrinae (Vespidae) vary between species from entirely mud nests as constructed by *Liostenogaster flavolineata*, through *Stenogaster* species with 60% and around 90% organic material (Hansell and Turillazzi 1991), to pure pulped vegetation nests of *Parischnogaster* species (Hansell 1981). One consistent characteristic of stenogastrine nests, regardless of the materials of which they are made is the absence of a fine nest suspension created out of the nest material itself. I concluded (Hansell 1987*a*) that it was the weakness of their nest material in tension that made the construction of nest suspension petioles impossible in Stenogastrinae. Their mud nests are too heavy, and the composition of their paper nests is of small plant fragments, some evidently rotted at the time of collection (Hansell 1987*a*; Hansell and Turillazzi 1995).

The poor quality of Stenogastrinae paper is indicated by the thicker paper walls of brood cells of these species compared with those of polistine wasps of a similar size (Hansell 1987*a*). In the nests of *Anischnogaster laticeps* the weakness of the walls of rotted plant fragments is apparently counteracted by a rich growth of fungal mycelium that permeates it (Hansell and Turillazzi 1995), a reinforcing fibre that at the same time consumes the plant component of the nest material. This living constituent of building material is also reported by Krombein (1991) in the nest material of *Eustenagaster eximia*.

The Polistinae, although termed *paper wasps*, have varied and species-specific nest materials. Wenzel (1991) recognises five different types, three distinguished on the basis of their plant material content: long woody fibres, plant hairs, and short vegetable chips (Fig. 2.4). The other two are characterised as mud, and pure glandular secretion. *Polistes* species typically build their small open-combed nests of long woody fibres bound with secretion (Fig. 2.4). *Ropalidia* species together use all five types of material recognised by Wenzel (1991).

Wenzel's (1991) categories of material are broad and on closer examination reveal a great deal of variation that is in need of study and explanation. Woody stem fibres may be collected from sound or rotted wood, plant hairs may be simple as in *Apoica*, boat-shaped as in *Protopolybia sedula* (Fig. 2.4), or stellate as in *Ropalidia guttatipennis* (Hansell 1996*b*). Amounts of glandular secretion in wasp nest papers also show great variation. *Ropalidia opifex* is unusual in having an envelope of secretion alone, while the papers of *Polybia* and *Chartergus* contain very little secretion from the wasps (Wenzel 1991), similar in that respect to the papers of Stenogastrinae (Kudo and Yamane 1996).

The nature and origins of the glandular secretions found in polistine papers may vary between species but seem predominantly to be proteinaceous. Schremmer *et al.* (1985) claimed to have identified a high chitin content in the glandular secretion of the paper of *Pseudochartergus chartegoides*. Maschwitz *et al.* (1990), however, found that the transparent envelope of *R. opifex* contained no chitin and was predominantly protein, with serine, glycine, and alanine making up more than two-thirds, consistent with the β-keratin structure of fibroin silk although not in crystalline form (See Section 2.4.3). The glandular component of the paper of *Polistes chinensis* is 50% of the nest material dry weight, and its composition is 70% protein. Kudo *et al.* (2000) identified more than 20 amino acids in the oral secretion, but five in quantities that ranged from 9% to 24% (in ascending order: proline, valine, alanine, serine, and glycine).

Differences between the comb and envelope papers of wasp nests was investigated by Cole *et al.* (2001) in a detailed examination of the paper of two *Dolichovespula* species and *Vespula vulgaris*. They found that comb fibres were significantly shorter than envelope fibres in all three species, but the same width. *V. vulgaris* comb and envelope fibres were, however, much shorter and

Fig. 2.4. Variety in the composition of wasp nest paper. (a) Boat-shaped plant hairs of *Protopolybia sedula* paper, showing predominant alignment probably in the direction of application; (b) long woody fibres in paper of *Polistes bernhardi*; (c) shattered woody fragments of *Polybia occidentalis* paper. Three pulp loads can be distinguished aligned from bottom left to top right. Scales: separation of two white bars—a) = 10 μm, b) & c) = 100 μm. (Photos Margaret Mullen.)

much wider than those of both *Dolichovespula* species. In *Dolichovespula norwegica* complete fibres were significantly longer in the envelope than in the comb paper, whereas there was no difference in the proportion of complete fibres. This gave some support to the view that not only are comb and envelope papers different but that this is partly as a consequence of differences in the material collected rather than simply how it is processed. The manufacturing of comb and envelope paper does, however, also appear to differ. Comb paper was thinner than envelope in all three species. The strength of *Dolichovespula* envelope paper was significantly greater that that of *V. vulgaris* when tested in the direction of the pulp fibres, its axis of greatest strength, although there were no differences when the tensile stress was applied at right angles to the direction of the fibres.

Differences between *Dolichovespula* and *Vespula* species in the fibre type and strength of the papers result from pulp being collected from different places. *Dolichovespula* species collect long fibres from weathered timbers such as fence posts, whereas *V. vulgaris* collects from more varied sources but typically rotted wood (Edwards 1980). The functional significance of these differences can be understood in terms of climatic demands of their respective nest sites. The stronger paper of *Dolichovespula* nests is consistent with the more exposed sites in which they nest. The sites of *V. vulgaris* nests, more sheltered from the wind and rain, may permit weaker materials to be used, suggestive of a trade-off between nest strength and paper processing costs.

The resistance of the various polistine papers to weakening by rain appears not to have been investigated experimentally. However, Yamane *et al.* (1999) interpret the greater amount of oral secretion in the nest material of *P. chinensis* compared with the related *Polistes riparius* as due in part to the wetter climate experienced by the former. Also, the proportion of oral secretion in the paper of nests built by foundress *P. chinensis* is greater in sites exposed to direct rainfall, compared with nests in sheltered sites (Kudo *et al.* 1998).

The generally tropical distribution of the Polstinae almost certainly means that the composition of the paper of some species is under selection pressure from predators as well as climate (Jeanne 1975), however, we lack the evidence on this. In the Vespinae, where the majority of species occupy northern temperate latitudes (Wilson 1971), climate is more likely to be the dominant influence on the properties of paper.

2.4 Self-secreted materials

The predictions of self-secreted materials made at the start of this chapter were that, compared with collected materials, they will be more uniform in composition and performance. It was also predicted of self-secreted materials that they would be a more prominent feature of traps than of houses, and that those used in traps would exhibit a greater complexity than those used in houses. These predictions can be evaluated through the examination of self-secreted wax, mucus, and silk building materials, and in particular the last two which are used in the construction of both houses and traps.

2.4.1 Wax and associated secretions

Waxes are complex mixtures of long-chain carboxylic acids and long-chain esters (Campbell 1999). They occur as a water-insoluble coating on leaves and on terrestrial arthropods, protecting them against desiccation. In certain insects waxy secretions are produced in much larger quantities and used as a construction material. The most conspicuous examples are the construction of nests by social bees of the family Apidae. However, in one of the four sub-orders of Hemiptera, the Sternorrhyncha, a variety of structures are produced from glandular excretions, some of which contain waxes, although some are of other types or of unknown composition. These insects range across three superfamilies Aphidoidea, Coccoidea, and Psylloidea. All these insects have a plant-sucking mode of nutrition so, in order to obtain sufficient protein in their sap diet they may ingest an excess of sugars. Some of these bugs excrete the excess as sugary honeydew, but in others it is used in the synthesis of building materials, including wax.

The shelters constructed by nymphs of a number of the Psylloidea turn out to be varied in their composition and do not include wax. These structures are referred to as *'lerps'*, a name originating in Australia, the centre of their distribution. They mostly take the form of a fixed shallow dome or shield under which the nymph is able to feed on the host plant. The lerp of *Celtisapis usubai* has a rough fibrous appearance (Takagi and Miyatake 1993), while that of *Cardiaspina* species are characterised by a delicate open lace-like appearance with spreading fan-like rays, linked by multiple fine bridges (Gilby *et al.* 1976) (Fig. 2.5).

These varied lerps are made entirely from secretions of the animals themselves, the material itself dependent upon species, although mostly reflecting the use of surplus carbohydrate in the diet as a substrate for building material manufacture. The lerp of *Cardiaspina albitextura* and *Lasiopshylla rotundipennis* appear to be over 80% starch with small amounts of protein and lipid, with the rest water. However, the lerp of *L. rotundipennis* contains proteinaceous filaments originating from glands ventrally on the abdomen (Gilby *et al.* 1976), while the lerp of the African species *Aritaina mupane* has about equal quantities of water soluble and insoluble components. The former contains significant amounts of fructose and glucose, while 30–40% of the water-insoluble carbohydrate is uronic acid (Ernst and Sekhwela 1987).

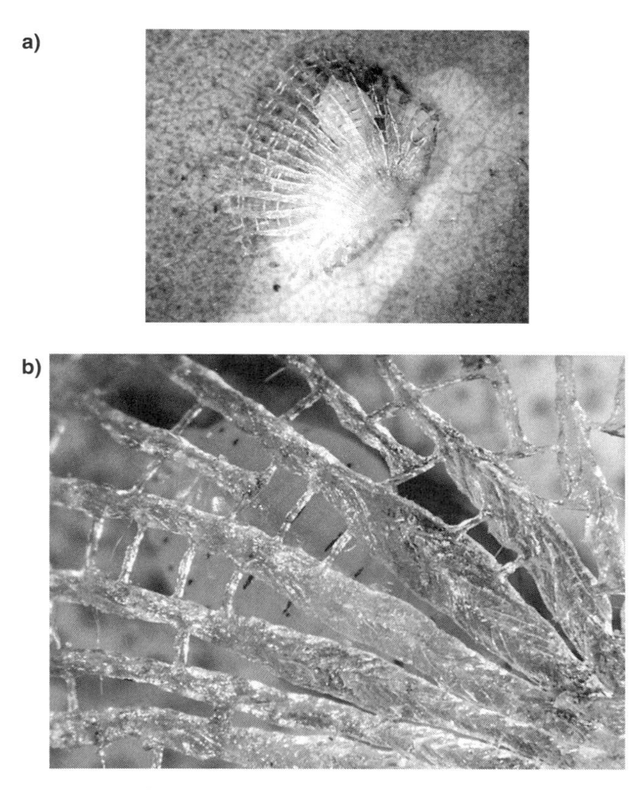

Fig. 2.5. The intricate lerp of a *Cardiaspina* species: (a) whole lerp with the outline of the bug's abdomen discernable through it, (b) detail of the lerp. The 'herring bone' pattern on each radial spine is built up by repeated additions of secretion. Each bridge between radii is a single application of material. (Photos Gary Taylor.)

In the Coccoidea, the fixed coverings over the bodies of the insects are generally referred to as 'tests' or 'scales'. In the family Draspididae, tests are created from fine filaments of material described as 'wax', in some species forming a loose mass, although in others more dense and fused. In the Conchaspididae, the test may be formed either from an anal secretion, or from an anal secretion in which are embedded layers of broken wax filaments secreted from glands on the ventral side of the abdomen (Takagi 1992). Analysis of the waxy threads covering the body of the scale insect *Drosicha corpulenta* (Margarodidae), showed 82% wax esters and 18% hydrocarbon (Hashimoto and Kitaoka 1982). The majority of the wax ester component had a chain length of C52 formed by 1-hexacosanol esterified by hexacosanoic acid (Hashimoto and Kitaoka 1982). Coverings of profuse wax threads over the body have convergently evolved in a small number of species of mites, causing them to resemble minute versions of their insect equivalents (Manson and Gerson 1996).

The comb building material of honeybees (*Apis mellifera*) can more strictly be referred to as wax. It is characterised by 67% monester (Milborrow *et al.* 1987) plus other esters and small amounts of a variety of other organic compounds (Tulloch 1980). Wax produced by the stingless bee *Trigona australis* (Meliponini) is, however, markedly different with nearly 90% hydrocarbons and only around 6% esters (Milborrow *et al.* 1987). Honeybee wax originates as flakes or 'scales' from specialised glands beneath the abdominal sternites which are active in workers for a short period early in their adult life (Michener 1974). The secretion of the wax apparently requires higher than normal abdominal temperatures because wax secreting honeybees congregate at sites of comb expansion in special groups called *festoons* (Hepburn and Muller 1988). This scale wax is then processed in the mouth parts to become comb wax (Hepburn 1986), a process that alters it structurally. *Scale wax* is crystalline, with the crystallites vertically inclined to the plane of the scale. Mandibulation, and the addition of salivary proteinaceous secretions, transforms this to *comb wax*, a material that lacks texture and has a random crystallographic signature (Kurstjens *et al.* 1985; Kurstjens and McClain 1990).

These differences between scale and comb wax are also reflected in their mechanical properties. The tensile strength of virgin scale wax is greater than that of comb wax over a temperature range of 25–45°C, where nest temperature is always maintained by bees close to 35°C. The extensibility at 35°C before breaking under tension is 70% for scale wax but about 10% for comb wax. The stiffness of comb wax is about twice as high as that for scale wax. Finally, the work of fracture (toughness) is constant for scale wax over the whole temperature range of 25–45°C, while that for comb wax declines sharply with temperature, being significantly lower than scale wax at nest temperature beyond 30°C) (Kurstjens *et al.* 1985; Hepburn 1986).

From the constructional point of view, the mandibulation converts the wax into a form which requires less work to mould at nest temperature. In the completed structure it has greater stiffness, greater isotropy due to the randomisation of the crystal orientation, but less strength. Mature honeycomb, however, has greatly enhanced mechanical properties due to the inclusion of fragments of larval cocoon silk. After adults emerge from their cells, the cocoon can be broken up and incorporated into the walls of other cells. This makes mature comb a composite material, with protein fibres embedded in a wax matrix. This can increase the tensile strength and toughness about fourfold, and nearly doubles the stiffness depending upon the amount of silk included (Hepburn and Kurstjens 1988).

2.4.2 Mucus

Mucus is a term generally applied in biology to extracellular secretions that have various sticky or slippery properties. It is likely to describe a mixture of organic molecules but, in published studies on animal-built structures only a few, if any of these components are identified; even when they are, comparisons are difficult because of confusing terminology.

An essential ingredient of mammalian mucus secretions is polysaccharide, an important class of which are linear disaccharides called *glycosaminoglycans*, sometimes termed mucopolysaccharides. These are generally found arranged as a series of long unbranched chains attached to both sides of the length of a core protein molecule. These complexes, which are termed *proteoglycans*, have a tendency to bind water, giving them a high viscosity and low compressibility, making them suitable, for example, as lubricants for mammalian joints (Zubay *et al.* 1995). Complexes of carbohydrate chains attached to amino acid residues of a protein may also be referred to as *glycoprotein*, however, this term is generally confined to complexes with a low carbohydrate content, while proteoglycan is reserved to describe such molecules with a very high (85–95%) carbohydrate content.

Mucoid materials are associated with a number of structures, both homes and traps. As traps, they occur as sticky strands, generally in the form of a net. In houses, they may act as cement binding together other materials, as complete walls, or in the coating of foam bubble structures.

Animals build foam structures in one of two ways: either agitating water to which the secretion has been added, or 'blowing' bubbles one at a time. The whisking process is exemplified by the manufacture of foam nests by certain frog species: Hylidae (Hadad *et al.* 1990), Leptodactylidae (Heyer and Rand 1977), Rhacophoridae (Fukuyama 1991). Processing of the secretion to produce the foam involves both sexes since the female produces both foam precursor secretion and eggs, while the male whips the secretion into a froth. In *Physalaemus pustulosus* the male first collects the eggs and secretion by drawing his hind feet over the female's vent. He then whips the secretion into a foam with alternate thrusts of the hind feet at the rapid rate of about one beat per 0.04 s, in bursts of about 15 (Heyer and Rand 1977). In *Polypedates leucomystax* the female secretion has been identified as substantially mucopolysaccharide (Kabisch *et al.* 1998).

Bubble nest building in fish is a characteristic of males of a number of species among the anabantoid families, for example, the siamese fighting fish *Betta splendens* and a number of species of gouramis. A male of the catfish *Hoplosternum littorale* (Cattichthyidae) gulps air and water, which pass over the gills where mucus is added. This mixture, on emerging is stirred by movements of the pelvic fins to create the foam in which the fertilised eggs develop (Andrade and Abe 1997).

In the hemipteran suborder Auchenorrhyncha, nymphal bugs of the family and Cercopidae (froghoppers and spittlebugs) (Hemiptera) blow bubbles to create protective shelters. *Philaenus spumarius*, for example, expels air from abdominal spiracles into a cavity where it can be blown through a film of hind gut secretion to create bubbles that accumulate over the nymph while it sucks the sap of the host plant. Included in the mixture of components reported from the froth of a *Deois* species of spittlebug were both glycoproteins, and at least 10 different polypeptides (Mello *et al.* 1987).

Burrowing species use mucous secretions, probably to stabilise the walls of their dwelling tubes and faclitate subterranean locomotion. Polysaccharides have been identified as components the secretory materials of eight tube-dwelling polychaete families (Vovelle 1997). The mucus secretion that emerges as strands from dispersed glands on the appendages of some mud shrimps *Callianassa* apparently serves the same function (Dworschak 1998). The kidney secretion used by male three-spined sticklebacks (*Gasterosteus aculeatus*) to stick together nest material was found to be dominated by one kind of glycoprotein molecule (Jakobsson *et al.* 1999). It is noteworthy that the role of glue in the prey capture droplets of the webs of araneid spiders, is also played by a glycoprotein (Section 2.4.3).

The role of mucus in home building is therefore fairly limited, including only some bubble structures and as glue to help stabilise burrow walls. As to their molecular complexity, we have essentially no evidence to compare them with the mucus secretions used in the construction of feeding devices. Such a structure is the mast built by the filter feeding amphipod crustacean *Dyopedos monacanthus*, to place it beyond the boundary layer of water flowing over the substrate. This mast is about 1 mm in diameter and between 50 and 80 mm high (Fig. 2.6) (Mattson and Cedhagen 1989). It has four components; two are self-secreted glandular materials, the third the organisms own faecal pellets, the remainder collected materials. One of the glandular secretions is described simply as 'spinning thread', the other as 'mucus'; they emerge from different structures and are applied in different ways. This superficially simple feeding structure is, at the material level an interesting composite.

The majority of mucus, feeding devices are nets, and it is significant that this is a self-secreted material and forms the whole of the structure. Such feeding nets are found within the tubes of certain polychaetes of the family Chaetopteridae. Typically these worms live in tubes open only at the inhalant end. *Mesochaetopterus taylori*, for example, filters suspended fine particles drawn into the burrow by posteriorly directed peristaltic waves passing along its body. Its net is consumed and replaced about once every 5.5 min (Sendall *et al.* 1995). In the polychaete *Praxillura maculata* the mucus net is external to the tube and operates like a spider web capturing particles drifting in the current. To deploy the web, the worm first constructs a tube of coarse sand and shell fragments which projects 20–70 mm above the substrate, at the top of which are added 6–12 radiating spokes (Fig. 2.7). What cement is used to

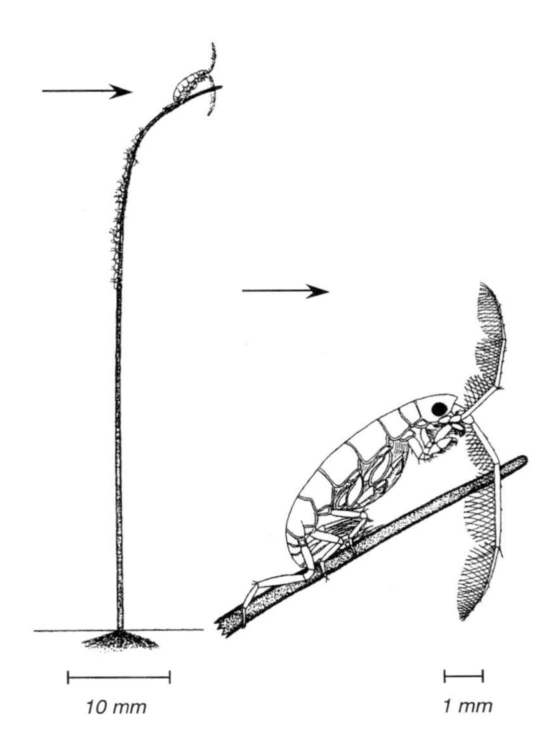

Fig. 2.6. *Dyopedos monacanthus* female in filter feeding position on its constructed mast accompanied by young. (Adapted from Mattson and Cedhagen 1989 with permission from Elsevier.)

Fig. 2.7. The mucus capture net of the polychaete *P. maculata*, is suspended between radiating spokes made of cemented-together sand grains. (Photo McDaniel and Banse 1979.)

hold this structure together is unclear, but the spokes provide a frame which supports the mucus net, which once more is repeatedly consumed to be replaced by another (McDaniel and Banse 1979).

A similar capture principle, without the supporting architecture has evolved in the specialised family of sessile gastropod molluscs, the Vermetidae. They secrete strands of sticky mucus, rich in protein apparently for added toughness, that can extend 100 mm around them. These, when stirred by the nearby wave action, capture fine particles, which are gathered in and consumed once or twice an hour (Kappner *et al*. 2000).

The ability of mucus to trap fine particles suspended in water has led to the use of feeding webs in open water, planktonic habitats by both polychaetes and molluscs. *Poeobius meseres* is an unusual, unsegmented polychaete found in midwater plankton, that can feed either by directly grabbing particles with cilliated tentacles or by the deployment of a mucus net (Uttal and Buck 1996). The pteropod mollusc *Gleba cordata* also occupies this midwater habitat in which it is able to deploy a feeding web much larger than itself (Gilmer 1972). This captures fine particles drifting past in ocean currents or falling slowly through the water column.

Exploiting the same planktonic resources are organisms that construct the most complex of all mucus architecture, the Larvacea (Appendicularia, Chordata). These tadpole-like organisms live in houses equipped with fine filters to trap microscopic organisms suspended in the water current driven through the house by the beating of the organism's tail. A larvacean house is a differentiated structure with intricate, species-specific detail. In this case we do know something not only of the construction behaviour but also of the mucus material itself.

Among the most complex are those built by *Oikopleura* species. In *Oikopleura dioica*, water enters the house through a pair of inlets within which lies a mucus mesh of $30 \times 100\ \mu m^2$ (Fenaux 1986). The water is then drawn past the head of the organism and into the chamber containing the beating tail and through the pair of filters before leaving through an exhalent aperture (Flood and Deibel 1998).

Each of the filters is in the form of a sandwich of upper and lower net filters for trapping food particles with a coarser scaffolding net in between (Flood 1991; Flood and Deibel 1998). In *Oikopleura labradoriensis* the mesh of the inhalant filter net is $80 \times 10\ \mu m^2$ and that of the food filter $0.3 \times 0.3\ \mu m^2$ (Flood 1991) (Fig. 2.8).

In spite of the complexity and precision of a house, an *Oikopleura* can create up to 16 per day and 50 or more in one lifetime (Sato *et al*. 2001). The construction behaviour is also very simple. The organism essentially just lashes its tail to expand a capsule of mucus it secretes from glands covering its body (Alldredge 1976*b*; Flood and Deibel 1998) (Fig. 2.9).

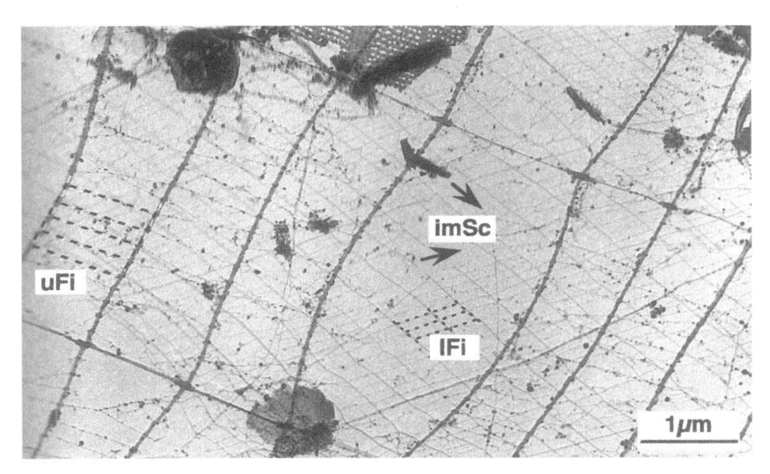

Fig. 2.8. The regular mesh of the mucus food capture net of the larvacean *O. dioica* derives from the properties of the material itself. Upper filter net = uFi; lower filter net = lFi; intermediary screen = imSc. (Adapted from Flood 1991 with permission from Springer-Verlag GmbH.)

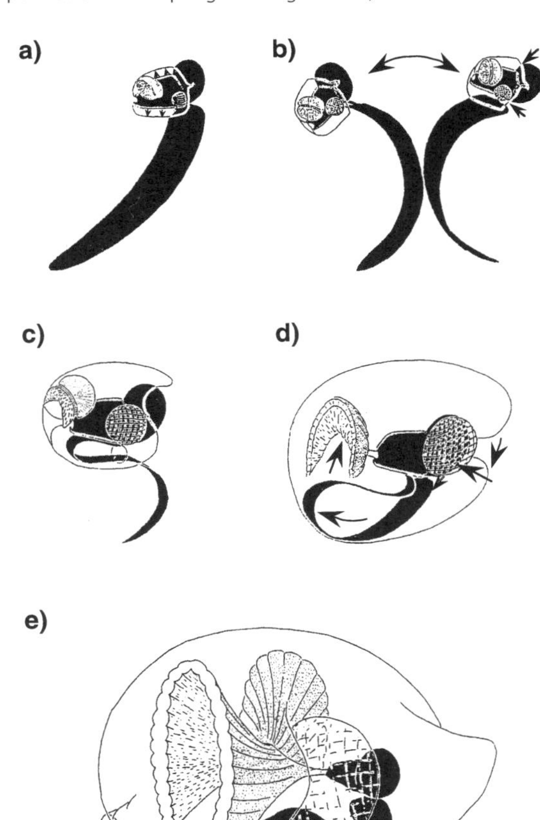

Fig. 2.9. The house of *Oikopleura* species can be built with very simple behaviour. The tadpole-like organism is shown in solid black. (a) and (b) The mucus capsule is secreted by the head and initially enlarged by up-and-down and back-and-forth movements of it. (c) and (d) The organism inserts its tail into the capsule and, with lashing movements, enlarges it to full size. (e) Organism in feeding position in the completed house. (Adapted from Flood and Deibel 1998.)

The ability of Larvacea to build their houses so rapidly and with such simple behaviour, results from the spatial arrangement of the mucus secreting glands on the body of the organism and the type of mucus each secretes. A monolayer of cells covering the trunk of the animal secretes the initially diminutive house. The positions of the inlet filters and food concentrating filters, for example, reflect the relative positions and patterns of the mucus cells that secrete them. The nets of these filters are not however spun by the cells themselves, but results from an as yet unknown self-assembly process (Flood and Deibel 1998).

Spada *et al.* (2001) found that the house of *O. dioica* incorporated at least 20 polypeptides, a number of them highly glycosylated, that is resembling proteoglycans. Three particular proteins, named *oikosin 1, 2,* and *3*, were matched to identified groups of cells on the epithelium. The arrangement of these cells matches some of the structures emerging in the expanded house. Oikosin 2 and 3, for example, are synthesised in regions of the epithelium that produce the mesh filters, although both are also secreted in other areas. This shows an architecture dependent upon a complex array of specialised secretions.

2.4.3 Silk

Silk is composed of polypeptides in the form of long unbranched chains, but what defines it as silk is that it is stored internally as a viscous liquid and spun into threads (Craig 1997). The extrusion process by which threads are created contributes towards the conformation of the protein molecules, which in turn has an important effect on the mechanical properties of the threads. Silk threads are produced by a wide range of arthropods (Figs. 8.5 and 8.6). The amino acid composition of silk secretions is variable across taxa but glycine, alanine, and serine predominate. Craig *et al.* (1999) identify three broadly different silk types, according to the proportions of these and some other amino acids. However, the distribution of these silk types across the arthropods has no clear phylogenetic pattern, although the majority use for them if for the building of various kinds of shelters or dwellings.

The long-chain polypeptides of silk adopt secondary and tertiary conformations of the protein molecules, which are crucial to their mechanical properties. Four different conformations are possible: α-helix, cross-β, parallel-β, and irregular. The α-helix is the form of the keratin molecule found in the hair and feathers of higher vertebrates, but the polypeptide chains can also be folded back on themselves, either in a direction across the axis of the fibre (cross-β), or in the direction of the axis (parallel-β). In the latter, the protein molecules may be aligned side-by-side in a parallel or antiparallel direction to form sheets which, due to the small size of the amino acid residues, can fit closely together to form highly organised crystalline domains.

All of these conformations are known from examples of insect silks, however, the most specialised silks are produced by areneoid spiders. Some of their threads combine α-helical, parallel-β, and amorphous domains in the same thread (Craig 1997). The spider thread is therefore a composite material with crystalline and amorphous components, giving the threads particular advantages in prey capture.

The labial silk glands of Trichoptera and Lepidoptera, which are homologous structures, are paired tubular glands in which two distinct secretions are produced; silk fibroin and sericin (Sehnal and Akai 1990). In Lepidoptera the fibroin forms the core of the thread, while the sericin forms a complex coating which contains lipids and polysaccharides, as well as other polypeptide molecules (Magoshi *et al.* 1994). The most common function of silk produced by these two insect orders is in the construction of larval and pupal shelters.

The cocoon silk of the silkworm *Antheraea pernyi* is rich in polyalanine sequences, and some other unique repeated sequences. This is, however, quite different from that of *Bombyx mori*, a member of another 'silkmoth' family. It is largely composed of glycine and alanine, which occur in multiple repeats of the hexapeptides GAGAGS and GAGAGX, where G = gyycine, A = alanine, S = serine, and X varies between tyrosine and valine. These sequences spontaneously form crystalline β-pleated sheets, which together make up 40–50% of the volume of the *B. mori* silk fibre. These crystalline blocks appear to be linked by short chains (30 residue) non-repetitive sequences, in an unstructured or amorphous arrangement. This model of the silk thread composed of the two domains, crystalline and amorphous, is referred to as the *network model* (Gosline *et al.* 1999).

In the advanced spiders (Araneomorphae), the anterior median pair of spinnerets may form the cribellum, which produces the capture threads typical of the cribellate spiders, or as in the Araneidae be absent (Opell 1995; Bond and Opell 1997). The spinnerets in Araneid orb makers are linked to seven glands each producing a secretion with a distinctive amino acid composition, five concerned largely or exclusively with prey capture, one concerned exclusively with home building (Vollrath and Knight 2001) (Fig. 2.10). The product of the major ampullate (MA) gland is used for dragline or frame threads, that of the minor ampullate (MI) gland for radii and other structural roles, while flagelliform gland (Fl) silk produces the core of the capture spiral thread (Vollrath and Knight 2001).

The strength of MA silk is very high for a biological material, and comparable with, although less than, that of high tensile steel. However, this comparison is misleading because, in other fundamental respects the mechanical properties of MA silk are quite unlike that of high tensile steel (Table 2.1). The steel wire has a stiffness about 20 times that of *Araneus* MA silk thread. MA silk has a maximum extensibility of 27%, whereas that of high tensile steel is

Gland	Function	Composition
Major ampullate	Dragline and frame threads	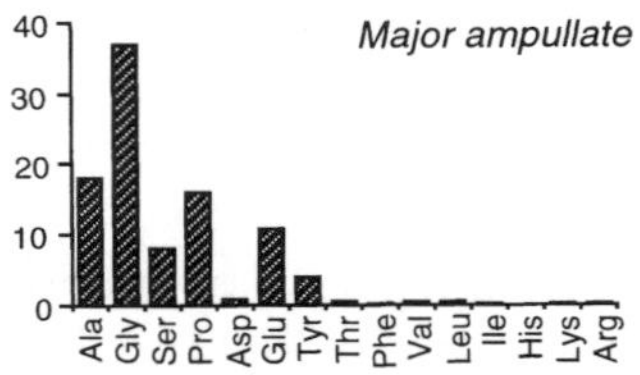 *Major ampullate* 0 10 20 30 40 Ala Gly Ser Pro Asp Glu Tyr Thr Phe Val Leu Ile His Lys Arg
Minor ampullate	Additional structural threads and radii	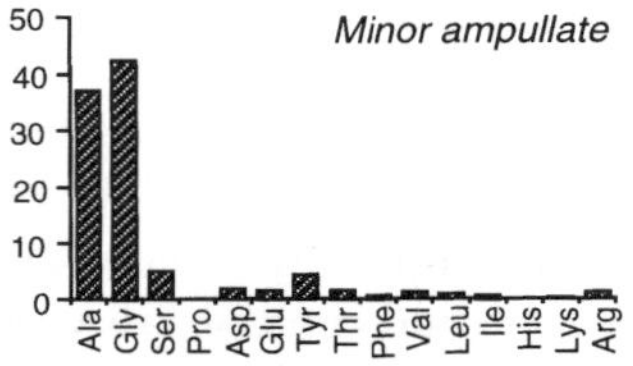 *Minor ampullate* 0 10 20 30 40 50 Ala Gly Ser Pro Asp Glu Tyr Thr Phe Val Leu Ile His Lys Arg
Flagelliform	Core of the capture spiral thread	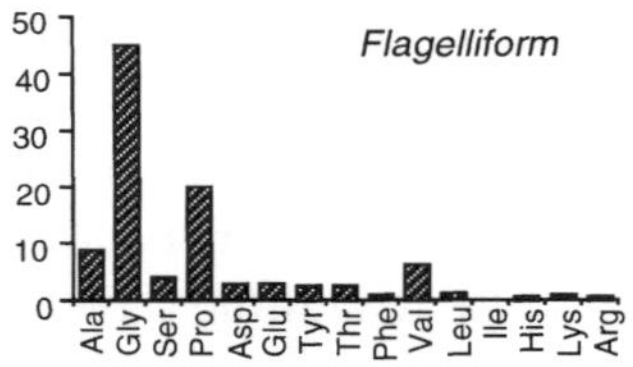 *Flagelliform* 0 10 20 30 40 50 Ala Gly Ser Pro Asp Glu Tyr Thr Phe Val Leu Ile His Lys Arg
Aggregate	Coating of capture spiral thread	Significant amounts of 15 different amino acids. Other substances include glycoproteins for stickiness and hygroscopic organic compounds
Pyriform	Attachment or joining of threads	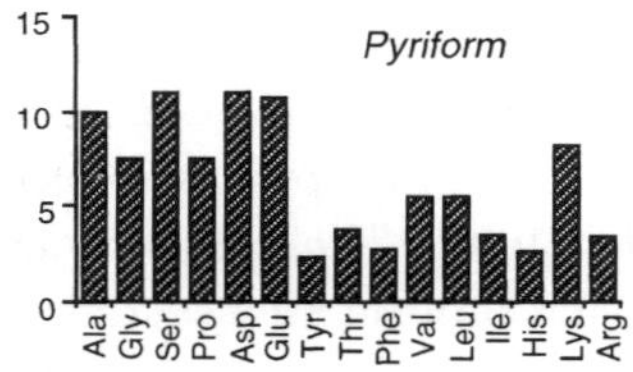 *Pyriform* 0 5 10 15 Ala Gly Ser Pro Asp Glu Tyr Thr Phe Val Leu Ile His Lys Arg
Aciniform	Wrapping prey	Significant amounts of 15 different amino acids
Cylindriform	Egg sac covering silk	Significant amounts of 13 amino acids with alanine, glycine, serine, and glutamic acid predominating

Fig. 2.10. The amino acid compositions of the specialised secretions produced from the silk glands of a typical Araneid orb web spider. (Adapted from Vollrath and Knight 2001.)

Table 2.1. Mechanical properties of spider and lepidopteran silks in comparison to high tensile steel

Organism	*Silk type*	*Stiffness,* E_{init} *(GPa)*	*Strengths* σ_{max} *(GPa)*	*Extensibility* ε_{max}	*Toughness* *(MJm^{-3})*	*Hysteresis* *(%)*
Araneus	MA	10	1.1	0.27	160	65
Araneus	FL{viscid spiral}	0.003	0.5	2.70	150	65
Araneus	MI		1.2	0.40		56
Nephila	MA	22	1.3	0.12	80	
Bombyx	Cocoon	7	0.6	0.18	70	
Steel	High tensile	200	1.5	0.008	6	

Data from Kohler and Vollrath (1995) and Gosline *et al.* (1999).

less than 1%. High tensile steel, with its low extensibility, is very suited to the construction of suspension bridges, but the extensibility of MA silk gives it a toughness much greater than that of the steel, and well adapted to task of absorbing the kinetic energy of flying or struggling prey.

In comparison to MA silk, the viscid silk produced by the flagelliform gland has only half the strength. Flagelliform gland silk also differs in having a much lower stiffness, with an extensibility of 270% before failure; the consequence of this is that the area under the stress/strain curve, the measure of its toughness, is comparable to that of MA silk.

The measures of strength given in Table 2.1 are obtained by the application of a load in the direction of the axis of the thread. However, the loads that orb web threads have to bear are commonly normal (at right angles) to the axis of the thread. This makes a substantial difference to the maximum load that can be borne by the thread, which depends upon both the load and the deflection angle. If the load is imagined to be a spider hanging from a thread that when unloaded would be horizontal, then the force acting upon the silk is given by $F\,silk = l/2\,\sin\theta$, where l is the load induced by the hanging spider. This shows that where θ is small because of low thread extension, $F\,silk$ will be largest, but that, as the thread extends, the force on the thread decreases as the direction of the force approaches more closely to that of the axis of the load (Fig. 2.11) (Gosline *et al.* 1999).

All other things being equal, this would mean that the more the thread was deflected, the greater the load it could bear. However, extension of the thread also causes it to narrow, so reducing its strength. Denny (1976) showed that load bearing would be maximised by fibres having an extensibility of 42%. MA silk, with an extensibility of 27%, is less than but similar in magnitude to that figure (Gosline *et al.* 1999). The orb webs of *Araneus* are oriented vertically to intercept flying prey, which will therefore tend to impact web threads at right angles to their axis (Fig. 2.11), so again the extensibility of the web threads helps to maximise the size of prey that can be captured.

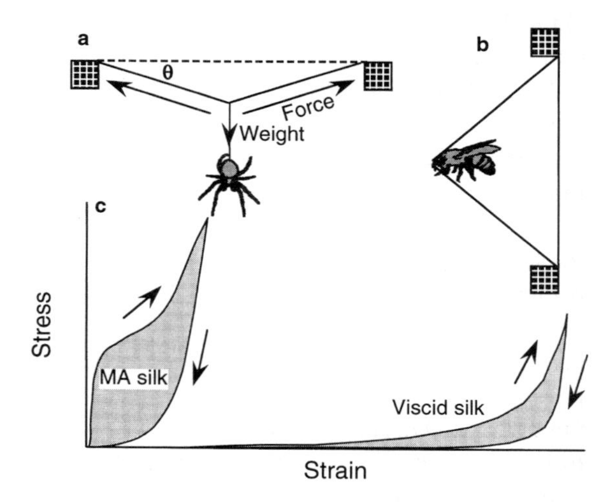

Fig. 2.11. Load bearing and energy absorption by major ampullate (MA) gland and viscid spiral silk of the orb web spider *Araneus diadematus*. (a) When a static load is applied at right angles to a fibre, a large angle θ due to extension, allows a larger load to be supported. (b) Thread extension and deflection on the impact of flying prey. (c) The stress–strain curves of MA and viscid capture silk threads. Shaded areas show energy lost in the load–unload cycle due to hysteresis. (Redrawn from Gosline *et al.* 1999.)

The silks of orb webs have a further property that helps them to fulfil their prey capture function, their viscoelasticity. Both MA and viscid spiral threads extend elastically almost to the point of failure under stress. If all the energy of impact of flying prey was stored during the elastic extension of the threads, there would be a danger that on recoil, the prey would be flung back off the web. But both MA and viscid spiral threads show 65% hysteresis (Fig. 2.11), dissipating the majority of impact energy as heat. This is achieved by the viscoelasticity of the thread material, a property derived from its molecular structure. The radial threads of *Araneus diadematus* produced by the minor ampullate glands (MI silk) have similar properties (Kohler and Vollrath 1995) (Table 2.1).

It is also a feature of the viscoelastic extension of these threads that their effectiveness in the mechanical properties they exhibit depends upon the rate of application of strain. The figures given in Table 2.1 show values for static application of loads, but for *Araneus* MA thread, the values of stiffness, strength, extension at failure, and toughness can all be enhanced under conditions of rapid extension such as the impact of flying prey (Gosline *et al.* 1999).

Two fibroin molecules have been identified from the MA silk of both *A. diadematus* and *Nephila clavipes*. In *A. diadematus*, both molecules have a repeat sequence of over 30 amino acid residues in which the β-sheet, crystalline element is the product of polyalanine sequences of 6–8 residues and not, as in *B. mori*, of GAGA sequences. The remainder of the repeat of both molecules contains short sequences in which glycine and proline predominate as GPX, and GPGXX/GPGQQ, where X is one of a number of possible amino acids. These proline-rich regions do not take up the β-sheet conformation, but the pentapeptide sequences may adopt a β-spiral form (Guerette *et al.* 1996;

Hayashi *et al.* 1999; Winkler and Kaplan 2000). These molecular data provide a method of calculating the proportions of crystalline and amorphous regions in MA silk based on mechanical data. The prediction from the biomechanical data for a crystalline volume in *A. diadematus* MA silk of 20–25%, and amorphous region sequences of 16–20 amino acid residues, is reasonably closely matched by the molecular data (Gosline *et al.* 1999).

How exactly this relates to the mechanical properties of MA threads is unclear as the exact nature of the secondary and tertiary conformation of the protein chains themselves remains uncertain. Gosline *et al.* (1999) envisage that, in terms of the network model, an MA web frame thread under normal conditions of tension will have its crystalline regions aligned with their axes parallel to the fibre axis and linked by a network of amorphous chains, but Simmons *et al.* (1996) interpret nuclear magnetic resonance data from MA threads of *Nephila madagascarensis* as showing not only highly oriented crystalline regions that account for about 40% of the alanine-rich material, but also poorly oriented β-sheets that account for the remaining 60% of the alanine component. Both these elements, according to this model, are linked by a network of amorphous regions. However, the 'amorphous' regions may possess significant structure. For example, the GPGGX sequence, to which Hayashi and Lewis (2001) attribute the ability to form a β-spiral conformation in flagelliform silk (see below), is also found in *A. diadematus* and *N. clavipes* MA silk.

The prediction of specialisation of capture silks is supported by the distinct differences between MA and MI silk in both *Araneus* and *Nephila*. MI silk of *N. clavipes* is composed of two protein molecules, both dominated by glycine and alanine, which occur in GAGAGA repeats typical of *B. mori* silk, and of GGX. These repeat units are combined in repetitive regions separated by spacer regions of about 140 residues. Neither the spacer nor the GGX regions can form β-sheets so in the MI silk, the GAGA domains seem largely responsible for the formation of the crystalline component. In *A. diadematus* this is estimated at 68% of the volume, with short amorphous regions contributing to the remaining 32%, with similar proportions shown in *N. clavipes* (Colgin and Lewis 1998; Gosline *et al.* 1999). Proline, such a prominent feature of MA silk, is absent from the MI fibroin molecules of *N. clavipes* and *A. diadematus.* The relatively large volume of the crystalline regions of MI silk embedded in the GGX and spacer regions result in a thread which is similar in strength to MA dragline thread (Colgin and Lewis 1998).

The third major thread of the orb web is flagelliform gland silk, the exceptional extensibility of which is also accompanied by special molecular features. The fibroin molecule identified in *N. clavipes* flagelliform silk is composed largely of repeated regions, each with the same subrepeat. In *N. clavipes* the repeated region begins with about 65 repeats of the pentapeptide GPGGX, and about eight repeats of the glycine-rich 'motif', GGX. Inserted in this there is also a spacer sequence of 28 amino acids, quite unlike other typical silk proteins (Hayashi and Lewis 2001). Also, peculiar among spider silks, there is an absence of poly-GA or AA sequences, required to create crystalline regions. This means that an explanation other than the network model needs to be invoked to account for the special qualities of extensibility, high hysteresis, and toughness shown by the capture spiral. Hayashi and Lewis (2001) develop the model outlined in Hayashi *et al.* (1999), that GPGGX repeats can take up a spiral conformation, with intermolecular bonds to stabilise the turns of and layers of the configuration, and with the fibre axis in the direction of the spiral progression. This β-spiral could account for the mechanical properties of the thread, that is, initial elongation with little change in stress, then steep rise before fracture (Fig. 2.11).

The double-stranded, flagelliform gland thread of the capture spiral of ecribellate spiders is coated with a water soluble secretion from the four aggregate glands (Peters 1987; Vollrath and Knight 2001). This coating of the core fibre seems to perform a major role in keeping it hydrated. The coat is 80% water content (Edmonds and Vollrath 1992) and in its absence, the elasticity of the capture thread protein is greatly reduced. Thread coated with aggregate gland secretion immediately assumes its new equilibrium value under tension, whereas dry thread shows a greater stiffness, reaching equilibrium only gradually. In *Araneus* and *Argiope* hygroscopic compounds such as choline chloride and *N*-acetyltaurine are in solution in the thread coating, helping to keep the core threads moist (Vollrath *et al.* 1990; Townley *et al.* 1991).

This coating, in addition to preserving the elasticity of the capture thread, serves the function of keeping the thread sticky. The coating is unstable and, through surface tension, forms into droplets (Edmonds and Vollrath 1992), and each droplet protects the all-important adhesive glue, a glycoprotein ring which envelopes the core thread within the droplet (Vollrath 1991; Vollrath and Tillingast 1991; Peters 1995).

The capture material of cribellate spiders is dry and restrains prey in a quite different manner from the sticky thread of ecribellate species. The material is in the form of a mass of fine threads each emerging from a separate tiny spigot located on the cribellum. This cloud of fine silk fibrils is collected by a comb device, one on each metatarsus of the fourth leg, the *calamistrum* and laid onto a paired core thread. Probably due to these combing movements, the cribellar threads are laid down as a regular series of bulging masses sometimes referred to as *cribellar puffs*. The amount of cribellar silk on the capture threads varies with the web type, being greatest in species building simple webs but least in orb web builders (Opell 1994; Craig 2003).

The nature of the cribellar silk capture principle and its effectiveness was investigated in the orb web of *Uloborus glomosus* and the simpler triangular web of *Hyptiotes cavatus*. This showed as expected, that the threads were least effective at holding onto moth wings because of their detachable scales, but could ensnare the setae of other flying insects. However, another capture principle was also found to be involved, since the threads attached strongly to smooth surfaces such as lady beetle elytra. The breaking energy of cribellar silk is similar to that of the viscid capture thread of ecribellate spiders, but is achieved in a different manner. It has a higher tensile strength, but has a higher Young's modulus and shows less extension before breaking than its ecribellate equivalent (Opell and Bond 2001).

2.5 Assessment of predictions

This review of materials reveals problems in assessing predictions made at the start of the chapter. First, it reveals a sampling bias, weighted strongly towards self-secreted trap materials, particularly silk. Second, apart from research on spider orb web silk, no detailed connection has been made between material and function so, although some self-secreted components of waxes or mucus have been identified, we have little idea of what they are designed to do. An exception is paper composition in relation to mechanical demands in the nests of the wasps *Vespula* and *Dolichovespula* (Cole *et al.* 2001).

Of materials in compression, very little is known. The materials of mound-building termites, creators of the largest structures in compression by non-humans, are known to use a combination of collected and self-secreted materials, but we know little beyond that.

In spite of these reservations it is still possible to make some evaluation of all the predictions. The first prediction, that self-secreted materials will exhibit greater quality control and be more uniform than collected materials, is not convincingly born out, at least not without qualification. There is evidence of a degree of uniformity in both kinds of building material, while more surprisingly, both also provide examples of flexibility.

The apparent differences in the composition of mud between cliff and barn swallows (Kilgore and Knudsen 1977), show that what might appear to be a crude category of material, may for the builder have rather narrowly specified optima. When unable to locate optimal materials, there is evidence among collectors of relaxation of selection criteria, as early experiments on the 'plasticity of building instinct' in caddis larvae showed (Copeland and Crowell 1937).

For self-secreted materials, it is evident that there is adaptive detail at the molecular level. Even when there is uncertainty in the identity of the materials themselves, there may be evidence of specialised secretory cells, for example, in the different regions of the trunk of the appendicularian *Oikopleura* (Spada *et al.* 2001). Even so there is also evidence of flexibility in the synthesis of secreted materials. *A. diadematus* individuals may exhibit differences in the amino acid composition of dragline thread on different occasions.

Vollrath (1999) uses this in support of the argument that this is due to adaptive responsiveness to a range of mechanical demands, indicating that dragline is more of a generalist than a specialist thread. Craig *et al.* (2000), however, provides experimental evidence that within-individual variation in silk composition in *Argiope* species can be due to current diet. *A. keyserlingi* has higher levels of serine to glycine in dragline silk when fed on crickets (*Acheta*) than when fed on blowflies (*Lucillia*). This could indicate a trade off in which silk quality is compromised in order to capture a more abundant prey, or that these components are non-vital.

It is interesting to note in this context, the virtual absence of several 'essential' amino acids (e.g. Leu, Lys, Phe, Thr) from the much used secretions of the MA and flagelliform glands in *A. diadematus*, particularly when compared with significant presence of these amino acids in less utilised pyriform gland secretion (M.W. Kennedy, personal communication) (Fig. 2.10). This might be construed as either conservation of scarce amino acids or adaptive design at the molecular level. Craig (2003) takes a different view on what is an 'essential' amino acid. In her view, if an amino acid is 'essential', then the animal will pay the price of synthesising it; ones that it does not synthesise are not essential. By this reasoning, those amino acids that are found to vary in an individual's silk are the components that are not critical to its effective operation.

Craig *et al.* (1999, 2000) explain the variation on the composition of spider silk compared with lepidopteran silk on differences in their diet. They argue that the predictable leaf diet of silkworms *B. mori* permits the production of a silk of uniform composition across the species, while spiders may be constrained by the proteins available from their most recent prey. The Trichoptera could provide a good test of this hypothesis since they, like spiders, continue to spin silk throughout their larval life (building either cases or nets), while some of these are herbivores, some detritivores, and some carnivores.

The prediction at the start of the chapter that self-secreted materials will be more prevalent in traps than in houses, while remaining tentative, looks valid. Animal traps are confined to invertebrates, and are composed almost universally of self-secreted capture threads, often organised into a net. Homes of invertebrates frequently incorporate self-secreted materials, but generally in combination with collected ones. Vertebrate home builders rarely use self-secreted materials, although many are builders.

Of the third prediction, it seems premature to conclude whether or not self-secreted trap materials exhibit greater complexity than do those used in making homes. The main problem here is the sampling bias. Craig (1997) points out that any silk type from webs of an araneid spider has greater complexity than the cocoon silk thread of a lepidopteran. Advanced spiders may also have up to nine different types of silk, the majority of which have a use in web construction; on the other hand, spider egg cocoons are not always simple structures and may contain as many as three different kinds of silk thread.

The basis of the prediction that self-secreted trap materials would be more complex than their home building counterparts was based on the dynamic nature of traps compared with houses. I still feel that this premise is generally true. However, the construction of the mucus houses of *Oikopleura* shows that this generality needs qualification. These houses have both a dwelling and a trapping role (Fig. 2.9) inviting the prediction that the trap materials will be more complex. But, in this instance the creation of the house, through a self-organising, expansion process, requires all the materials to respond dynamically at least during the construction phase, and for that reason to have rather similar levels of complexity. So, thinking of secreted materials only in their final role, neglecting the constructional one, may miss an important aspect of their adaptation.

3 Construction: behaviour and anatomy

3.1 Introduction

Construction involves the coordination of appropriate anatomy in effective action; this is the subject of this chapter. This effectiveness will be evident in movements of precision and of power, although the importance of each of these will vary greatly depending upon the method of construction. Many types of construction involve assembly of materials but some, notably burrowing, require the removal of material. This chapter characterises different methods or techniques of construction, using the schemes developed in Hansell (1984, 2000). Particular species tend to use exclusively or predominantly one of these techniques, but their full building repertoire may cover two or more. This chapter also examines predictions on the nature of specialisation in anatomy and behaviour associated with the different building methods.

Comparing predicted relationships with available evidence is not a strong test, but this is a field where it is difficult to apply an experimental approach. Accepting that there are limitations in the evidence, the aim here is to use the predictions to clarify important issues and point the way to future research. Three predictions relate particularly to the anatomy of construction, two to the behaviour. The predictions concerning anatomy are:

1. Morphological adaptations for construction will generally also be used in other functional contexts. This prediction is based on the broad evolutionary assumption that construction behaviour evolved from the use of anatomy already evolved for other purposes. It is a weak prediction, but it is important to examine, in order to judge whether there is likely to be a trade-off between the efficiency of construction and other behaviours.

2. Where construction anatomy serves other functions, its degree of specialisation for building will in general depend upon the relative proportion of its time spent in the other behavioural role. This was the argument put forward by Hansell (2000) for the lack of specialisation of bird beaks for nest building compared with feeding.
3. Anatomy associated with power will be more characteristic of construction by the removal of material rather than assembly. This was argued by Hansell (2000), who drew attention to the extreme morphological adaptations seen in some burrowing species.

The testable predictions made regarding behavioural specialisation for construction are:

4. Building behaviour should be characterised by elements that are discrete, simple, and repetitive. This prediction follows from the conclusion of Chapter 2 that materials are generally chosen to have narrowly defined qualities, and that this will allow simplification and repetition of the behavioural actions.
5. 'Getting started' behaviour will show greater variability than the behaviour of adding to an established structure. This argument was touched upon in Hansell (2000) in the construction of bird nests, where some architectural variability is evident where nests are fitted to the individual topography of a nest site, while the composition and construction of the nest as a whole is rather invariant and species-typical. At the behavioural level, this flexibility might be gained in more than one way. For example, by a larger repertoire of simple actions, or by greater variation within each action, possibly accompanied by modification through experience.

The relationship between specialised behaviour and specialised anatomy for construction is an important issue, but harder to predict. Specialisation in anatomy should constrain behavioural repertoire. Less specialised anatomy should permit a greater behavioural range, but there is no obvious reason why some of these behaviours should not also be highly specialised in their execution. Another important issue for which there is no clear prediction is overall repertoire size. If there is selection for standardised materials and simple actions, there should also be selection for a small behavioural repertoire. In the absence of any precise definition of 'small', what this chapter can do is note repertoire sizes used in construction, where clearly given.

3.2 Piling up: fetch and drop

The behaviour sequence *fetch and drop* can result in the building of a structure by the accumulation of objects deposited at the same location or in a particular relationship to one another. Crater-like rings are created round the nest entrances of some ants, for example, *Myrmecocystus* species by virtue of excavated soil particles being carried a short distance from the entrance in whatever direction and then dropped; cone-shaped mounds that form the nests of wood ants (*Formica rufa*) are accumulated plant fragments repeatedly dropped by returning workers (Wheeler 1910). The anatomy of the legs and mandibles that they use are identical to, and the behaviour similar to that used in foraging for food.

How such a simple procedure as fetch and drop might gradually create architecture, is examined by Franks *et al.* (1992) in the creation of an essentially two-dimensional nest space by the minute ants *Leptothorax tuberointerruptus* that live in narrow cracks in rocks. When introduced into a new cavity, each ant forager will bring in a particle in its jaws to the centre of the space, where its colony companions initially cluster. It then turns through 180°, moves outwards about one body length, and then drops the particle. As particles accumulate, a forager will use the grain held in its mandibles as a bulldozer, pushing other grains before it. Additional rules are needed to complete the enclosure, and site the entrance. These decision rules are the subject of Chapter 4, but this example shows that simple, repeated elements of behaviour using generalised anatomy can build up structure. The horned coot (*Fulica cornuta*), nesting in Andean lakes creates a small island by the process of repeatedly dropping stones into the shallow water, on top of which a vegetation nest is built. Ripley (1957) reports an island estimated at more than 1.3 t of stones.

The yellow-head jawfish (*Opisthognathus aurifrons*), uses a combination of digging (Section 3.8) and piling up to create a masonry-lined vertical shaft in the substrate. A crater is created by digging out sediment with the mouth. Rocks retrieved from around the crater are then piled up within it to create a shaft, at the bottom of which a lower chamber is excavated. Finally, sand is distributed around the burrow with the mouth to level the surface (Fig. 3.1) (Colin 1973), an example of the mouth, an organ with other obvious uses, performing at least two distinct construction behaviours.

A number of species of ground-nesting birds use a piling up technique to create a nest (Hansell 2000). Others make nests in the form of stick platforms or cups in trees, for example, storks (Cicioniidae), pelicans (Pelicanidae), herons (Ardeidae), and eagles (Accipitridae). In at least some of these, the construction behaviour does seem to be strictly fetch and drop, as, for example, in the little sparrowhawk (*Accipiler minullus*) (Liversidge 1962). However, black-billed magpies (*Pica pica*) and wood pigeons (*Columba palumbus*) both let many twigs fall in the early stages of constructing such nests (*personal observation*), evidence that getting the nest established does pose special problems. Detailed examination of this getting started behaviour might reveal a degree of variability in this stage of the construction, where careful twig placement is required in a non-standardised configuration of branches.

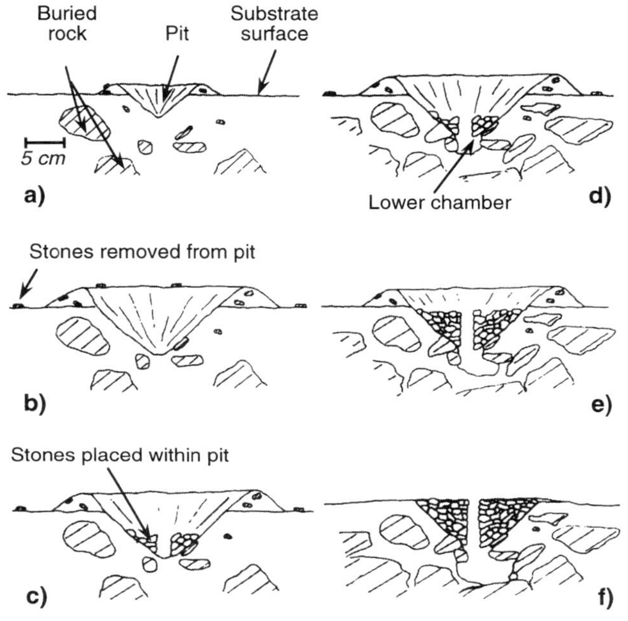

Fig. 3.1. The yellow-head jawfish (*O. aurifrons*) creates a retreat by a combination of excavation and assembly using its mouth. The initial crater is filled with rocks to produce a masonry-lined vertical shaft, at the bottom of which a cavity is finally created. (Adapted from Colin 1973.)

Some degree of interlocking may also occur in the construction of stick platform nests to help hold the structure together. The 300 or so species of pigeons and doves (Columbidae) show rather invariant nest architecture, and more than half of them build a platform nest in trees (Goodwin 1983). The nest of the spotted dove (*Streptopelia chinensis*) is a platform of twigs that is built up by the laying of twigs across one another. However, in the nest of the wood pigeon (*C. palumbus*), the projecting ends of the twigs are bent back into the body of the platform, probably giving it additional rigidity (Hansell 2000).

The building of the stick nests by larger Corvidae apparently involves behaviour that is more than simply dropping the twig in the required position. The black-billed magpie *P. pica* tests a twig in one position then another, accompanied by ventral or lateral shuddering movements of the body apparently in an attempt to entangle the twig more securely (Goodwin 1976). The nest of the white-tipped brown jay (*Psilorhinus mexicanus*), which is a platform of twigs with the finer ones located at the top, has a depression for the eggs in the middle. This is created, not simply by placement of the twigs with the beak, but also by shaping movements of the breast and scrabbling with the feet (Skutch 1960). The breast is in no obvious way modified for this purpose, while the feet and beak have predominantly non-building functions.

The conical-shaped house of the wood rat (*Neotoma micropus*) is created by the piling up of a variety of materials including pieces of *Opuntia* cactus, cattle dung, sticks, and other plant materials, under which the rat excavates a system of short tunnels and chambers (Thies *et al.* 1996). The building of dams and lodges by the beaver *Castor fiber* also includes elements of piling up but performed in rather different ways depending upon the material concerned. Wilsson (1976) describes beavers placing pieces of branch on top of the dam at the lowest point of its profile, in order to curtail the water flow at that point. However, construction of the dam also involves other placing behaviours such as the insertion of sticks, end first into the streambed to act as buttresses on the downstream side (Richard 1955).

On the upstream, side of the dam, beavers place mud to create a water resistant seal. This also can be said to employ a fetch and drop technique but in this case the material is collected from the floor of the lagoon by being scooped up on the backs of the 'hands' (Wilsson 1976). The anatomy of these front feet is unusual in being adapted not simply for carrying mud but also small branches held between the back of the hands and the chin while walking bipedally. Beavers are also able to employ a form of opposable grip for manipulating sticks. Teeth and hands are therefore adapted to the roles of collecting and placing of building materials, however, it should be remembered that beavers cut down and collect branches on which to feed, and handle small branches in order to feed on the bark. Feeding and building therefore utilise essentially the same anatomy, and share common behaviours (Richard 1958).

3.3 Interlocking and weaving

Bringing building materials together may not be sufficient to ensure that the structure remains intact, fastening mechanisms may also be required. This section describes construction processes by which pieces of building material become fastened together without the use of an adhesive. The nature of the fastening depends upon a combination of the actions of the builder and the nature of the material, the contribution of one being essentially the inverse of the other. At one extreme are materials that join so readily that they require little more behaviour than being brought into contact with one another, while at the other, the pieces of material can only be made to stay together by means of elaborate fastenings created by the builder.

Hansell (2000) divides *interlocking* into three separate techniques. The first is *Velcro* fastening, a title which implies correctly that the fastening is in the material rather than the behaviour; the other two are *stitches, and pop-rivets* and *entangle*, which place more stress on the creation of a fastening by the builder. To these is added *weaving*, which exemplifies the most complex fastening behaviour. Overwhelmingly, examples of these techniques are provided by birds, where the dominant construction tool is the beak, an instrument that shows morphological adaptation predominantly to feeding rather than building (Hansell 2000).

3.3.1 Velcro fastening

The principle of the Velcro fastening, invented by the Swiss engineer George de Mestral, is that the hooks covering one surface readily become ensnared in the loops covering the other surface when the two are pressed together. Two features of this fastening have made it popular with humans. First, the materials can be repeatedly fastened and unfastened with minimal decline in their effectiveness, a feature that is very convenient for clothes, and second the fastening process is very easy to perform, easy enough for children, for example. Both these features would make Velcro fastenings very suitable for animal building both during initial construction and subsequent repair and maintenance, provided appropriate hooked and looped materials were available. They prove to be, and are respectively plant materials and spider silk, a combination used only, it seems, by birds (Hansell 2000).

Arthropod silk is a strong material (Chapter 2), strong enough that it can be a suitable building material for birds building nests under about 30 g (Hansell 1996*a*). Silk is used as a structural material in the nests of some species of at least 25 of the 45 (= 56%) of passerine families. The way in which this silk is used is overwhelmingly to provide the loops in a Velcro fabric. As a result of this, spider egg cocoon silk, which is in many species a loose tangle of coarse silk, is popular, and lepidopteran silk is also effective (Hansell 2000).

The hooked material in animal Velcro fabrics can be any material that has plenty of small projecting processes to become ensnared. In the case of the long-tailed tit (*Aegithalos caudatus*), it is sprigs of small-leaved mosses, for the bushtit (*Psaltiparus minimus*) it includes pieces of a distinctive lichen bearing numerous stiff wiry projections (Hansell 2000).

Evidence that construction behaviour is indeed simple when using Velcro as the interlocking principle is indicated by the surprised conclusion of Tinbergen (1953) that the nest of the long-tailed tit, in spite of its complex appearance, is built using rather few and evidently stereotyped movements. 'Few' appears from this account to be about 15 distinct actions, leaving aside those concerned with gathering the different materials. Skutch (1954) describes the female of the white-collared seedeater (*Sporophila torqueola*), which in spite of having a beak that seemed little suited to delicate manipulation, was able to secure the initial nest attachments to supporting twigs simply by wrapping silk around their surface. The goldcrest (*Regulus regulus*) begins its nest by creating a framework of silk strands, which are carried in the beak between neighbouring twigs and wrapped round them. Within this is then built a hanging cup of moss and silk, finally shaped by vigorous breast movements and scraping with the feet (Thaler 1976).

3.3.2 Stitches and pop-rivets

These are related, discrete fastenings made with fine fibres. The common tailor-bird (*Orthotomus sutorius*) forms its purse nest by linking together the margins of green leaves with a yarn of arthropod silk and plant down. The stitches are made by driving the thread through the leaf, grasping it on the other side, and drawing it through again. As only a handful of stitches are required to secure each leaf margin, relocating the thread on the other side of the leaf may not be difficult. Certainly an almost identical fastening with a similarly mixed yarn is made by species of cisticola and prinia, for example, the golden-headed cisticola (*Cisticola exilis*) (Hansell 2000).

The grey-headed camaroptera (*Camaroptera brevicondata*) uses a combination of stitches and scores of plant down pop rivets driven through the leaf surface, to hold the structure together. The nest of the little spider hunter (*Arachnothera longirostra*), which is in the form of an egg chamber and entrance tunnel of skeletonised leaves, is suspended underneath a large leaf by about 150 fine silk threads. These fastenings have been driven through the leaf from below, presumably by the fine, curved beak. Each thread ends in a knob of silk so that, after passing through the leaf membrane it is trapped by the recoil of the leaf tissue (Madge 1970).

The snapping shrimp *Alpheus pachychirus* is described as stitching together mats of filamentous algae by thrusting through them an algal thread held in the claws of the second pereiopods. The thread can then apparently be grasped again and pulled back through (Schmitt 1975), although how the shrimp is able to relocate the position of the end of the thread is not clear. These limbs seem highly suited to the task, but it should be noted that the second pereiopods of this *Alpheus* species are not in any obvious way different from those of non-stitching species. Read *et al.* (1991) describe in detail the second pair of thoracic limbs in *Alpheus heterochelis* as not only being long and fine but also having for crustacea a specialised carpus joint that gives exceptional mobility for use in probing the substrate, retrieving small food items and grooming (Fig. 3.2). This species shows no stitching behaviour.

3.3.3 Entangle

The modification of the piling up technique by the wood pigeon (*C. palumbus*) to bend twig ends back into the core of the nest, illustrates the basic element of this technique. With more flexible lengths of plant material the entangle technique is easier, and a survey of the nests of several hundred species of birds (Hansell 2000) reveals its widespread occurrence.

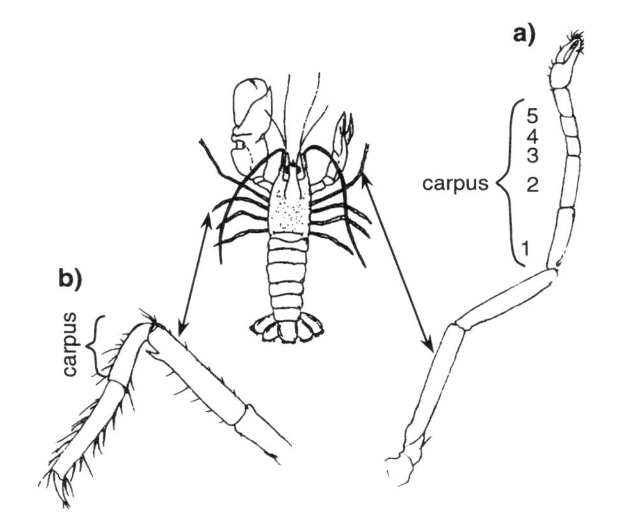

Fig. 3.2. A comparison of the carpus of the second and third pair of thoracic appendages of the snapping shrimp *A. pachychirus*. The former (a) has five segments, suitable for delicate manipulation, while the latter (b), which is adapted for walking, has one. (Adapted from Read *et al.* 1991.)

The primary nest building instrument for these birds is again the beak, and evidence also supports the view that behavioural elements are simple and repetitive. Aichorn (1989) emphasises the simplicity of the building behaviour shown by the female white-winged snowfinch (*Montifringilla nivalis*) in the construction of its grass and rootlet cup nest. She inserts materials into the nest with vibrations of the beak and shapes the cup with movements of the beak and feet. Having formed a loosely tangled hollow ball of material, the nest wall is thickened as she tucks in the material with rapid pushes of the beak.

Construction of nests by birds using woody vine tendrils provides an example where, unlike more flexible strands of vegetation, building elements brought together entangle readily and securely. Skutch (1969) described the rufous piha (*Lipaugus unirufus*) creating a nest platform of stiff, wiry vine tendrils by standing in the middle of an incipient nest, pulling the projecting ends of the tendrils towards her and tucking them in. These had a tendency to spring out again but, with the persistence of this simple behaviour, the bird created the nest platform. Using the entangling properties inherent in vine tendrils a number of bird species use this as their nest building material; for example, the wompoo fruit dove (*Ptilinopus magnificus*) constructs a skeletal platform nest, while the metallic starling (*Aplonis metallica*) creates a massive hanging bag of around 450 g (Hansell 2000).

3.3.4 Weaving

Collias and Collias (1964*b*) describe a variety of fastenings used by the village weaver (*Ploceus cucullatus*) to attach strips of vegetation to one another and to secure the nest in position. These include loops, half-hitches, hitches, bindings, slip knots, and overhand knots, as well as a more regular over-and-under pattern of weft through warp that defines human weaving. This broad characterisation of weaving by animals is the one adopted here. To effect these various fastenings requires the same thread to be placed around or through some existing structure, picked up again, pushed through and gathered possibly several times until incorporated into the structure.

The morphological requirement for a weaver is a bodkin-like structure that can grasp the thread and pierce the fabric. This needs to be accompanied by fine control of movement and acute sense organs to maintain the progress of a thread and so build up the fabric. All birds have this combination in the form of sharp beak, flexible neck, and acute vision, although only a small minority are weavers. There is currently no evidence that bird species which show weaving have any additional adaptation of beak, eye, or neck, but these may have so far gone unremarked because they are quite subtle.

There is, however, some evidence that the behaviour of weaving is not only complex, but also differs from that of foraging or grooming. This has not been analysed at the level of the basic elements of movement, although it could be argued that creating knots and hitches is likely to involve movements that would be unusual in foraging. Starting a nest, by attaching the first few strands to a twig, poses particular difficulties. Spiral wrapping around a twig may give temporary stability but it must be secured with a hitch or knot before being released. The red-billed weaver (*Quelea quelea*) achieves this by holding the strip of material in position on a branch with both feet while securing it with a hitch (Chapin 1954). Howman and Begg (1995) report that the formation of the initial attachment to a fine twig and the formation of an inverted-V of strands for the attachment of the hanging nest, demands great dexterity and may require almost a day to complete.

Experimental evidence of a difference in quality of nests completed by experienced and naive village weavers (*P. cucullatus*) shows that learning is required to perfect the weaving technique (Collias and Collias 1964*a*). An indication that this requires complex skill development would be a demonstration that mature birds, even though they acquired a similar range of weaving techniques, exhibited some individuality in the manner of their execution.

Weaving has arisen independently in two taxa of birds, the Old World weaver birds (Ploceinae) and the New World, oropendolas, orioles, and caciques (Icterini, Fringillidae). In the crested oropendola (*Psarocolius decumanus*), and the yellow oriole (*Icterus nigrogularis*), single strands of plant material are woven into the nest wall following a complex path. Examination of the fabric of these nests shows that weavers of both groups use a similar range of attachments (loops, loop tucks, hitches, and fine spiral coils; Heath and Hansell 2002). But this does not tell us how many distinct actions are required to accomplish this, nor how similar they are between individuals.

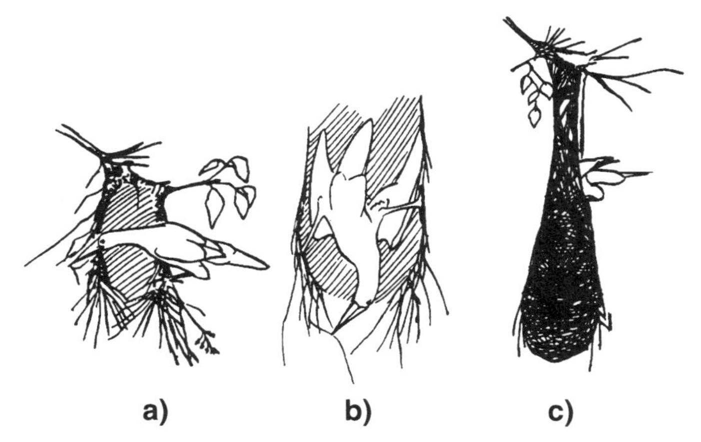

Fig. 3.3. A female crested oropendola (*P. decumanus*) uses the fine point of its large beak to weave vegetation strands to build its nest. (a) *Pull around and tuck* movement being used in the construction of the nest entrance. (b) Female inside the sleeve of the nest, weaving together strands to close the bottom. (c) Female perched on the rim of the completed nest, feeding the young. (Adapted from Drury 1962.)

Psarocolius decumanus during the initial stages of nest construction will use its feet to anchor newly gathered material, but employs distinct, repeated, short behavioural sequences using the beak to build up the nest fabric. These are described by Drury (1962) as *peck-pull-poke*, and *peck-pull around-tuck*. The fine control of movement in these sequences is revealed in Drury's description of the binding on the lower rim of the nest entrance as being *'buttonhole like'*, resulting from a vertical peck through the fabric, pulling of the strand over the rim and pushing it through again (Fig. 3.3). The level of motor control exhibited here is impressive, but as remarkable is the apparently small behavioural repertoire employed to achieve it.

3.4 Sticking together

In this construction method, fastening is achieved by means of an adhesive or glue, so there are two different sorts of materials, building units (bricks) and glue. The latter may be secreted by the builder or collected while initially plastic, and used before hardening. Two self-secreted materials are characteristic of this method of construction, mucus (in the broadly defined sense of Chapter 2) and silk, while mud provides an example of such a collected material.

Tube making by the polychaete *Pectinaria* (*Lagis*) *koreni* involves applying a mucus secretion of the cementing organ to mineral particles collected by the tentacles, and placing them on the anterior tube rim with the building organ

(Vovelle 1973). This combination of structure for collection of particles, dedicated cement producing gland, and construction organ is evident in at least three other families of polychaetes. These therefore show very simple kinds of construction behaviour but some dedicated construction anatomy.

A few vertebrates also use self-secreted adhesives, including fish and birds (Hansell 1984). Some swifts and swiftlets (Apodidae) stick together collected plant materials with an oral mucus secretion. The white throated swift (*Aeronautes saxatilis*) sticks together plant down and feathers, the chimney swift (*Chaetura pelagica*), cements regular straight pieces of twig (Pearson 1936, Hansell 2000); the cliff flycatcher (*Hirundinea ferruginea*) (Tyrannidae) has convergently evolved the same technique (Sick 1993). In all these species, manipulation of the collected material and application of the secreted material must involve that generalised organ of manipulation the beak, but I have been unable to find any detailed record of their building behaviour.

Silk is characteristically extruded as a thread, consequently the behaviour of sticking together must involve carrying the thread from one brick to another by the builder. In the case of the weaver ants this is slightly more complex since the silk is not in fact secreted by the worker ants that use it to stick leaves together, but by their larvae. In *Oecophylla* species the larva remains rigid and motionless, while moved back and forth held in the mandibles of a worker (Hölldobler and Wilson 1990). In *Dendromyrmex* species, however, it is the worker ants that remain immobile while holding larvae to the site where silk is to be applied, while it is the larvae, using movements similar to those seen in cocoon spinning, that secure the building units together (Hölldobler and Wilson 1983). In the related genus *Polyrachis* it is again the behaviour of the larva that secures the collected plant materials together. So, although the building behaviour of ants or larvae in these species is specialised, it is simple. The anatomy employed is, in the case of the mandibles and legs, also simple although specialised in terms of the silk secreting apparatus itself.

The sticking together of particles with silk to create a case takes a similar form in a variety of species of caddis larvae. In *Silo pallipes* a selected sand grain is held to the anterior rim of the case in different orientations, and the mouth parts applied to it. It may be manipulated by the legs into several different positions and tried in different orientations, which involve the rotation around and flipping over of the particle using all legs. If accepted, the particle is held in a characteristic 'fitting position' with the meso and metathoracic legs holding the sides and the prothoracic legs pulling it towards the case. The head is then bent between the legs to apply strands of silk back and forth to secure the particle to the case (Hansell 1968*a*). The number of different actions observed in this sequence is quite low, but the ordering of the elements seems quite flexible and deserves more detailed observation.

Case building caddis larvae normally build their first case on emergence from the egg, thereafter adding to the front end and periodically cutting a section off the back (Hansell 1968*b*, 1972). If taken from a case at any stage, larvae can rebuild, but this involves behaviours other than normal case extension. Bierens de Haan (1922), in studying *Limnophilus marmoratus*, regarded this as sufficiently different to be a separate mechanism (*covering instinct*) to rapidly fabricate a characteristic *provisional case*, while the '*building instinct*' was concerned with the construction of the *definitive case*. For the leaf panel builder *Lepidostoma hirtum*, the rebuilt case is initially of sand grains, a material typical of the first two instars, but of larger and more irregular size, before adding the characteristic leaf panel case after its body is covered with stones (Hansell 1972). So, in some caddis species there may be alternative building routines, although the number of behaviour patterns in each may be small.

3.5 Modelling

This describes the working of a collected or secreted plastic material into the desired shape. Hansell (1984) grouped this with the *extrusion moulding* of silk threads to create a broader category of *moulding*. Here, these are treated separately because in the present, more detailed consideration of building, extrusion is seen as a part of the secretion of silk material (Fig. 2.1). It leads to construction by deploying linear filaments to build up a three-dimensional structure, and is treated under the heading of *spinning* in Section 3.7.

A balance of great sophistication in the biochemistry of the building material coupled with simplicity of construction behaviour and anatomy was described in Chapter 2 (Section 2.4.2) for case construction by appendicularia. The complex case architecture of this organism derives from the pattern of secretory glands on the body surface of the organism. The construction itself amounts to little more than expansion of the structure by repeated thrashing of the larval tail (Flood and Deibel 1998) (Fig. 2.9), which has the more enduring task of maintaining filter feeding currents in the completed case (Alldredge 1976*b*).

Salivary mucus is used to construct the nest of the edible-nest swiftlet (*Collocalia fuciphaga*) (Medway 1962), but the more typical modelling material of birds is mud. Swallows and martins (Hirundinidae) build up the nest shape by the placement of large numbers of mud pellets. The modelling in this case is characterised by a vibrating movement of the head as each load is applied (Emlen 1954). The vibration of each new load probably causes partial liquefaction of the mud through the phenomenon of thixotropy. This disperses the moisture more evenly and allows the fresh mud to overrun small air spaces.

The sphecid wasp *Trigonopsis cameronii* prepares mud by regurgitating water from its crop before gathering up the damp soil. In applying the earthen pellet to build up the brood cell it adopts a relatively constant procedure in which the lump is placed on the nest and manipulated with the mandibles, while more crop water is added and the head vibrated rapidly. At the same time the wasp spreads the mud by opening and closing her mandibles while swinging her head backwards and forwards and emitting a soft buzzing sound (Eberhard 1974). The eumenine wasp *Paraleptomenes mephitis* constructs tubular mud cells by the addition of small mud pellets but, while this is revealed in the rough texture of the cell exterior, the interior surface is smoothed using the mandibles (Krombein 1978).

Although mandibles are in wasps, the organs of prey capture and feeding, there is evidence of their modification for use as building instruments. This is seen in differences in mandible design in Vespidae that model mud and those that model paper. Mandibles of the mud building Eumeninae are relatively long and flattened (a spatula form that may be suited to modelling), compared with the shorter, broader design seen in the paper making Polistinae and Vespinae. The length of the mandibles is expressed as the angle formed between them when closed. This is relatively small (45–70°) for the Eumeninae compared with around 80° for the Polistinae (Fig. 3.4) (Hansell 1987*a*). The short, broad mandibles of the paper makers (high width/length ratio) are adaptations to rasp the plant fibres. This is evidence of anatomy adapted for power, but in this case for material processing to create paper rather than for construction.

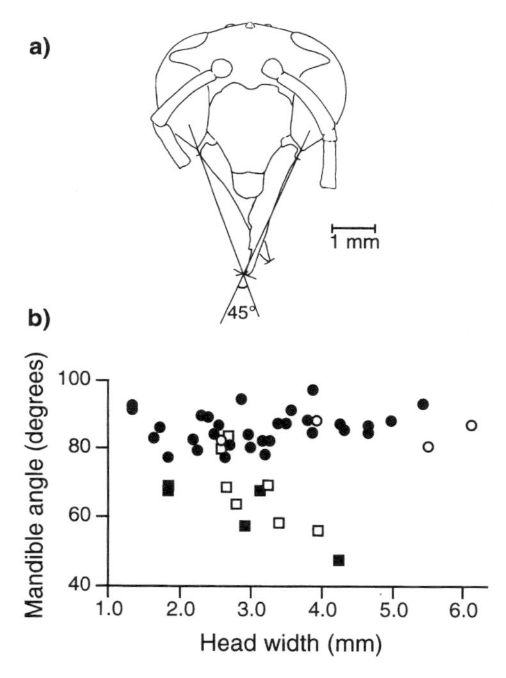

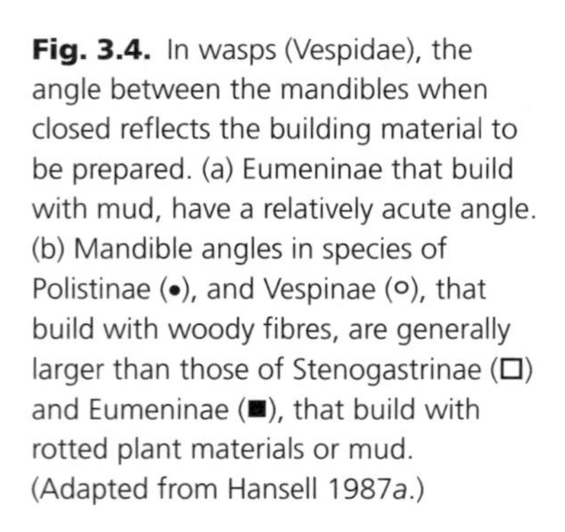
Fig. 3.4. In wasps (Vespidae), the angle between the mandibles when closed reflects the building material to be prepared. (a) Eumeninae that build with mud, have a relatively acute angle. (b) Mandible angles in species of Polistinae (•), and Vespinae (○), that build with woody fibres, are generally larger than those of Stenogastrinae (□) and Eumeninae (■), that build with rotted plant materials or mud. (Adapted from Hansell 1987*a*.)

The Stenogastrinae, a subfamily of the Vespidae, includes species such as *Stenogaster* sp. that build nests entirely or largely of mud (Hansell and Turillazzi 1991), and species that build with fragments of rotted wood, such as *Parischnogaster mellyi* and *Holischnogaster gracilipes* (Hansell 1981, 1986). This group has a mandibular anatomy intermediate between that of Eumeninae and Polistinae in terms of the mandible angle (Fig. 3.4), and a weaker mandible design (lower width/length ratio) than Polistinae and Vespinae (Hansell 1987*a*).

Wasps model paper into sheets to create combs of hexagonal brood cells or thin layers of envelope. The paper wasp *Polistes fuscatus* first places the pellet of macerated woody plant material fairly thickly along the top of the wall of the cell that is to be extended. This load is then worked to the required thickness with successive passes of the mandibles along the rim, each time reducing its thickness (West Eberhard 1969).

3.6 Folding and rolling

Sheets of material if treated in this way become three-dimensional structures, in which animals can live or use as a shelter for their offspring. Leaves provide such sheets of material and a few species, although of widely differing taxa, are able to utilise them in this way. Folding and rolling is not a technique suited to dry leaves as they would split and break up under the distortion. However, living leaves are specifically built to resist bending or crumpling, creating a problem for the builders to overcome. The rigidity of the leaves depends upon two factors, the strength of the veins and the turgor pressure in the cells. The builder must either weaken these, overcome them with a greater force, or use a combination of the two.

So-called tent-making bats use the behaviourally simple technique of biting into the veins of large leaves to create a perforated line along which the leaf folds under the force of gravity. The neotropical bat (*Aritebus watsoni*) creates a day roost in this way, using leaves of a variety of understorey plant species including *Carludovica palmata* (Cyclanthaceae) (Timm 1987). The basic elements of the behaviour range from localised biting to completely or partially chewing through leaf veins, however, the arrangement of perforations varies with plant species. The bifurcated frond of the palm *Astrocaryum standleyanum* is modified to the shape of an inverted boat by two rows of perforations in a V-shape, the angle of which is at the leaf base (Choe 1997). These basic tent-making techniques are shown by at least three other genera of New World bats of the Stenoderminae (Phyllostomidae) including *Ectophylla* and *Vampyressa* (Timm and Mortimer 1976; Timm 1984) and *Uroderma* (Choe 1994), and is also reported in two species of Megachiroptera (Timm 1987; Kunz *et al.* 1994). Whether the teeth in any of these species show any adaptations for tent

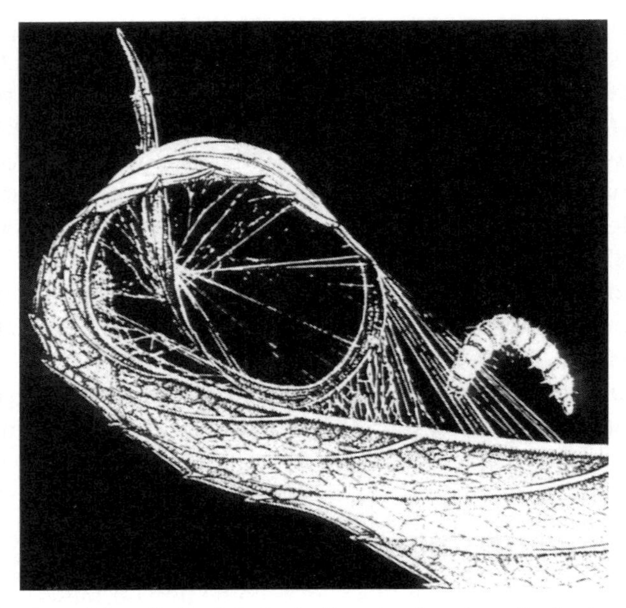

Fig. 3.5. Construction of the rolled leaf shelter by the caterpillar *C. serotinella*, showing the small size of the insect relative to the leaf. The contraction of successive silk threads gradually rolls up the leaf. (Adapted from Fitzgerald and Clark 1994 with permission of Kluwer Academic/Plenum Publishers.)

making is unrecorded, however, in *Cyanopterus sphinx*, only males seem to be the builders (Balasingh *et al.* 1995), so differences in tooth morphology between the sexes related to tent building may await description.

The folding and rolling of leaves by arthropods generally involves a combination of wilting the leaf, and bending it. The former is achieved with cuts and perforations, while bending is either achieved by muscular force or the use of contractile silk. Both spider and lepidoptera species show the ability to use the contraction of silk strands to roll leaves.

The lepidopteran larva, *Caloptilia serotinella* (Gracillariidae) rolls leaves to create a shelter and secure a supply of leaf tissue on which to feed, however, in this case the leaf is rolled from its tip against the force exerted by the midrib. To do this the larva spins a sequence of several hundred silk strands linking the outer surface of the leaf roll to the underside of the leaf (Fig. 3.5). Each thread is stretched by 13–14% by the spinning movement of the caterpillar, this creates traction forces of greater than 0.1 N. A feature of the spinning behaviour of the larva is a repetitive, rhythmic swinging of the body between the two anchor points of the thread (Fitzgerald and Clark 1994). This movement resembles the *'stretch-bend'* movement shown by the silkworm *Hyalophora cecropia* as a basic, repeated behavioural element, when building up a silken cone in which to pupate (Van der Kloot and Williams 1953).

The *C. serotinella* larva periodically interrupts the spinning behaviour to enter the leaf roll to bite into the leaf midrib, so facilitating the rolling process. As the leaf rolls up, threads that have previously been in tension slacken as the gap between the two attachment points shortens under the force exerted by new threads. The potential of the silk threads to supercontract if wetted does not seem to be involved in this leaf rolling process, although in mature leaf rolls most of the threads do seem to be in tension. This suggests that supercontraction does ultimately contribute to the rigidity of the leaf roll (Fitzgerald and Clark 1994). Using the same principle of accumulation of small forces, the aspen leaf rolling caterpillar *Pseudosciaphila duplex* generates forces of around 0.15 N, with an additional 25% increase in the force when the silk is wetted (Fitzgerald *et al.* 1991).

Rolling leaves to create a shelter is also shown by certain species of spider. The orbweb araneid spider *Phonognatha graeffei* spins silk between the edges of a leaf, using the contractile properties of the silk to create a roll. Thirunavukarasu *et al.* (1996) attribute the contraction to the drying out of the threads, although it seems more likely that it is supercontraction by water absorption.

The force to roll leaves generated by females of certain weevils of the family Attelabidae is not dependent upon silk, but their own muscle power. The purpose of their leaf rolling is to create a shelter and food supply for larvae emerging from eggs laid within the roll. Without the aid of silk, these beetles show quite elaborate behaviour aimed at minimising the forces needed to effect the rolling. Various genera are capable of this: *Hybolabus*, *Attelabus*, *Euscelus*, *Apoderus*, *Deporaus*, *Byctiscus*, and *Chonostropheus* (Sakurai 1988*a*,*b*; Zuppa *et al.* 1994).

A female of *Apoderus coryli*, which lays its eggs in hazel leaves, effects wilting by making a cut from the side of the leaf through the mid-vein to a point just beyond it. She then notches the secondary veins on the side of the leaf away from the cut to complete the wilting. Leaf rolling starts with the bending of the leaf along the axis of the mid-vein by pressing the two halves together at several places along the line of the fold. The double layer so created is then rolled upwards from apex to petiole, until it reaches the transverse cut. The eggs are laid in the leaf tissue just before rolling begins (Daanje 1957).

More complicated rolling procedures risk the creation of creases or crumples that obstruct progress. Some species exhibit rolling routines designed to overcome this problem. In *Deporaus betuliae* this is achieved by the path of the initial cut. After leaf inspection, the female cuts with her mandibles right across the leaf, leaving a diagonal perforated line that partially severs the mid-vein as the only connection between upper and lower parts of the leaf. Wilting is then accelerated by perforations using the claws over the whole leaf surface below the cut. The control of the rolling is assisted by the trajectory of the cut before the mid-vein. It closely follows the line of the *evolute* to the *evolvent* provided

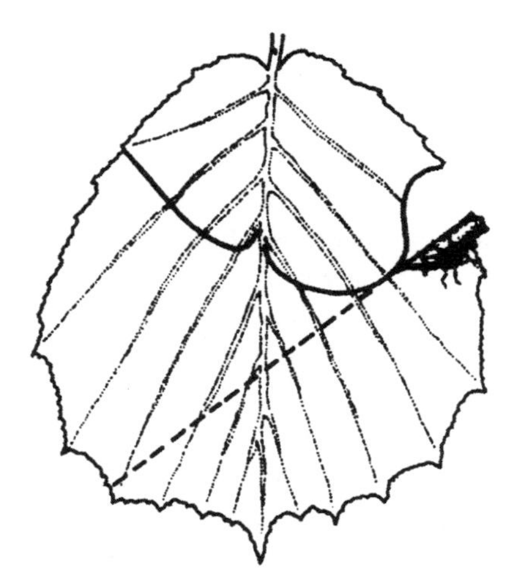

Fig. 3.6. Leaf rolling by the beetle *D. betulae* to create an oviposition site. The solid line shows the trajectory of the cut and the broken line shows the axis of the roll forming a tangent to the arc of the cut. (Adapted from Daanje 1964.)

by the shape of the leaf margin. The effect of this is that, as the axis of the roll sweeps round the edge of the leaf, the changing axes at the proximal end of the roll do not cross over and so cause crumpling (Fig. 3.6) (Daanje 1964).

These attelabids are all leaf-eaters, with mandibles equipped for the purpose, while the legs obviously have functions other than to bend leaves, so these are not anatomical features dedicated to leaf rolling. However, since females roll leaves and males do not, comparison of male and female exoskeleton could be undertaken to look for evidence of dimorphism indicative of adaptations for leaf rolling in the former. It is at least interesting to note that in European members of the genus *Apoderus*, females have wider heads than males and also have two distinct terminal claws to the anterior tibiae compared to only one in males (Daanje 1964).

3.7 Spinning

Silk spinners essentially produce a one-dimensional fine strand. It is then the deployment of these strands that builds up a two- or three-dimensional structure. This is illustrated particularly by the web spinning of spiders, and cocoon construction by the insect order Lepidoptera. The latter typically lay down silk threads, bending and stretching while standing at the same spot or while moving forward slowly; spiders, however, have evolved the ability to carry a single thread from one point of attachment to another some distance away, so building up a large array.

The caterpillar of the cecropia silkworm (*H. cecropia*) begins its preparation for pupation by spinning a cone-shaped silk frame, inside which the cocoon will be created. To do this the larva uses only two basic movements, described by Van der Kloot and Williams (1953) as *stretch–bend* and *swing–swing*. In the former, the caterpillar stretches up and attaches a thread to a point above it, then bends its body out to the side, securing the other end of the thread at the furthest point to the side that it can reach. Repetition of this at different angles to the side, builds up the cone. The base of the cone is created when the caterpillar adopts a head down posture within the cone, and performs the second movement to build up a flat sheet of silk.

Tent construction by the eastern tent caterpillar (*Malacosoma americanum*) (Lasiocampidae) has what appears to be a very similar basic spinning movement to create the walls and the sheets that divide the internal space into levels. A sheet starts off thin and transparent but becomes thicker and more opaque as it is added to by caterpillars moving over its surface laying strands in a repeated and regular pattern of spinning movements. A caterpillar swings the anterior half of its length to one side, returns to a straight body position and then advances, all the time spinning a silk strand. As this is repeated, sometimes with a swing to the left, sometimes to the right, with the caterpillar frequently changing the direction of its forward movement, a fairly even distribution of silk results (Fitzgerald and Willer 1983) (Fig. 3.7).

The swallowtail butterfly (*Papilio machaon*) pupates head-up on a plant stem, the posterior end of its abdomen secured in a silken mat, the body supported around the upper part of the body by a silk sling attached to a second mat. The sling and the mats are created by simple repetitive movements. The sling is formed by the larva carrying a thread from the upper mat over the side of its body and between the first and second pair of thoracic legs to attach it back where it started, so creating a large loop. The larva repeats this movement in the reverse direction, then back and forth about 25 times before inserting its body into the sling and pupating (Wojtusiak and Raczka 1983), in total a very small repertoire of behaviour.

The construction of a hibernaculum by the lepidopteran larva *Lemenitis populi* is clearly more complicated, and illustrates how the construction behaviour of one species may range over more than one category of construction technique, in this case, sticking together, folding and rolling, and spinning. In preparation for the winter, it performs a sequence of eight construction stages: (1) spinning a silken mat over a leaf, (2) cutting off a large piece of the silk-covered leaf, (3) moving the cut piece to the adjacent stem, (4) fixing the piece of leaf by one end to the stem, (5) folding the leaf to form a cavity, (6) adjusting the shelter to the larval body length, (7) aligning the case in an upright position, and (8) silking the whole length of case to the stem (Bink 1985). This may not convey the full behavioural range, because it appears that each of these stages may contain additional distinct actions. This is instructive since the sequence is a once in a lifetime performance, unlikely to depend upon more than a small amount of acquired skills.

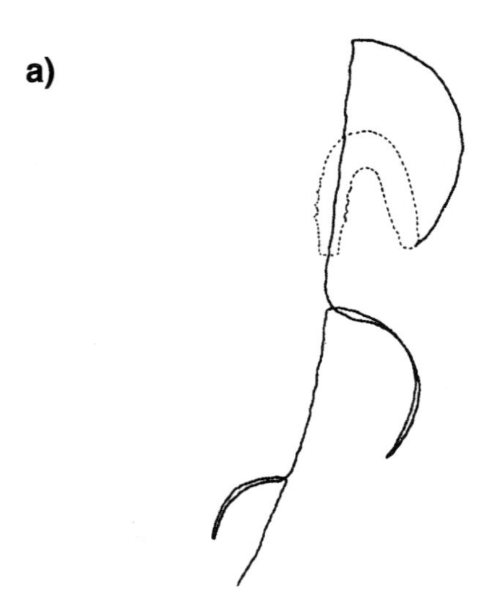

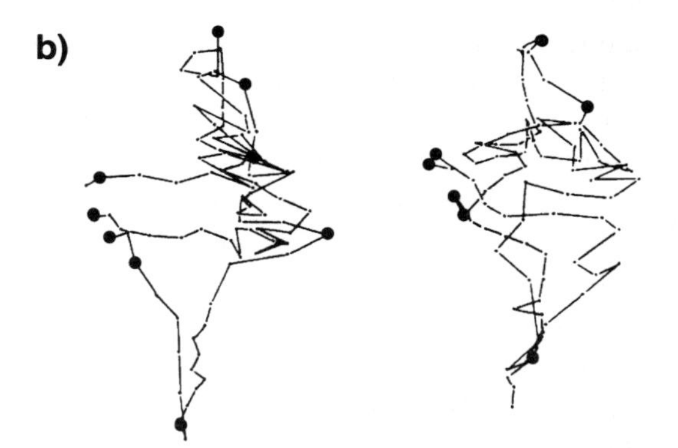

Fig. 3.7. Pattern of silk application by the eastern tent caterpillar (*M. americanum*). (a) Line of silk thread laid down by a single caterpillar as it moves over the tent wall. The dashed line represents the current position of the caterpillar as it spins the thread. (b) Pathways taken by two caterpillars while spinning silk to strengthen the wall. Dots indicate one second intervals; solid circles indicate the edges of the tent face. (Adapted from Fitzgerald and Willer 1983.)

A repetitive movement used by a number of silk spinning insects to build up a two- or three-dimensional structure is the *figure-of-eight tour*. For example, both swallowtail and peacock larvae apply silk to the pupal attachment site by repeated figure-of-eight movements of the head accompanied by small systematic body movement. In this way they build up a surface covering of silk (Wojtusiak and Raczka 1983). This same behavioural pattern is also reported for the construction of an enveloping cocoon by the caterpillar of *Antherea pernyi* (Lounibos 1975), by the larvae of ants of the genus *Formica* (Schmidt and Gürsch 1971), and by mantispid species (Order Neuroptera) (Bissett and Moran 1967).

It was pointed out by Streng (1974) that there are two different ways to execute the figure-of-eight movement, which he describes as the *positive* and *negative tours*. The difference is shown if you repeatedly draw the letter S without taking your pen from the pad. Moving your hand to the right while doing this, creates a positive tour, to the left the negative tour. There is a difference in outcome between these since, in the positive tour, the threads cross the x-axis of the progression at a more acute angle, bringing about an increased possibility of meshing. Streng (1974) therefore concluded that the positive tour ought to be more favoured by natural selection. The pupal cocoon of *Hydropsyche* is positive, as are the cocoons of *Formica* species (Wallis 1960; Schmidt and Gürsch 1971).

The figure-of-eight tour is not universal in the cocoon spinning of insects. The pupal cocoon of the aquatic dipteran larva *Simulium vittatum* is built up in a series of six characteristic stages, but the application of threads in each is from point to point, frequently in direct line or simple V-shapes (Stuart and Hunter 1995). These building stages show little variation across species and are preceded by a further six behaviours for cleaning and preparing the pupation surface.

The distance between the hub of the web of the New Guinea ladder web spider to its bottom edge is in the region of 900 mm, while the spider is only about 10 mm across (Robinson and Robinson 1972). Radii of this modified orb web span the distance from hub to web base as single continuous threads, illustrating the ability of spiders to deploy straight threads much longer than their own body dimensions. Spiders accomplish this in essentially three ways: by dropping vertically down, paying out a thread from the attachment point as in the above example, by drifting a thread on the air stream till it catches on something, or by using a detour technique.

The orb web spider *Araneus diadematus*, given a U-shaped frame in still air to preclude two of the options for thread deployment, will adopt the detour solution. It attaches a thread to the top of one side of the frame, walks down, across and up the other side paying out thread along the way. Finally tightening and attaching the thread, it has spanned the frame. To create an array of threads radiating from a hub, a further series of detours are employed, using the growing array of threads and frame uprights (Fig. 3.8) (Zschokke 1996). The sequence observed in the placement of radii is quite variable, even including the removal of some threads as well as their addition. This may reflect the getting started problem of having to adapt the web arrangements to available topography (Hansell 2000).

This detour routine can equally be applied to the building up of a three-dimensional web, provided a pathway between the two desired attachment points can be found. This ability has been confirmed for the construction of the domed sheet web of *Modisimus* sp., which includes a three-dimensional array of threads to support the sheet from above. This scaffolding of non-sticky dragline is built up and extended into a complex array by a fairly stereotyped sequence of attach, detour, tighten, attach (Eberhard 1992).

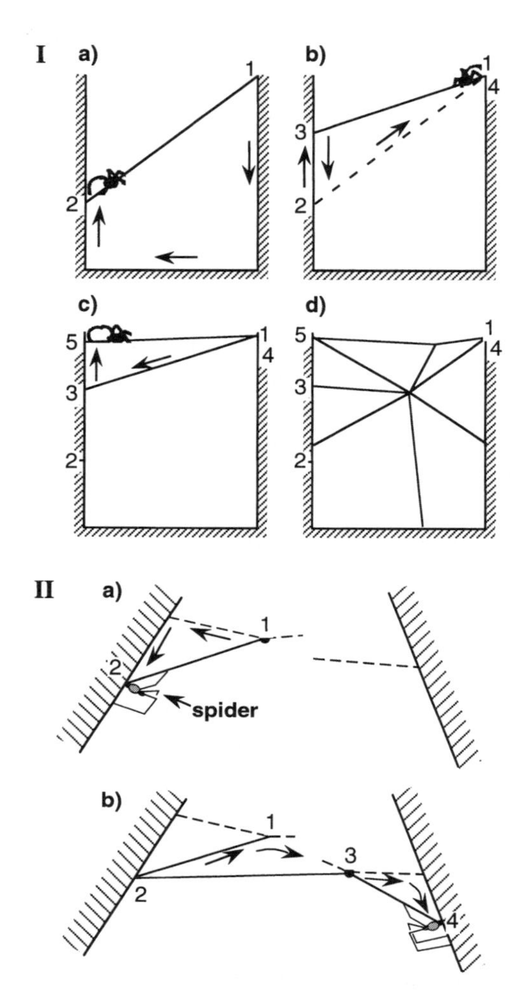

Fig. 3.8. The use of detour behaviour by spiders to deploy web threads. (Numbers denote successive points of attachment, and arrows, the path taken between the two points.) I. *A. diadematus*, constructing the radii of an orb web. (Adapted from Zschokke 1996.) II. *Modisimus* sp. building up a three-dimensional web by a series of detours. (Adapted from Eberhard 1992.)

3.8 Sculpting

This describes the formation of a structure by the removal or displacement of material from the initial mass. For the most part this is represented by species that make burrows. *Burrowing* in this context is therefore distinguished from *burying* where species dig themselves into a soft substratum that closes behind them as they progress (Atkinson and Taylor 1991). The latter is outside the concerns of this book.

Burrow digging species can be found among both vertebrates and invertebrates that have close relatives that do little or no burrowing. This provides an opportunity of testing prediction 3 of *Section 3.1*, that the degree of anatomical specialisation will depend upon the proportion of the animal's life that is spent in digging behaviour. If prediction 2 is correct, we should expect on the basis of prediction 3, that these differences will be particularly marked because the morphological specialisations of diggers include those for powerful movement. Another essential component of creating a burrow is the removal of spoil dislodged at the workface. In the introduction no prediction was made about anatomical specialisation for this, but pushing or carrying are two of the more obvious methods of achieving it.

Arthropods with their small size, hard exoskeletons, and jointed appendages should be well suited to burrowing. This is confirmed by the variety of aquatic and terrestrial arthropod burrowers. The small grapsid crab *Brachynotus gemmellari* excavates a simple blind, or U-shaped burrow in fine sediment using its chelipeds, plus second and third pereiopods. The spoil is then removed from the burrow by being carried between folded chelipeds, with the second and third pereiopods of the trailing side of the body assisting by forming the bottom of the carrying basket. Sediment dropped at the burrow entrance is bulldozed away with sweeps of the leading cheliped (Atkinson *et al.* 1998). The leading side of the crab may change in the digging sequence, so neither side is apparently specialised for carrying compared with the other.

In some crustacea, the large chelipeds have become sculpting instruments to excavate in solid substrates. Members of the thalassinid shrimp genus *Upogebia* typically excavate burrows in the mud (Nickell and Atkinson 1995), however, *Upogebia operculata* lives as a filter feeder in galleries it excavates in massive corals apparently using its first chelipeds to 'bite' through the coraline skeleton (Kleeman 1984). *Alpheus saxidomus* lives in cavities it excavates in rock. Shrimps of this genus are particularly known for the shock waves they use offensively and defensively by the snapping action of the major cheliped. However, abrasion on the dactylus of the claws of this rock-boring species appears to indicate that the snapping action is used to drive this 'hammer' to fragment the rock (Fischer and Meyer 1985).

Alpheid shrimps are more typically sediment burrowers, and can apparently show a degree of versatility in the use of their appendages. Karplus *et al.* (1972) describes sediment removal in *Alpheus djibontensis*. Digging is mainly carried out by inserting the first pair of chelae into the sand wall and twisting them to loosen the particles. Removal of dislodged particles is then achieved by rapid movements of the pleopods, with the posterior end of the abdomen directed towards the burrow entrance. Alternatively sand grains and small stones can be bulldozed with the first pair of chelae held together, their pushing surface apparently broadened for this purpose by rows of hairs fringing their dorsal and ventral margins. Larger stones and shell fragments are grasped by the first chelae and carried out. A similar variety of spoil removal techniques is also seen in other *Alpheus* species (Karplus 1987).

A number of orders of aquatic insects create burrows in the sediment. In some of these there is evidence of anatomy adapted for digging and/or spoil removal. In the Ephemeroptera, for example, large, tusk-like protruberances are clearly associated with burrowing in some genera. Nymphs of *Pentagenia vittigera* burrow within compacted clay while those of *Hexagenia limbata* inhabit soft silt. Both species show quite stereotyped and repeated cycles of movement, each adapted to their respective burrowing substrates. In *H. limbata* the head and prothoracic legs are the principle organs of substrate displacement, while the meso and metathoracic legs drive the body forward. The prothoracic legs also have the principal role of clearing the sediment from in front of the animal, while upward thrusts of the broad flattened head have the effect of anchoring the body. The tusks, which probably supplement the role of the head, are slender, smooth, and not heavily sclerotised (Fig. 3.9) (Keltner and McCafferty 1986).

Pentagenia vittigera uses upward sweeps of the heavily sclerotised, short, and serrated tusks, as the main instrument to dislodge the compacted clay. Its broad prothoracic legs push the loosened substrate back, at the same time widening the burrow principally with the spurs located on the dorsal margins of the tibiae and stout setae on the tibiae and tarsi of the forelegs (Fig. 3.9).

Burrowers in some softer aquatic sediments bear no marked or at least remarked upon evidence of anatomical specialisation. The maggot-like larva of the African horsefly *Tabanus biguttatus*, when pupating in the mud of a drying pool, avoids exposure through the cracking of the mud by the ingenious creation of a mud cocoon. Starting at the mud surface, it burrows downwards in a tight spiral, largely separating a cylinder of mud from its surroundings. It then burrows into the cylinder, seals up the hole, and creates a cavity within it before pupating. When the mud finally dries, any cracks pass round and not through the pupal cocoon (Oldroyd 1954).

Excavation using the mouth is the most common technique shown by fish, however, the aquatic medium allows vigorous fanning with selected fins to be effective in the removal and even excavation of finer sediments. The shell-breeding cichlid *Lamprologus* (*Neolamprologus*) *ocellatus* digs a hole beside an empty snail shell using its mouth and vigorous beating of the pectoral and caudal fins. After pushing the shell into the hole, it then drives sand back around the shell with its fins. When only the hole of the shell is visible, it is ready for occupation and in the case of females, for spawning (Haussknecht and Kuenzer 1990).

The red band-fish (*Cepola rubescens*), a species that digs a substantial vertical shaft as much as 90 cm into the mud, intersperses powerful thrusts of the mouth into the substrate with a regular pattern of carrying out mouthfuls of mud and dumping them (James *et al.* 1996). Periodically the fish shows 'coughing' spasms to clean the gills, a behaviour seen more widely in mouth

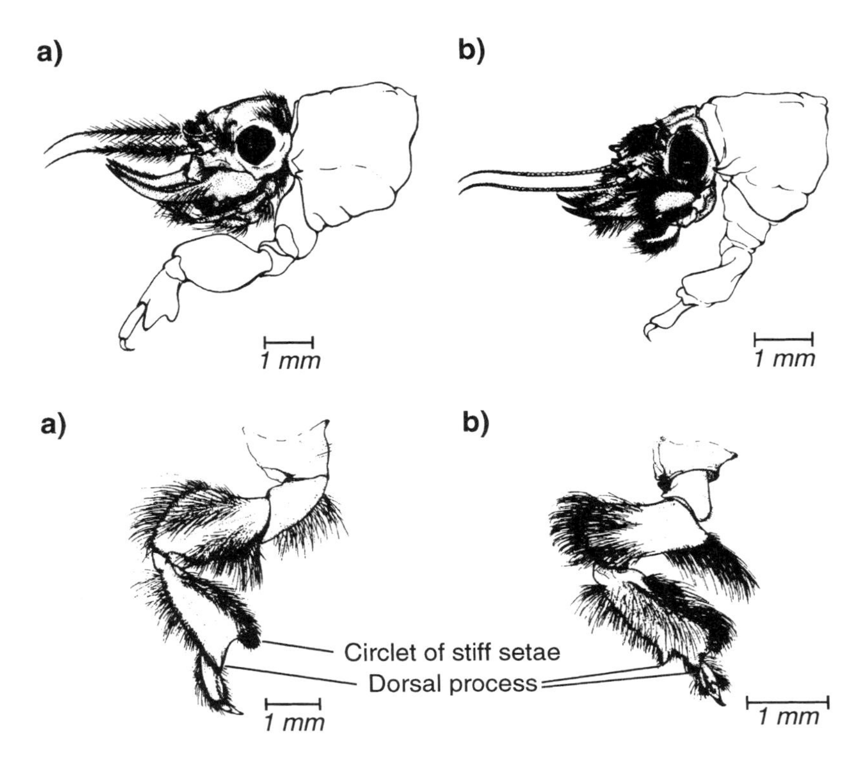

Fig. 3.9. Specialist burrowing anatomy of two mayfly nymphs living in two different substrates. (a) *H. limbata* uses its tusks to anchor itself in soft silt, and its front tibiae to sweep the sediment backwards. (b) *P. vittigera* burrows in compacted clay using its tusks to dislodge the clay, and dorsal processes on the margins of its powerful front tibiae to widen the burrow and clear the spoil. (Adapted from Keltner and McCafferty 1986.)

digging fish although the swordtail jawfish (*Lonchopisthus micrognathus*) shows the unusual adaptation of inserting its caudal fin into the gill chambers through the operculae (James *et al.* 1996). In some species of Gobiidae, sediment taken into the mouth is discharged directly through the opercula, so avoiding the carrying of mouthfuls out of the burrow (Atkinson and Taylor 1991).

Burrow digging by wasps of the families Pompilidae, Sphecidae, and Vespidae illustrate anatomical specialisations in three body areas: robust mandibles, spines on the legs, and a flattened pygidial plate on the abdomen (Evans and West Eberhard 1970). Olberg (1959) divided these wasps into four different categories depending upon their excavation behaviour: *rakers*, *pushers*, *pullers*, and *carriers*. The majority of species are rakers. They scrape the soil beneath the body with vigorous strokes of the front legs, either alternate as in Pompilidae or synchronous as in most Sphecidae. This method is particularly suited to species burrowing in sandy substrates and that burrow obliquely and at a fairly shallow angle (like *Bembix,* Sphecidae). In *Hoplisoides nebulosus* (Sphecidae) mandibles are initially employed to break up the hard ground before the spoil is ejected with rapid synchronous backward kicks with the front legs (Evans 1966).

Pushers back out of the burrow pushing the soil behind them using the front or even all legs and, in some species aided by the pygidial plate on the upper abdominal surface. This is a technique well suited to vertical shafts. *Tachetes* (Sphecidae) uses the forelegs (Kurczewski and Spofford 1986) which, with *Cerceris,* has an especially strong pygidial plate (Evans and West Eberhard 1970). In *Oxybelus aztecus* (Sphecidae) mandibles and forelegs loosen the soil before the hindlegs and abdomen are used to push spoil up the vertical burrow (Peckham 1985). Pullers back out of the entrance bearing a mass of soil between the head and forelegs. This technique is suited to soil that is a little moist and so adheres in lumps.

Sphex ichneumoneus shows that there can be flexibility in the choice of excavating technique within one species; it behaves as a raker when digging in dry sandy soil but, when reaching down to moister soil layers, adopts the puller pattern (Evans and West Eberhand 1970). *Philanthus* behaves as a puller, but will also bulldoze material forward with its head.

Carriers are like an efficient form of puller. They too hold material between the head and front legs, but, once out of their burrows, walk forward or even fly away from the burrow entrance before dropping the spoil. Mandibles are again frequently employed in dislodging material at the workface. *Ammophila dysmica* bites at the soil while buzzing, apparently to break up hard earth before carrying away the spoil (Rosenhein 1987). In some genera there is a specialised carrying structure, the *psammophore* that forms a basket between the ventral surface of the head and the anterior surface of the thorax (e.g. *Belomicrus*), or other equivalent structure, for example, *Pterocheilus* (Fig. 3.10) (Evans and West-Eberhard 1970), *Anacrabro,* and *Moniaecera* (Evans 1964).

Birds burrow in a variety of substrates always involving the beak as the principal instrument of excavation, but showing no characteristic beak anatomy for it. Some species such as trogons (Trogonidae) and parrots bite into the substrate and have short beaks, others such as Kingfishers (Alcedinidae) and jacamars (*Galbula*—Galbulidae) have longer, pointed beaks and excavate with a chiselling action (Hansell 2000).

In mammals, the fossorial habit has arisen a number of times independently, with the result that differences can be looked for between near-relatives that are fossorial and those that are not, while similarities arising through convergence can be sought between unrelated burrowing species. Two digging solutions have repeatedly emerged, enlarged forelimbs (the mole-like solution) and enlarged incisors (the biting solution).

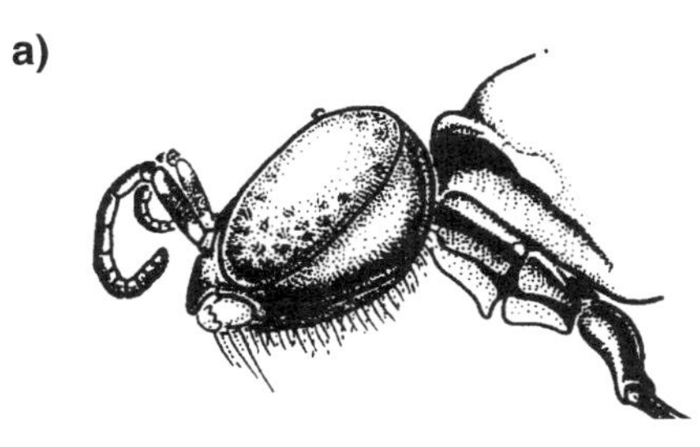

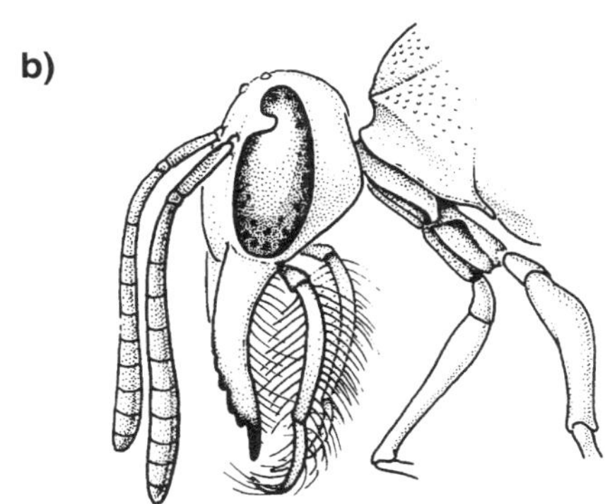

Fig. 3.10. Specialist spoilcarrying anatomy in solitary wasps. (a) The sculpted surface of the psammophore of *Belomicrus forbesii* (Sphecidae) creates a device for carrying a pellet of earth away from the burrow, between the head and the thorax. (b) Hair-fringed palps and elongate mandibles form a basket for carrying earth in females of *Pterocheilus* (Eumeninae). (Adapted from Evans and West Eberhard 1970; Permission University of Michigan Press.)

Convergent adaptations to a mole-like habit are, evident between the insectivore, European mole (*Talpa europea*) and the marsupial mole (*Notoryctes typhlops*). Both have short, laterally rotated anterior limbs with a massive humerus, a rounded digging paw with greatly enlarged claws. Among the Insectivores, the fossorial *Scapanus latimanus* has a humerus as broad as that of *Talpa*; the humerus of the semifossorial talpid, *Neurotrichus gibbsii* is rather less broad, while that of the shrew *Sorex trowbridgii*, unspecialised for digging, is relatively slender (Dubost 1968) (Fig. 3.11).

The largest number of burrowing species in mammals is found in the rodents, with fossorial species in at least six families: Geomyidae, Cricetidae, Spalacidae, Rhizomyidae, Ctenomyidae, and Bathyergidae (Dubost 1968). Between them they have evolved highly modified digging instruments from both the front limbs and the incisors. In *Myospalax* (Cricetidae) digging is done primarily with the large-clawed forelimbs, in *Ellobius* (Cricetidae) and *Spalax* (Bathyergidae) by prominent, rapidly growing curved incisors, and in Geomyidae and Ctenomyidae generally by a combination of teeth and forelimbs, with both showing moderate development (Dubost 1968; Nevo 1979). The mole rats *Tachyoryctes* (Rhizomyidae) and *Heliophobius* (Bathyergidae) dig into the workface with upward sweeps of the incisors, bracing their bodies by gripping the burrow sides with laterally rotated hind feet. In specialist limb-digging species such as *Notoryctes* or *Myospalax*, the forelimbs and pectoral

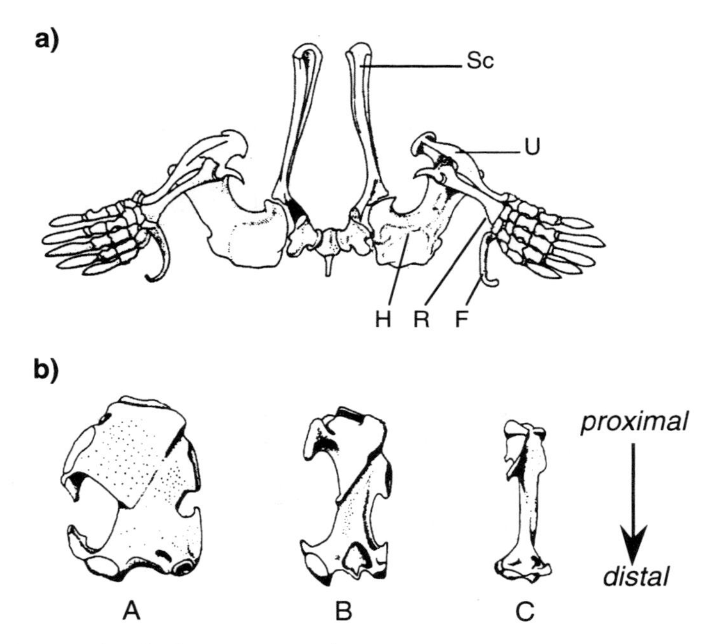

Fig. 3.11. A comparison in the insectivores of the evolution of a short, broad humerus for use as a powerful digging instrument. (a) Fore-limb of the mole *T. europea* (U = ulna; F = os falciforme; H = humerus; R = radius; Sc = scapula). (b) Anterior view of the humerus of (A) *S. latimanus*, (B) *N. gibbsii*, and (C) the relatively unspecialized *S. trowbridgii*. (Adapted fom Dubost 1968.)

girdle show substantial modification for powerful sculpting of the workface. In the less specialised limb-digging species such as *Ctenomys*, specialisation is evident in strong claws on the hands and, in *Ctenomys rionegrensis*, a supernumerary digit (the *thenar*) supported by a pre-pollex and a cartilaginous rod (Dubost 1968; Ubilla and Altuna 1990).

Ctenomys is a South American genus, while, *Geomys* is the North American equivalent. Both loosen the soil with a sharp downward motion of the wrist of the forelimb followed by forelimb flexion. Both genera have also individually evolved similar anatomical adaptations for digging, for example, a ventral extension of the scapular border which may be extended into a thin prominence, the *postscapular prominence*, which is absent in non-burrowing species such as the rat *Rattus* sp. This scapular modification is the site of attachment of the teres major muscle used in the flexion of the humerus in digging (Gorman and Stone 1990).

Removing excavated soil in rodents is effected using a variety of techniques: with the hindfeet (some Bathyergidae), by turning round and pushing with the head and forefeet (*Tachoryctes*, Rhizomyidae), or by bulldozing with the head alone (*Spalax*) (Nevo 1979). In some burrowing specialists, the tunnel space is not made entirely by the excavation of the soil but by compacting it

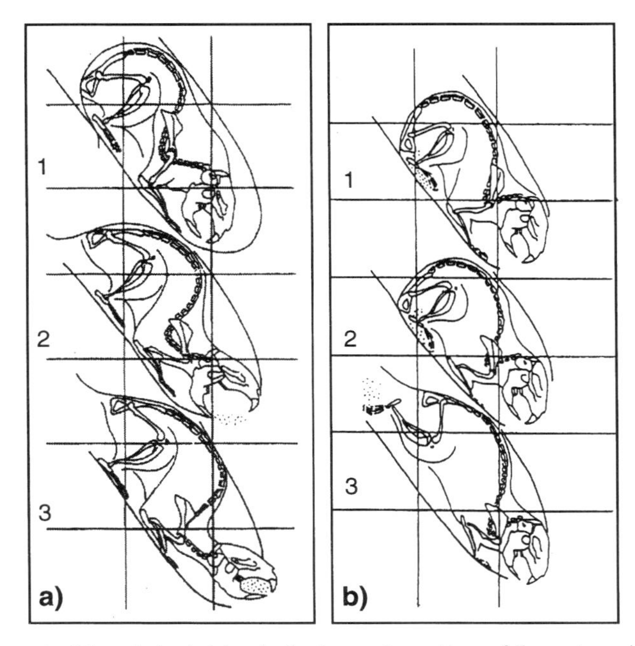

Fig. 3.12. Involvement of the whole skeleton in the burrowing actions of the water vole (*A. terrestris*). (a) Pressing its arched back against the roof of the burrow, the rodent thrusts its head forward to grasp a mouth-full of substrate. (b) The loosened spoil is passed back by the fore limbs and kicked down the burrow with the hind limbs. (Adapted from Laville 1989.)

into the walls of the tunnel. This appears to be the explanation of the combination of powerful front limbs and massive dorsally flattened face in the Oligocene fossil mammal *Xenocranium*, which probably allowed it to compact or displace the roof of its burrow (Rose and Emry 1983).

Evidence that digging should not be regarded as confined to fore limbs or incisors but may involve the whole skeleton in stereotyped cycles of behaviour, was obtained by Laville (1989) for the water vole (*Arvicola terrestris*) (Arvicolidae). This bites into the workface with the back arched and pressed against the tunnel roof to anchor the body as the head is thrust forward (Fig. 3.12). The soil is then spat out and passed back with rapid alternating movements of the fore limbs and finally kicked back with the hind limbs.

3.9 Conclusions

Table 3.1 shows that the parts of the anatomy used in construction are overwhelmingly legs and mouthparts (mandibles, beaks, and jaws), supporting prediction 1 of Section 3.1, that construction anatomy would generally be used in other contexts. The only other significant anatomical feature is spinnerets or

Table 3.1. A summary of the examples given in this chapter to illustrate the parts of the anatomy used in the construction process

Class	*Construction anatomy*					*Main construction forces*		
	Jaws Teeth	*Walking legs*	*Other*	*Specialised?*		*Builder*	*Building materials*	*External*
				No	*Yes*			
Piling up								
Insects	2			2		2		
Fish	1			1		1		
Birds	6			6		6		
Mammals	2			1		2		
Interlocking and Weaving								
Crustacea			I (2nd pereiopods)	I		I		
Birds	15			15		15		
Sticking Together								
Annelid			I (building organ) (mucus)		I	I		
Insects	5	2	3 (larval body)	5		5		
Birds	3			3		3		
Modelling								
Inspects	6			6		6		
Birds	2		I (mucus)	2		2		
Folding & Rolling								
Spiders		1		1			I (contractile silk)	
Insects	2	2	2 (larval body)	3	I	2	2 (contractile silk)	
Spinning								
Spilders		2		2		2		
Insects	3	I	3 (abdomen)	3		3		
Sculpting								
Crustacea			Chelae & pereiopods	3		3		
Insectra	6	5	I (psammo phore)	5	3	6		
Fish	2		I (abdomen)	2		2		
Birds	I			1		1		
Mammals	7	6	I (head)		13	13		

Notes: The table shows the dominance in the use of jaws and to a lesser extent the legs. There is a general lack of specialized anatomy, with the exception of that used in digging or carrying spoil created by digging.

other structures concerned with the secretion of building materials. These glandular structures, however, have no behaviour of their own, depending during construction upon movements generated in other parts of the body. They are therefore not strictly structures used in construction, since it is not possible to ask questions about their specialisation for manipulation or strength.

Legs and mouthparts may be more or less specialised for their non-building roles, however, this chapter reveals a general lack of specialised anatomy for building. There are nonetheless exceptions, the conspicuous example being for construction by sculpting, where both limbs and mouth-parts may show marked anatomical specialisation for their construction role. As between power and skill, it is evident in this case that the modifications are predominantly for the former. This can be seen clearly in rodents through comparison of fossorial and non-fossorial species, the nature of the adaptation depending upon whether the limbs or teeth predominate in the dislodging of materials at the workface (Figure 3.11).

What was not predicted was that burrowing species may have another anatomical specialisation, that needed for spoil removal. In the rodents this is expressed in further modification of the head and hands, but in some wasp species (Sphecidae and Vespidae) highly specialised structures, for example, the pygidial plate, and the psammophore (Fig. 3.10) have evolved for spoil transport. So, sculpting does support prediction 2 of Section 3.1 that where the construction role predominates, anatomy with other roles will show greatest modification to construction. It also supports prediction 3 that the anatomy of power is particularly expressed in construction that involves the removal of material.

The association of anatomical modification with power, rather than with both power and manipulative skill, was not predicted. One possible reason for this is that the movement of jaws (including associated head movements) and limbs in contexts other than construction, also requires fine control of movements. The second pereiopod of Alpheus species is evidently designed for great mobility and delicacy of movement (Fig. 3.2), but only one or two species have been recorded as using it for construction. The design of the beaks of birds can readily be associated with feeding; good vision and a mobile neck can be seen as associated adaptations, yet all assist effective nest construction. Even those birds that show the greatest manipulative skill, the weavers, show no obvious anatomical adaptation of the beak, or indeed the feet to the task. But anatomical modification for delicacy of manipulation may simply be not that obvious, at least in comparison to the anatomy of power where additional muscle and robust skeleton are a prerequisite. The relationship between specialised behaviour and specialised anatomy is therefore, as anticipated in Section 3.1, complicated.

Detailed comparative studies might reveal more subtle anatomical specialisation for skilled building behaviour in birds, but they are currently lacking. In some species, insight might also be obtained where there is a difference in the building role of males and females. In the ploceine weavers it is predominantly males that are the builders (Collias and Collias 1984), while in the oropendolas and caciques it is the females. It is interesting that in attelabid weevils where only females roll leaves, small anatomical differences between the sexes are discernible (Daanje 1964).

There is a general methodological problem in demonstrating anatomical and behavioural trade-offs between rival functions, but examples from other contexts show how it can be tackled experimentally. Trade-offs between tail length and flight agility have been studied in birds by the addition or removal of feathers (Evans and Thomas 1997). The manipulation of leg length through forced leg regeneration in spiders (Vollrath 1987) provides a rare example in the field of experimental manipulation to study the role of construction anatomy (Fig. 4.6).

Prediction 4, that behavioural elements of construction should be small in number, discrete, simple, and repetitive does in general seem to be upheld by qualitative accounts in the literature. However, data are scarce and there is an absence of information on how stereotyped these patterns are on repetition. With such information, it would be possible to address questions on, for example, limitations placed by their nervous systems on the building behaviour of invertebrates, and on the utilisation of greater learning abilities in higher vertebrates. An obstacle to this is also methodological. What is the appropriate way of quantifying behavioural complexity? Byrne *et al.* (2001) explore this in a study of the repertoires used by mountain gorillas in the handling of a nutritious but prickly thistle species prior to ingestion.

Eating leaves may seem like a relatively trivial problem but, in processing this difficult food plant, gorillas used 222 separate behavioural *elements*. As many of these could be regarded as minor personal variants, this was reduced to 46 groups of *functionally distinct elements*. As sheer size of repertoire could be regarded as lack of organisation rather than the reverse, Byrne *et al.* (2001) calculated the number of selective arrangements of actions, or *techniques*, and identified 256. This is however, very much less than the number of techniques possible from the repertoire of behavioural elements, indicating a high degree of selectivity in the processing behaviour, evidence of a cognitive capacity at least comparable to that required in tool use by chimpanzees.

This approach is a possible model for the further study of animal building behaviour. Primary candidates for such a study would be species of weavers, where manipulation is apparently difficult and learning involved, however, also worth examination are invertebrate species with apparently variable building sequences. Here, case building caddis larvae might be suitable candidates, because manipulation of mineral particles is prolonged, delicate, and not obviously in any stereotyped sequence (personal observation).

In the absence of detailed behavioural studies, prediction 5 that 'getting started' behaviour will show greater variability than later building behaviour, cannot be resolved either. In some species there is clearly special behaviour that is necessary at the start of the building sequence; the spinning of silk attachment pads or scaffolding at the beginning of the pupation sequence of lepidoptera is an example. But the prediction relates not to this, but specifically to the unpredictability and variability of the site of the cocoon, nest, or whatever, compared with the predictability of the established structure, the parameters of which are defined by the builder itself. Is this difference in predictability reflected in a comparison of behavioural variability between these two phases? Some evidence, from birds, for example, seems to support this (Hansell 2000), but more data are needed. Certainly selection could favour flexibility in a habitat where nest sites are scarce. Limiting the acceptable configuration of twigs for a nest site, limits choice.

4 Work organisation and building complexity

4.1 Introduction

A 7 m high termite mound is a challenge to the argument developed through Chapters 2 and 3 that standard materials allow simple actions and a limited behaviour repertoire to generate architecture. Even the sheer scale of the mound relative to a termite would seem to preclude the individual builder from any concept of the extent of the structure, let alone its architectural complexity. This chapter tries to assess how demanding on the decision-making capacity of the organism building behaviour is, it looks at how organisation of work forces can be achieved without leadership, and examines the evidence for learned skills and innovation among non-human builders.

4.2 Evidence of spatial knowledge

To build requires feedback. For many builders, however, the completed structure extends well beyond the organisms sensory range. Such a structure could be built-up by repeated responses to local stimuli, or alternatively the organism might be able to acquire a spatial memory of the structure. This chapter shows that both these are possible, and the latter is considered here.

The problem facing females of leaf rolling weevils (Attelabidae) is whether a leaf is of a sufficient size to provide protection and food for growing larvae. As the leaf is much bigger than the beetle, this decision can only be made after a beetle has learned through exploration something about a given leaf's size. The pattern of leaf inspection differs between species but in each is an orderly process of exploration. In *Apoderus balteatus*, it consists of four distinctive and stereotyped paths over the leaf surface (Sakurai 1988*a*).

Using model leaves Sakurai (1988*a*) showed that in *A.balteatus* it was leaf length that determined its acceptance and then leaf area that determined whether one or two eggs were laid on it. The beetle can clearly obtain a direct measure of leaf length from walking, as it does, the length of the mid-vein; however, it appears able to compute an indirect measurement of leaf area, possibly as a ratio between leaf circumference and length.

Chonostropheus chujoi rolls oak leaves again showing four behavioural elements in its exploration, similar to those of *A. balteatus* (Sakurai 1988*b*). Experiments with leaves of different shapes indicate that the female starts her measurement of the leaf at its most protrusive point, judged not by the tip of the mid-vein but by the leaf's angularity. She then calculates the size of the leaf from its circumference, not length or leaf area. The position of the cut across the leaf before rolling begins is determined by walking a designated distance during the phase of *upward circumference walk*. If the leaf is small, so that the leaf base is reached before this, cutting position is determined by a second method provided the leaf is above a certain minimum size. In this method, the beetle calculates the position of the cut by a *backward walk* from the leaf base for about 5 mm. The number of eggs laid in the roll, which can vary from one to four, is however related to the leaf area, so this can apparently be determined as well.

Experiments on ants and bees show that similar abilities to assess spatial relationships beyond their immediate sensory range are possessed by some social insects, allowing them to determine the suitability of a cavity as a nest site. *Leptothorax albipennis* is a species of ant that lives in narrow crevices between rocks, so the suitability of a nest site is determined essentially by the nest site floor area. New sites are discovered by individual scouts, which then recruit the rest of the colony to it. Mallon and Franks (2000) tested three rival hypotheses on how an ant might assess the cavity area through exploration. These were: (1) it measures the length of the internal perimeter; (2) it uses a mean path length algorithm (i.e. it measures the mean path length between collisions with the wall; (3) it uses the algorithm *Buffon's needle*, a solution proposed by Comte George de Buffon in the eighteenth century for estimating the area of a plane from the frequency of intersection of two sets of randomly arranged lines of known length.

By offering scout ants cavities of different design, hypothesis (1) was ruled out, because scouts were found to discriminate between cavities of different area but the same perimeter. Hypothesis (2) was rejected, when ants found equally acceptable two cavities of the same area, one of which had a partial barrier down the middle. For hypothesis (3) (Buffon's needle) to be supported, it is necessary for an ant to visit the potential nest cavity more than once, laying a pheromonal trail in the first visit with which encounter frequency can be measured in the second visit. In the experiments it was observed that ants did visit a cavity more than once and, on the second visit, did pause at points of intersection with their path of the first visit. Two additional features of the ant's behaviour ensured that this was a remarkably accurate method for estimating cavity area. First, an ant was able to recognise its own individual pheromone trail, ensuring that on its second visit it was not confused by visits from other individuals. Second the ants were found to standardise the length of the trail of the first visit regardless of the size of the cavity (Mugford *et al.* 2001).

Honeybees, before founding a new colony by swarming, send out scout bees in search of a new nest site, and report back on its suitability and location using dance language. Cavities of 40 L are preferred over ones of 10 L or 100 L, indicating that there is both an upper and lower preferred limit. Cavity inspection lasts about 40 min and involves walking around the interior wall of the cavity. Seeley (1977), by making the walls of a cylindrical test cavity rotate either in the direction of or contrary to the direction of a walking scout bee, showed that the perceived walking distance round the walls and back to the entrance, was important in the estimation of cavity volume.

4.3 Organisation of workforces

Significant research has been directed towards how colony activities of social insects are organised and, in view of the ability of these insects to construct elaborate nests very much larger than themselves, the organisation of their workforces in building behaviour has been an important feature of such work.

4.3.1 Morphological castes and temporal polyethism

When a solitary bee such as *Augochlora pura* (Halictidae) builds its nest, it has to undertake a *series* of acts, excavation of burrow, then of brood cells, finally lining the cells with secretion (Michener 1974). Where there is a workforce, as in the eusocial bees of the Apidae, individuals could attempt to carry out building acts in parallel, the higher the number of bees, the greater the chance that the job will be completed. This process is described as *parallel–series* (Oster and Wilson 1978) (Fig 4.1). However, if each of the series of tasks is undertaken by its own specialists, then the failure of one individual, results in that task alone being affected, and another member of that task group may still complete it. This kind of operation is termed *series–parallel*, and can be shown to be more likely to complete the task than one in which individuals are organised in parallel, provided that the competence level of individuals is above a threshold value of efficiency (i.e. at a level which exceeds the average for individuals working on their own).

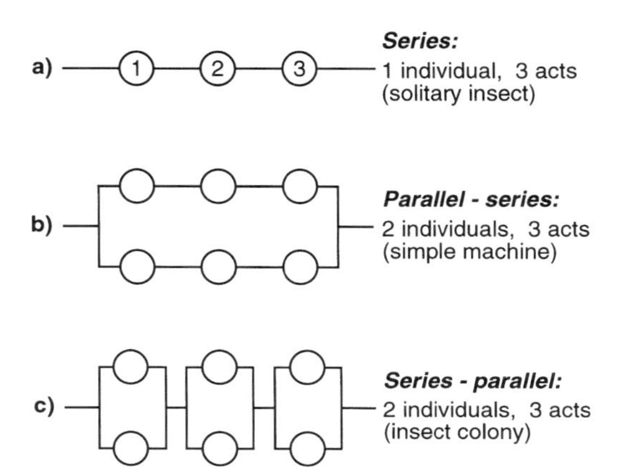

Fig. 4.1. Organisation of workforces, comparing the probability of job completion where organisation is in: (a) series, (b) parallel–series, and (c) series–parallel. This shows that, for colonial species, series–parallel is the most reliable. This is because, with each individual having a specialised task, its failure is overcome by the success of a fellow task group member. (Adapted from Hölldobler and Wilson 1990.)

The large colony sizes, frequency of social interactions, and complex nest architecture of social insects give the impression of great behavioural diversity compared to that of solitary species. However, comparisons between the two show that, while the number of behavioural acts in species living in large colonies may exceed that of solitary species, the difference is in fact quite small (Oster and Wilson 1978). Therefore where social insects have evolved job specialisation, the explanation seems more likely to be due to enhanced efficiency in tasks compared with solitary species, than an adaptation to a wider range of tasks.

By definition, task specialisation is a specialisation in behaviour, so it may occur with or without specialisation in morphology. Morphological casts are, however, found only in a minority of social insects. In ants, for example, which show the greatest extremes of morphological cast differentiation of any social insects, they are present in just 44 out of 263 genera (Gordon 1996), and very few species have more than three clearly distinct morphs (Oster and Wilson 1978).

The development of morphological castes in ants has been shown to generally result from rather simple allometric relationships. The colonies of leaf cutter ants *Atta sexdens*, at an estimated 5–8 million individuals (Hölldobler and Wilson 1990), are among the largest of any of species insect or otherwise. They are also known to have among the highest number of identified behavioural tasks of any ant, 29 being listed by Wilson (1980). The allometry that underlies their varied body proportions is nevertheless very simple; all castes are generated by a single allometric curve for any pair of body measurements (Fig. 4.2) (Wilson 1980). Any particular task is characteristically undertaken by individuals with a particular morphology, as measured by their head width.

These head width morphs form four role clusters centred on head widths of 1.0, 1.4, 2.2, and 3.0 mm, categorised respectively as (1) *gardeners and nurses*,

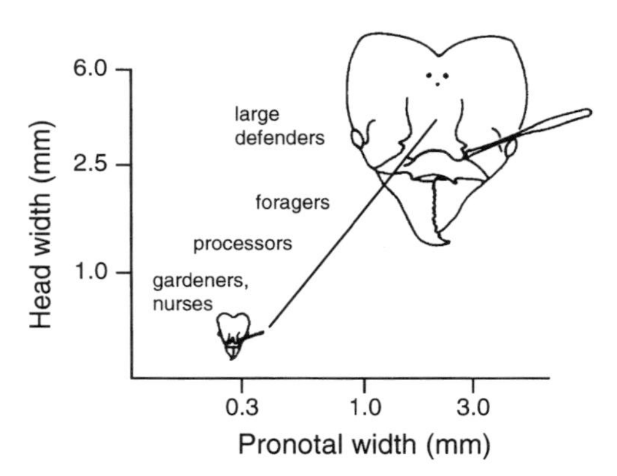

Fig. 4.2. Morphological caste differentiation in the leaf cutter ant *A. sexdens*, although apparently complex for social insects, can be generated from a single allometric curve for any given pair of measurements, shown here as head width against pronotal width. (Adapted from Wilson 1980.)

(2) *within-nest generalists*, (3) *foragers and excavators*, and (4) *defenders*. The 29 tasks are therefore divided between only four not entirely distinct morphological castes. Plotting task diversity against worker size shows that, although the smallest size is more abundant, it is the intermediate size individuals (head width between 1.2 and 2.2 mm) that perform the greatest number of tasks. So morphological caste creates a degree of specialisation but retains a significant measure of flexibility. One further factor contributes to the polyethism in *A. sexdens*—and this is age. Callow workers, that is, young adults with exoskeletons not fully darkened and tanned, engage more frequently in activities within the nest than do older workers of the same size class. In this species, age polyethism (task specialisation in relation to age) does not specifically concern nest building but is part of the range of simple mechanisms that can create job specialisation of which morphological caste division is only one.

A combination of morphological and temporal polyethism is also found in some termites, including specialisation in building behaviour. In *Macrotermes bellicosus* four morphological castes are recognised: *major* and *minor workers*, and *major* and *minor soldiers*. When nest building occurs in the colony, it is undertaken almost entirely by minor workers, although this caste also predominates in the execution of some other nest activities. Foraging is largely undertaken by major workers. Damage to the nest induces immediate repair largely by minor workers, although this is the only task in which both major and minor soldiers participate at a high rate, contributing in spite of their small numbers, about 17% of the participants (Gerber *et al.* 1988). *M. bellicosus* also shows evidence of age polyethism with a general trend from undertaking tasks inside, followed by outside the nest.

Age polyethism as an organising principle has some inherent weaknesses. In the view of Tofts and Franks (1992) one of these is that it is a 'highly static' method of allocating roles, as ageing is likely to be a slower processes than changes in task demand in a fluctuating environment. There is also the problem of the colony having a worker age distribution which reflects past opportunities for brood rearing rather than present need for specialist workers. Tofts and Franks (1992) propose a different and simpler principle of task organisation, which they term *foraging for work*. It assumes that ants (in this particular case) on emergence find themselves in a role, which may simply be related to their location in the nest. This is the role in which they continue until they can no longer find that work, at which time they seek another. As these individuals age, and new individuals eclose, age-related differences become apparent (e.g. younger workers tend to carry out tasks within the nest, then later take on external roles) not because of an organising principle of temporal polyethism, but as an emergent property of a simpler system of job allocation.

This is not seen as a powerful enough explanation by Robinson *et al*. (1994) to account for the widespread occurrence of age polyethism in social insects. They also question the description by Tofts and Franks (1992) of age polyethism as highly static, implying that age polyethism progresses according to a strict timetable. Evidence does now point to flexibility in age polyethism, including in the organisation of building behaviour (Gordon 1996).

Ropalidia marginata is a primitively eusocial wasp, where there is no morphological caste differentiation at all, with all adult females retaining the option of later reproduction. Nevertheless this wasp exhibits age-related polyethism, with extranidial activities increasing with age as intranidial activities decline. In this sequence, the probability that an individual will bring pulp to the nest reaches its peak at about the time of this transition. The behavioural ontogeny exhibits the sequence: *feed larva* precedes *build*; then the extranidial activities first appear as *bring pulp*, followed by *bring food* (Naug and Gardagkar 1998). This is discernible in the two measures of task allocation: frequency of task performance with age (FTP) and probability of task performance with age (PTP) (Fig. 4.3).

Interestingly, relative wasp age explains a greater proportion of the variance than does absolute age. This seems to indicate that wasps of whatever age respond to perceived task need by shifting from one task to another rather than adjusting rates or work. This will be advantageous to a colony because it can provide greater flexibility, since the task allocation profile of an individual will be influenced by that of other wasps. This has obvious advantages in a developing colony faced with environmental fluctuations.

Wakano *et al*. (1998) test the adaptiveness of age polyethism using a simplified dynamic model of colony behaviour, which permits only two activities *inside work* and *outside work* (foraging). It follows the typical insect pattern that inside work is carried out by younger workers. The model confirms that in a static environment, age polyethism is adaptive where job specialization is efficient. More interestingly, it also finds that in a fluctuating environment, the adaptiveness of age polyethism depends upon the degree of perturbation, with

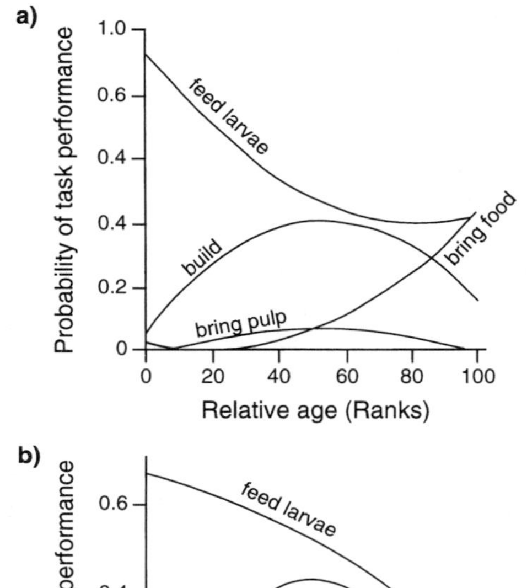

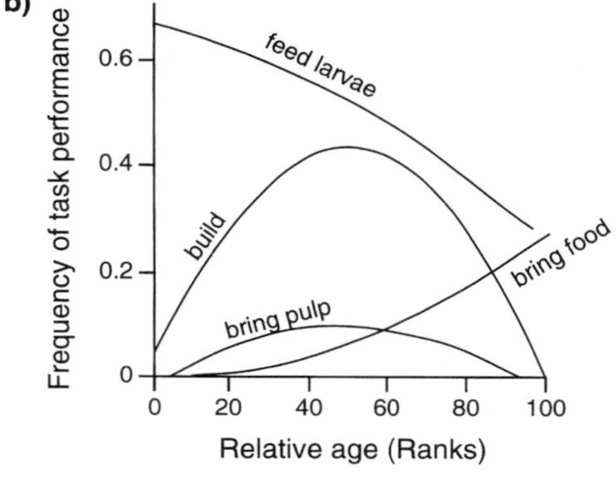

Fig. 4.3. Fitted polynomial regression lines for (a) probability of task performnce and for (b) frequency of task performance against relative age, in colonies of the wasp *Ropalidia marginata*. The fitted lines are for the four tasks plotted together on the same scale to show the overlap of performance of different tasks across age. (Adapted from Naug and Gadagkar 1998).

an absence of age polyethism becoming adaptive with greater fluctuations. However, if the strict conditions of age polyethism (only one task allowable at any age) are relaxed to one of shifting probability of performing one of the two tasks, age polyethism can remain adaptive in spite of some environmental fluctuation. This model therefore supports the view that age polyethism can be adaptive in a fluctuating environment particularly if task is not strictly tied to age but shows some flexibility.

This is borne out by the pattern of age polyethism observed in honeybees. The wax glands of honeybees grow over the first 3–4 days after adult emergence and then begin to regress, with a consequence that workers only engage directly in comb construction for a limited part of their adult life. In the few days prior to the maturation of the wax glands, workers remain in the hive engaged in tasks such as cleaning cells and feeding brood; after 15 days of age they are found increasingly outside the hive, engaged in foraging (Seeley 1982). The presence of two distinct, and a possible third age-related caste working inside the hive were recognized by Seeley and Kolmes (1991): (1) *cell cleaning*, (2) *brood care*, and (3) *food storage* (an activity which includes the task of comb building), after which the bees take to outside nest duties.

Experimental manipulation of the colony shows that the relationship between age and task is flexible. The addition of younger honeybees to a colony containing 8–13 day workers was observed to cause many older nest-bound bees to become foragers (Page *et al.* 1992). This is consistent with the forage for work hypothesis (Tofts and Franks 1992), however, it was also observed that the probability of changing role was greater for older than for younger bees and, when younger bees were removed, it was the least experienced foragers that were most likely to revert to within-nest activities. These findings are only explicable in terms of a flexible age polyethism.

4.3.2 Task partitioning

Defining a worker by its caste is rather a vague concept. Commonly used categories such as *soldier*, *forager*, or *builder* specify a 'trade' that can embrace a number of *tasks* (Jeanne 1986). Temporal castes are defined by changes that take place over a matter of days or weeks. The concept of task partitioning examines how individuals decide which of a number of alternative tasks they will perform at any moment. The implications of this for the behaviour of individuals have been extensively studied by Bob Jeanne and others on the nest building behaviour of wasps.

Polybia occidentalis is an advanced eusocial polistine wasp of the neotropics that founds new colonies by swarming. Within a few days of arriving at the new site, the colony constructs a new nest, consisting of several stacked combs contained within an envelope. Jeanne *et al.* (1988) found evidence of age-related specialisation with nest building activity predominating over foraging up to 20 or so days depending upon the colony, with individuals also showing some variation in the timing of their switch.

Nest building requires the performance of three tasks, *water foraging*, *pulp foraging*, and *nest building*. In *P. occidentalis* each task is performed by a separate group of workers; pulp foragers return to the nest with pulp, water collectors with crops full of water. At the nest, water is passed to pulp foragers and builders; pulp foragers pass their loads to builders, which share the load between themselves before adding the material to the growing nest (Fig. 4.4) (Jeanne 1986). This building sequence is therefore organised according to the series–parallel model (Oster and Wilson 1978). This system could be more efficient than parallel–series organisation in *P. occidentalis* simply because, as specialists, individuals are more efficient than generalists, although this is unproven. However, a pulp forager collects a load that is more than six times larger than can be handled by a builder. Specialisation therefore allows the number of foraging trips to be reduced by an estimated 2.6 times compared with parallel-series organisation (Jeanne 1986).

In small colonies, specialisation does cause some difficulties because encounter rates are low between individuals capable of linking the tasks in

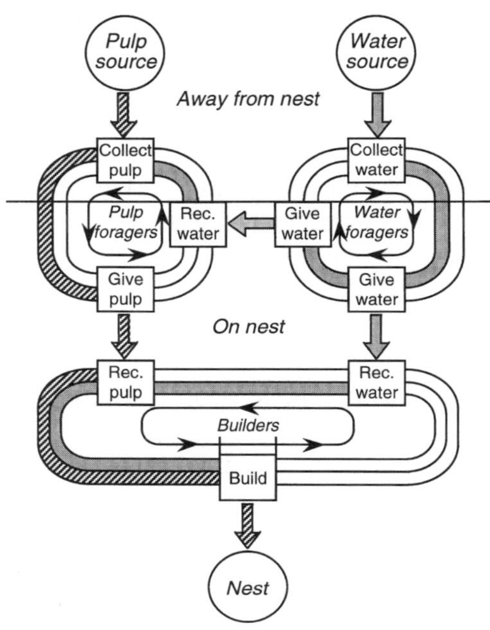

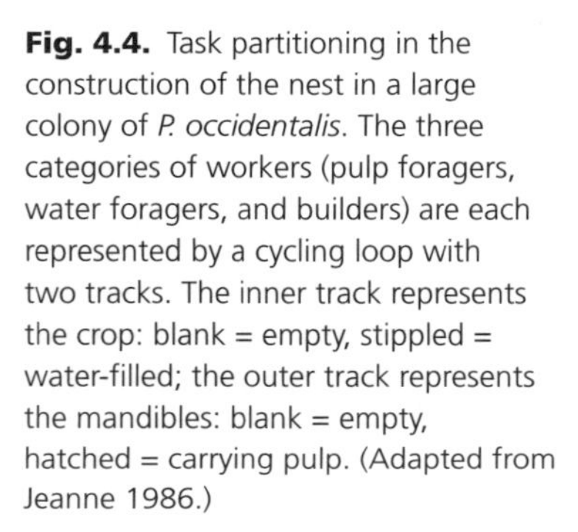
Fig. 4.4. Task partitioning in the construction of the nest in a large colony of *P. occidentalis*. The three categories of workers (pulp foragers, water foragers, and builders) are each represented by a cycling loop with two tracks. The inner track represents the crop: blank = empty, stippled = water-filled; the outer track represents the mandibles: blank = empty, hatched = carrying pulp. (Adapted from Jeanne 1986.)

sequence, compared to larger colonies. This leads to longer waiting times for the task specialists, however, the problem would be reduced if individuals were able to switch tasks. This proves to be the case with task switching, which is rare in large colonies, being enhanced in smaller ones. How task switching is regulated has subsequently been studied experimentally through the manipulation of the proportions of task specialists during nest building.

Jeanne (1987) found that in *P. occidentalis*, neither removal of active pulp foragers nor of a large proportion of the builders, had a very marked effect upon the rates at which the three tasks were performed. Removal of water carriers did have an effect but this was not replicated in a subsequent study (Jeanne 1996). Taken together, these results showed that when task specialists were removed, they could be replaced. Observations on marked individuals indicated that, while this might occur to a small extent through task switching, it was largely through recruitment of idle individuals to activity.

Maintaining a pool of inactive individuals that may from time to time be recruited to the workforce seems potentially an unnecessary cost, but alternatives also present problems. Recruiting individuals from other activities, not necessarily concerned with nest building, could itself be disruptive, while for workers simply to increase their work rate, might either be inefficient or even impossible. O'Donnell and Jeanne (1990) tested between these by observing the response of marked individuals on a nest of

P. occidentalis induced to additional building through experimental nest damage.

They found that foragers specialised in collecting one of four resources, water, pulp, nectar, and prey, the first two only concerned directly with nest building, the other two with feeding. When foragers switched task it was found to be generally within the function of nest building or food, and that stimulation of nest building did not reduce the rate of food collection. The observations reconfirmed that workers recruited into gathering water and pulp came from a pool of inactive individuals, and further that since they had neither been gathering food on the days preceding nest damage nor on subsequent days, they were specialist reservists.

Jeanne (1996) tested three hypotheses of how individual *P. occidentalis* workers might discover whether or not their task, at a particular moment, needed more or less participants. These were:

1. The nest feedback hypothesis, where each individual obtains information directly from inspection of the nest damage, with greater damage promoting greater participation.
2. The task-mates feedback hypothesis, where recruitment is regulated by the level of activity within an individual task group. Here, a high existing level of activity in the task group would inhibit recruitment, and vice versa.
3. The non-task-mates feedback hypothesis, where individuals decide on the basis of perceived supply or demand of members of other task groups. This predicts, for example, that a high level of activity in pulp gatherers could stimulate recruitment in the builder group, and vice versa.

In this experiment, manipulations involved not only removal of the majority of each of the task forces in turn, but also water supplementation by adding water near the site of nest repair and pulp supplementation (by placing an additional supply of pulp on a spatula just beside the nest).

The nest feedback hypothesis could only be partially supported, because repairs proceeded in spite of builders alone being regularly in contact with the nest damage. The number of builders that were active did, however, have a negative effect on others becoming active, supporting hypothesis (2). This negative feedback could be as simple as the level of crowding at the repair site (Jeanne 1996).

Supplementing pulp resulted initially in a reduction of the activity level of pulp foragers and then to the numbers participating, supporting hypothesis (3), that their level of activity was set by the level of demand for pulp. Spraying the nest with water to reduce its demand, or removal of pulp foragers both had the effect of reducing the water foraging. This indicated that water carriers too were regulated by feedback from the demand of pulp carriers or builders for their commodity. So it seems that only the builders are

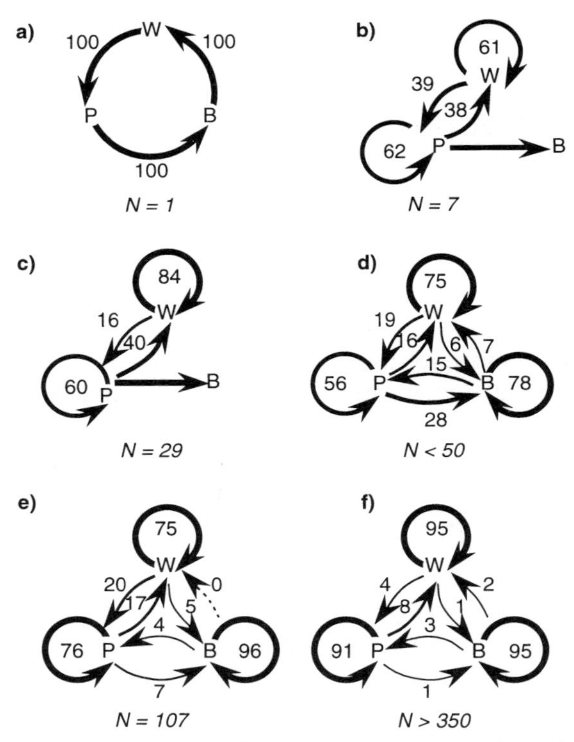

Fig. 4.5. All elements of the organisation of wasp nest construction evident in large swarm founding colonies, can also be seen in small colonies with single foundresses: what changes with colony size is the degree of individual specialisation. Diagrams (a)–(f) show the frequency of transitions between three construction tasks, pulp foraging (P), water foraging (W), and building (B), for each of six different species. The width of the arrows corresponds to the frequency indicated by the associated number. Below each species, *N* = the number of wasps on which the data are based. (a) *Polistes*, single foundress; (b) *Vespula sylvestris*, colony of seven individuals; (c) *Polistes fuscatus* colony of 29 individuals; (d) *P. occidentalis*, pooled data from four colonies; (e) *Metapolybia mesoamericana*, wasps from a single colony; (f) *P. occidentalis* (see also, Fig. 4.4), pooled data from three colonies. (Adapted from Karsai and Wenzel 1998.)

influenced by information directly from the nest damage, while the activity of pulp and water foragers is regulated by demand for their materials.

Every element of nest construction that is evident in large colonies of swarm founding wasps such as *P. occidentalis* exists also in primitive, single foundress and small colonies, however, what differs is the level of individual flexibility in their performance (Fig. 4.5) (Karsai and Wenzel 1998). In species with small colonies, individuals show flexibility in their capacity to take on any of the building tasks, while in large swarm founding species, parallel processing by specialists is possible, offering greater efficiency. This can be seen in the only qualitative difference between independent and swarm founding wasps; swarm founders never keep the whole pulp load but divide it up, indicating that load size is a secondary consequence of the more specialised organisation of the building sequence. As the colony gets bigger, a pulp forager is less likely to build and specialisation in all tasks is more evident. This supports the view that large colony size was a prerequisite for behavioural specialisation and not the reverse.

A *team* of workers is defined by Oster and Wilson (1978) 'as a group of non-interchangeable individuals who move around the colony as a group'. The inability under this definition of individuals to change roles necessitates them belonging to different castes (Hölldobler and Wilson 1990), creating a very narrow definition of team, which no social insects and only few vertebrate species can meet. Anderson and Franks (2001) seek to give recognition to intermediate levels of social coordination by proposing two new categories of cooperation: *group task* and *team task*. A group task requires multiple individuals to perform the same activity concurrently for the task to be fulfilled. An example of this related to building would be the formation in army ants of the structure of a bivouac by linking their legs together (Wilson 1971). This they distinguish from a team task, which requires different subtasks to be performed concurrently by individuals who might be interchangeable and not different castes. An example of this would be the construction of leaf purse nests by weaver ants *Oecophylla*, since some workers may form chains to pull neighbouring leaves together, while at the same time others must hold larvae, whose silk is used to bind the leaves together (Hölldobler and Wilson 1990).

In vertebrates the occurrence of team tasks seems to be particularly associated with group hunting and frequently involves individual recognition not least because, unlike members of social insect colonies, all participants have personal fitness and might benefit through cheating unless closely policed by their teammates. Anderson and Franks (2001) feel that within this definition more examples of team cooperation will be found. Nevertheless, there appears to be a virtual absence of team building among vertebrates other than humans.

4.4 Organisation of the building sequence

Possible paths towards the creation of pattern in living systems are reviewed by Camazine *et al.* (2001), in a book where the organisation of building behaviour features prominently. They are particularly interested in the power of *self-organisation* and related processes in building up complex structures but also consider the merits of four other mechanisms; the use of *leaders*, *templates*, *blueprints*, and *recipes*. *Leadership* is an aspect of the organisation of a workforce, which was covered in Section 4.3, although it does have consequences for the execution of the building work. In that section, it was apparent that although social insects commonly exhibit job specialisation, they show virtually no evidence of leadership.

The use of templates to guide building work is probably underestimated. The shape and size of the nest of the village weaver (*Ploceus cucullatus*) was claimed by Collias and Collias (1962) to be defined by the reach of the male bird as it stood in the initial vertical ring of material. However, the dimensions of bird cup nests may be defined in the same way, and virtually all portable case bearing or tube building invertebrates could and probably do use their own bodies or the reach of their limbs to create a similar template.

An architectural blueprint would give a builder a mental representation of the goal, leaving the means to achieve it open. Evidence, however, suggests that fairly constrained sequences of stimulus and response are more likely to guide building rather than innovative goal-directed ones (Hansell 1984). The building of a spider orb web (Section 4.4.1) is evidence of this.

A recipe requires rigid adherence to a sequence of actions. This has two severe limitations: first, it cannot respond to unpredictable circumstances, and second prevents the flexible sharing of building work between individuals. These limitations are avoided where a builder can move about the structure, engaging in a meaningful stimulus–response dialogue wherever building work is required. Such an ability was invoked by Grassé (1959) to explain how termites could coordinate their building activity, without the need to communicate directly with each other; this he called *stigmergy*.

A limitation of the stigmergy model has been its apparent inability to demonstrate that stimulus–response sequences using local cues alone could create such massive and complex structures as, for example, termite mounds. The principle of *self-organisation* has been invoked to close this gap. This is defined by Camazine *et al.* (2001) as: 'a process in which pattern at a global level of a system emerges solely from numerous interactions among the lower-level components of the system'. They add: 'Moreover, the rules specifying interactions among the system's components are executed using only local information, without reference to the global pattern.'

4.4.1 Local stimuli and linked decision rules

Section 4.2 demonstrates that some organisms in their construction behaviour are able to store sensory information in a way that gives them some appreciation of the extent and arrangement of the objects they manipulate. But, although the leaf being rolled by the attelabid beetle *Deporaus betulae* may be around ten times its own body length, this is trivial compared with the disparity in size between, for example, eusocial insects and their nests. So, do wasps and termites need to have special sensory capabilities and enhanced central nervous capacity to be able to do this, or are builders very little different from related, non-building organisms in these respects? Orb web spiders, which build large structures relative to themselves have shown that sequences of behavioural acts initiated by local stimuli can account for much of this building process.

A spider that looses its leg will regenerate it over the next few moults. Vollrath (1987) used this ability to create experimental *A. diadematus* with selected legs partially regenerated, and hence shorter than the rest. He then used these individuals to test whether spiders determine the spacing between turns of the auxiliary spiral and the capture spiral that replaces it, from leg positions. The auxiliary spiral, which a spider lays down while travelling from the hub to the periphery is a *logarithmic* spiral, that cuts the radii at a constant angle. The capture spiral, which the spider lays down as it returns from the edge of the web to its hub while picking up the auxiliary spiral, is an *arithmetic* spiral, that is, one that results in an equal distance between successive turns.

Vollrath (1987) found that the spacing of the threads of the auxiliary spiral was unaffected, even when all four legs on one side were shortened, whereas the distance between neighbouring threads of the capture spiral was altered from a normal to a bimodal distribution when only the front leg of one side was shortened (Fig. 4.6).

Krink and Vollrath (1999) used a computer simulation to test whether an organism like a spider could spin a web using only simple decision rules based on local spatial information. In this model they confined the simulation just to the construction of the capture spiral. This model was initially a development of one pioneered by Eberhard (1969) which used completed webs as the basis for generating the simulated behaviour.

The success of the model in reproducing a spiral path resembling that of *A. diadematus* gave support to the idea that the spiders relied heavily on local rules to guide their path (Gotts and Vollrath 1991). However, filming a spider during web construction showed that the actual path taken during the laying down of the capture spiral was in fact different from the path of the thread itself (Zschokke and Vollrath 1995). This required the modification of the simulation to reproduce a *way-finding* algorithm as opposed to a *thread placement* algorithm, to test how the spider behaviour might be organised (Krink and Vollrath 1999).

The reason for the difference between the two paths (spider and thread) is a consequence of the fact that the spider uses the auxiliary spiral for support as it moves from the web periphery to the hub, laying down the capture spiral. Being a logarithmic spiral the distance between its turns is greater at the edge than the hub. This means that particularly at the periphery, the spider may use the same turn of auxiliary spiral to lay down successive turns of capture spiral, varying its reach in order to maintain constant spacing of the capture spiral. As a consequence, the behavioural path of the spider laying

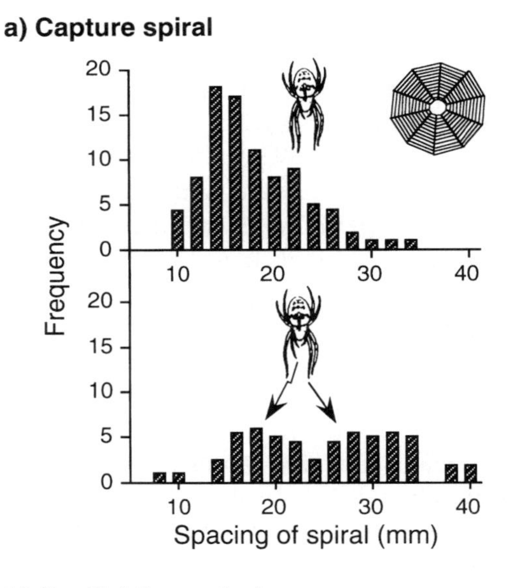

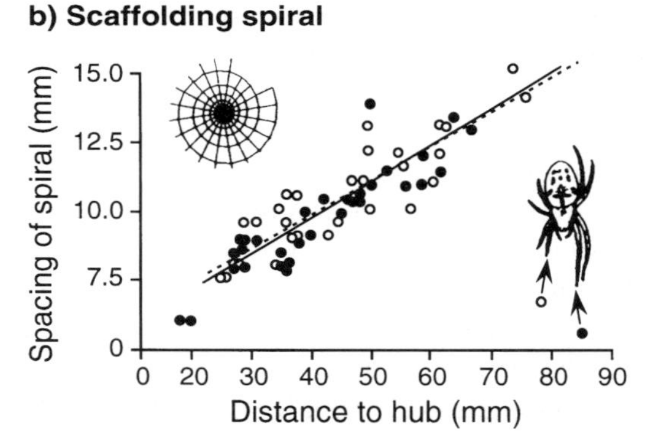

Fig. 4.6. Control of the spacing of successive turns of (a) the equidistant capture spiral, and (b) the equiangular auxiliary (scaffolding) spiral of the orb web of *A. diadematus*. In (a) is shown the frequency distribution of the spacing of the capture spiral in webs spun by intact individuals compared to those with one front leg shortened through forced regeneration. Note the bimodal distribution of the spacing in the latter. In (b) the positive correlation between capture spiral spacing and distance from the hub shown by intact individuals (•), is unaltered for spiders with all legs on one side regenerated (o). (Adapted from Vollrath 1987).

down an evenly spaced capture spiral tends to bunch around the track of the auxiliary spiral on which it depends for support.

The Krink and Vollrath (1999) model is an algorithm of decision rules to generate the track of the capture spiral, while mimicking the actual path of construction behaviour. It hypothesises that the spider determines the track of the capture spiral from two measurements: first the spacing between successive capture spirals and second the distance between the 'handhold' of the first inner leg on the auxiliary spiral and the outreach of the first outer leg on the innermost turn of the capture spiral (Krink and Vollrath 1998). This allowed the virtual spider to model the role of only the first pair of legs, while its success could be tested against the observed results of the leg regeneration experiment (Vollrath 1987).

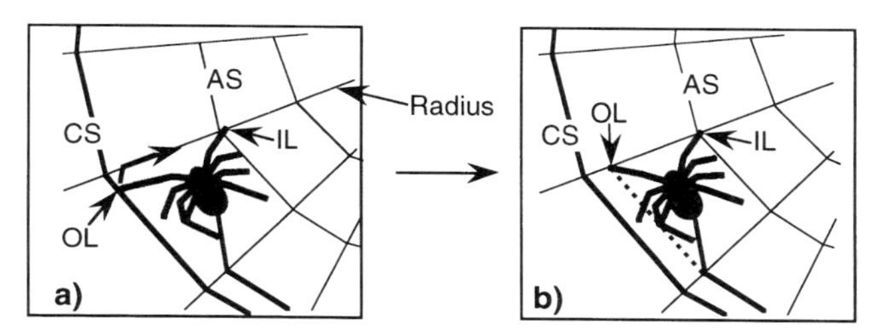

Fig. 4.7. The third rule of the basic orientation and construction group of the 'virtual spider' determines the position of attachment of the capture spiral to the radius (adapted from Krink and Vollrath 1999). (a) Holding the auxiliary spiral (AS) with the front right (inner) leg (IL), the virtual spider checks the position of the current capture spiral (CS) with the front left (outer) leg (OL). (b) Confirmation of the condition in (a) initiates first the movement of OL to the position shown on the radius, followed by attachment of the capture spiral at that point (dotted line).

It was found that for accurate web simulation, the virtual spider needed no more than 10 rules controlling simple local action patterns in combination with appropriate sensory and motor capacity. These rules were divided into three groups concerning respectively: the first group of four for basic orientation and construction (mainly in relation to radial, peripheral, and spiral threads, orientation along the auxiliary spiral, and attachment of new capture spiral threads). The second group of five rules were concerned with advanced orientation, in particular U-turn behaviour during the laying of the capture spiral, and a third class of rules (only one), for additional fine adjustment (Krink and Vollrath 1999).

Rules 3 and 4 of the basic orientation and construction group are the ones that determine the point of attachment of the capture spiral to the radius (Fig. 4.7); they are therefore a test of the virtual spider's decision rules against the observed output of experimental spiders with shortened front legs. In fact the model failed to mimic the regenerated leg results by still producing an essentially normal spacing. This led to re-examination of the actual behaviour of leg-regenerated *A. diadematus*. This showed that the spiders always kept the leg grasping the auxiliary spiral bent rather than straight as had been assumed in the virtual spider model. This has the effect of keeping the spider's body a fixed distance from the auxiliary thread. This explains why, when the shorter foreleg is gripping the auxiliary spiral in a bent posture, the normal leg is still able to reach the capture spiral (Fig. 4.8). However, when the spider reverses the direction of the capture spiral, as it does from time to time, the normal leg is now gripping the auxiliary spiral and the shortened leg is unable to reach the capture spiral. Adjustment of the model to mimic this effect, generated a bimodal output of the capture spiral closely resembling that of the leg regenerated spiders (Fig. 4.6) (Vollrath 1987).

The virtual spider web output still exhibits some discrepancies compared with real ones. This may be due to the elasticity of the silk itself, a feature not

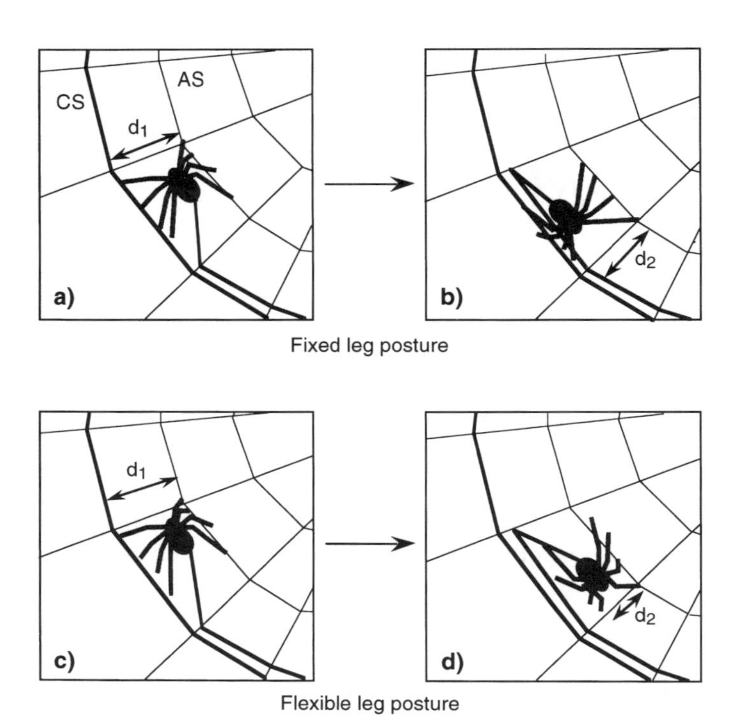

Fig. 4.8. Representation of the method of control of the spacing of the capture spiral by a spider with a regenerated foreleg that attempts to reach the previously constructed capture spiral after reversing its direction of movement. (Adapted from Krink and Vollrath 1999.) The original virtual spider model is shown (a)–(b), and the revised model (c)–(d). (CS = capture spiral, AS = auxiliary spiral.) In (a)–(b) both anterior legs are kept at a fixed posture after the reverse, thus the spider bridges the same distance in both ($d_1 = d_2$). This failed to produce the bimodal distribution of spacing shown by real spiders (Fig 4.6(a)). In (c)–(d) the model is modified to incorporate the observation that the leg grasping the auxiliary spiral is actually kept bent. The result is that after reversing direction $d_2 < d_1$, so the outer (shortened) leg is unable to reach the previously constructed capture spiral, conforming to the result observed in Fig. 4.6(a).

incorporated into the model. Nevertheless, the model gives strong support for the view that a spider could generate an ordered structure much larger than itself by applying simple rules based on local stimuli and needing a minimal capacity for memory storage. It also demonstrates that modelling linked to careful behavioural observation is a productive approach to understanding the control of building behaviour.

4.4.2 Stigmergy

Workers of the termite *Bellicositermes natalensis,* stimulated to construction behaviour, initially appear to place pieces of building material at random. Aggregations of material then begin to appear that coalesce into rather evenly spaced pillars, which are joined to neighbours to create arches and extended into walls and corridors (Grassé 1984). This focus of the termites on the building that is already formed led Grassé (1959) to propose the concept of stigmergy (meaning roughly *focus of work*). This is a building principle in which an individual is seen as responding with construction behaviour wherever the structure communicates to it a 'stimulating configuration'; the consequence of this is the creation of a new configuration (Theraulaz *et al.* 1998).

Deneubourg (1977) described a simple model for the pattern of building seen in termites based on responses to pheromones. A pheromone incorporated into lumps of building material by workers acts as a positive feedback loop, causing single deposits to grow into aggregations and then pillars. Theraulaz and Bonabeau (1995) and Theraulaz *et al.* (1998) describe the models simulating the building of pillars by termites as quantitative stigmergic models, because the organisms respond to quantitative stimuli, such as pheromone fields and gradients. This they contrast with stigmergy based on discrete or qualitative stimuli, as in the case of a wasp choosing where to add the next brood cell to the edge of a comb. These two stigmergic systems are not mutually exclusive since both could exist within one stimulus type (Bonabeau *et al.* 1999).

The nest of *Polistes fuscatus* is a simple comb suspended from an overhead attachment point by a fine petiole. The nest is initiated as a pad of paper pulp from which the downward-projecting petiole is created; at the end of this is constructed a flat sheet, which forms the foundation of the first and second cells. On these cells, successive cells are added to create the characteristic nest design. Downing and Jeanne (1988) ask whether stigmergy can account for this building sequence.

The behaviour sequence up to the initiation of the second cell does take the form of a linear series. However, Downing and Jeanne (1988) discovered that, rather than each stage forming the stimulus for the next, as is strictly required by the stigmergy hypothesis, nest foundresses continued to refer to stimuli used earlier in the sequence, combining that information with cues from the current stage of building. At a more advanced stage of building, *P. fuscatus* shows non-linear sequences of behaviour and alternative responses when faced, for example, with nest damage. This led Downing and Jeanne (1990) to suggest that, as the nest architecture became more elaborate, the wasps did need to compare stimuli in order to assess the appropriate response.

Downing (1994) tested four rival hypotheses of how the wasps could make their building decisions. These were:

1. The *hierarchical response* method, in which the wasp makes a random search of the nest, responding to the first cue that is greater than the insect's threshold value.

2. The *hierarchical search* method, where certain areas of the nest receive more inspection, enhancing the probability that the next response will be made there.
3. The *systematic analysis* method, which is a non-random search of all areas leading to simultaneous evaluation of information obtained from multiple sources.
4. The *hierarchical response and search*, which is a combination of 1 and 2.

Only the first of these is strictly stigmergic, although it could be argued that the second hypothesis (hierarchical search) still allows responses to be made simply on the basis of immediate local cues.

She tested these by attempting to predict the site of next building by a foundress occupying a 15 brood-cell nest of experimentally determined cell arrangement. None of these hypotheses proved to be fully correct, and she concluded that the site of next building during the non-linear phase of nest construction is based on the summation of multiple cues. Downing believes that this kind of evaluation process may be widespread in social insects where complex nests are built with non-linear construction sequences.

An alternative to testing for the presence of stigmergic control in live colonies, is to assess theoretically how effective a strictly stigmergic model is in reproducing the same structure. Karsai and Pénzes (1993) used this approach to test a model of wasp comb construction, where decisions are based on an algorithm in which responses are made only to local architectural configurations. This they describe as a *stigmergic script*. The building processes incorporated into the model are the collection of pulp, the lengthening of an existing cell and the placement of a new one. This was able to generate patterns resembling natural combs, demonstrating that creating a comb could be a *self-organising* process, and not dependant upon any higher order decision-making.

Returning to the study of the behaviour of the wasps themselves, Karsai and Theraulaz (1995) looked in detail at the nest construction behaviour of another *Polistes* species, *Polistes. dominulus*, starting from the foundation of the nest. They observed that in the early stages of building, the behaviour of the wasp is constrained by the few options open to it, whereas later there is more structure and therefore more opportunity for non-linear behaviour sequences. Nevertheless, even at the earliest stages of building, the behaviour is not entirely stereotyped.

Before the petiole is built, the *P. dominulus* female must stand parallel to the substrate. When the flat sheet is being added to the tip of the petiole, the wasp adopts a strongly bent position, hanging from the substrate by the two hind pairs of legs. Only when the flat sheet is completed, can the wasp stand upon it in the vertical position necessary for the initiation of the first cell. Only when the first cell reaches a certain size, is the wasp able to turn round on the nest to strengthen the attachment of the petiole to the substrate.

These observations demonstrate that the structure of the nest not only provides changing stimuli to the builder, but changing possibilities to build, the more options the more the nest develops. To Karsai and Theraulaz (1995) therefore, the emergence of non-linear building sequences as nest building progresses in *Polistes* does not preclude what they regard as a stigmergic method of building control. Indeed, they believe that this method of organisation of building decisions based on local cues has contributed to the evolution of building behaviour across the animal kingdom, without dependence upon advanced nervous systems. Their emphasis is therefore on the potential and actual simplicity of the individual decision-making process in social insect nest building, while that of Downing (1994) is on the evidence for and possible widespread occurrence of more complex decision-making processes.

Natural selection should operate on builders through the success of the structures they build and also the success of the building process compared with rival methods of construction. Jeanne (1975) hypothesised an ideal nest in terms of its economy in the use of material through maximal cell wall sharing. Comparing the pattern of comb growth in *P. dominulus* with that predicted by the *economically ideal* nest, Karsai and Pénzes (2000) found that fewer variations in cell arrangement occurred at any stage of comb development than this hypothesis would allow. They also found that certain cell arrangements that were non-optimal according to Jeanne's (1975) hypothesis, did occur. The functional, and possibly adaptive growth pattern shown by the combs were termed by them as *maximum cell compactness*, defined as the summed distance of every cell centre to the geometric centre of the nest. Karsai and Pénzes (2000) found that incorporating into the theoretical model additional stigmergic rules based in particular on the age of cells on the comb, the observed comb growth pattern could be reproduced. As with investigation of the organisation of spider web building, this progress in our understanding has come about through reference back and forth between the empirical and the theoretical.

4.4.3 Self-organisation

The use of self-organisation models in the study of building behaviour is a test of whether it is possible for organisms with a limited range of responses evoked only by local stimuli, and without direct communication with one another, to build a structure as elaborate as a social insect nest. Self-organisation relies upon four core characteristics. Two of these are *feedback*, positive and negative. The other two are *multiple interactions* and the *amplification* of random activities (e.g. locomotion or task switching), both inherent properties of social insect colonies (Bonabeau *et al.* 1997, 1999).

A queen of *B. natalensis*, taken from her cell and placed with her colony, stimulates workers to build pillars spaced about her. These coalesce, creating a wall that is built up over her to enclose a chamber larger than but shaped similarly to the elongated bulk of the queen herself (Grassé 1984). Grassé demonstrated that the stimulus for this building was a pheromone secreted from the queen's body, and concluded that the position of the queen's chamber wall was determined by a threshold value in the gradient of the pheromone cloud enveloping her. Bonabeau *et al.* (1997) test the ability of the theoretical model, which is based on stimuli generated by termites as they build, in producing a queen cell or similar nest feature. Their model is a development of the self-organisation model of Deneubourg (1977), which demonstrated how positive feedback of pheromone from the accumulating building material could give rise to pillars and then an arch.

The model of Bonabeau *et al.* (1997) includes dynamic elements of the movement of the termites, of the air surrounding them, trail laying by the termites, as well as the emission of a pheromone by the queen and by the pellets of building material. Using this model, the regular disposition of pillars can, under conditions of moving air, for example, generate walls. Similar, although weaker wall-producing effects can be created by assuming directional movement of the termites themselves. This model therefore does support the view that, even where building activity is dispersed and simultaneous, and consequently not strictly sequential, decisions based on immediate, local stimuli can produce a complex architecture. More importantly it shows that there is an interaction between the builders and the architecture that alters the system, which in turn alters the builder's own behaviour. The creation of walls, for example, comes to direct the movement of the termites, and therefore their application of trail pheromone which attracts further termites to the trail, leading to the accumulation of material along its edges. Inhibition by the trail pheromone of the deposition of material directly upon it, results in the formation of a corridor. This model therefore not only incorporates a template, but also the elements of positive and negative feedback (also involving pheromones), amplification, and multiple interactions. This simulated architecture is therefore not the outcome of a design rule but is an emergent property of the collective local responses, an example of self-organisation.

The nest construction of the ant *Leptothorax tuberointerruptus* provides a simplified system for testing aspects of this concept in a living system. These ants live in the narrow spaces between rocks, an almost two-dimensional world. In this, they defend the nest space by means of a circular wall of sand grains with one or more openings through it. When placed in an artificial nest space with a supply of sand grains, the ants show some typical, simple behaviours. Colony members group together while the external workers collect individual stones, approach, and touch their nest mates, before dropping the stone one or two body lengths from them. The colony members therefore create a physical template, in a similar way to the pheromonal one produced by a *Macrotermes* queen. A nest initiated by a larger colony, encloses a larger area.

It is also observed that the presence of stones influences the behaviour of builders. In a site where stones have accumulated, a new stone is more likely to be dropped; where stones are scarce, a stone is more likely to be picked up. These feedback effects amplify the effect already induced by the template. A mathematical model incorporating these simple activities and the probability that ants would perform them was able to predict the emergence of the wall, the relationship between colony size and nest area, and the occurrence of gaps in the wall (Franks and Deneubourg 1997).

4.4.4 Self-organisation and self-assembly

A single wasp comb, whether radially or bilaterally symmetrical is still some way short in terms of complexity when compared with the varied cell and comb arrangements found in some advanced eusocial wasp species. It can therefore be asked whether these can all be emergent designs based on the same sorts of local decision-making that have sufficed up to the level of complexity of a single comb, or whether beyond this level some more elaborate kinds of decision-making are essential.

Honeybee combs hang vertically, attached at their upper margins with back to back cells projecting laterally from each comb. Several combs, equally spaced, and parallel in arrangement, appear as the colony grows (Fig. 4.9(a)) (Skarka *et al.* 1990). Bees engaged in extending a comb are accompanied by chains or *festoons* of linked bees that are not directly engaged in the construction but which form a scaffolding that regulates its orientation and further extension (Darchen 1962; Hepburn 1998).

Deposition of wax acts as an attraction to the bees, creating positive feedback, and the orientation of the comb appears to emerge from initial competition between differently oriented groups of bees through amplification of some of the signals produced by building activities. In this case, however, there are also builder-to-builder interactions the complexity of which is not understood. Nevertheless, Belic *et al.* (1986), incorporating only the basic elements described above, were able to simulate the creation of parallel sided combs. Skarka *et al.* (1990), again concentrating on construction at the level of the comb, also demonstrated through a model that incorporated only these essential elements of bee–bee and bee–wax interactions, that parallel, equidistant combs could be a self-organising property of these simpler processes (Fig. 4.9(b)).

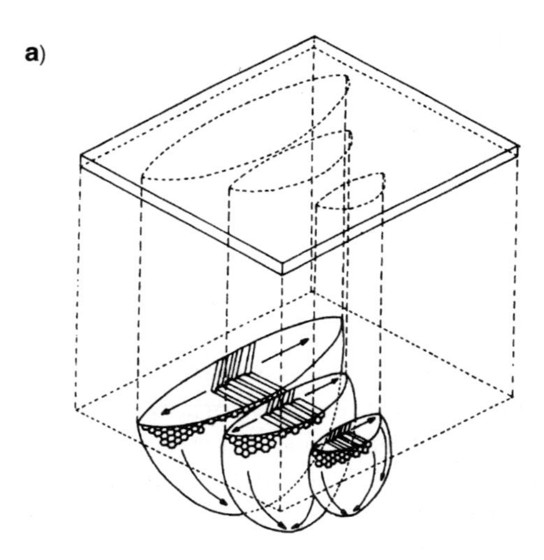

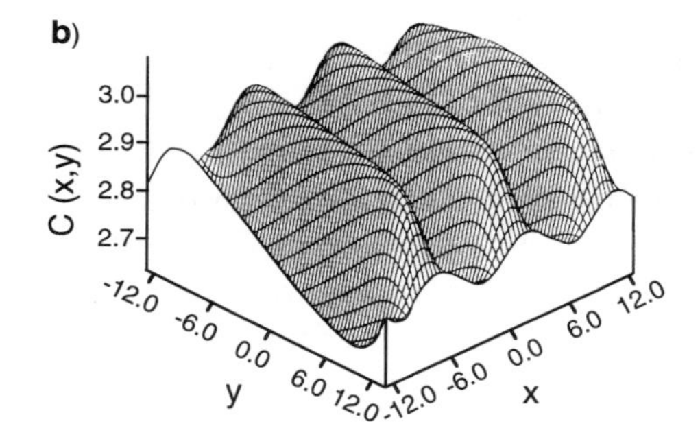

Fig. 4.9. Emergent design of comb arrangement in nests of honeybees. (a) Schematic view of the arrangement of parallel combs placed on the ceiling of a honeybee nest. Arrows indicate directions in which the building proceeds. (b) A model incorporating simply local bee–bee and bee–wax interactions was able to generate comb arrangements and growth patterns resembling those seen in a nest. (After Skarka *et al.* 1990.)

Theraulaz and Bonabeau (1995) and Theraulaz *et al.* (1998) express the view that quantitative stigmergic models, like those described here to simulate the building of termite nest walls or bee combs, are likely to become unmanageably complicated if called upon to model the full complexity of nest construction. Instead they propose a new class of model which is based on *qualitative* stigmergy, and incorporates the principle of *self-assembly*.

Some confusion exists in the literature of the distinction between self-organisation and self-assembly (Sendova-Franks and Franks 1999). An extreme form of self-assembly would be for each component of a structure to carry explicit coding for the type of component with which it could interact. This would ensure that each found its proper position in the architectural whole. This, as these authors point out, carries the risk that a single error may damage the integrity of the whole structure. However, where self-assembly occurs with self-organisation, the self-organising properties of the system ensure that appropriate specified configurations arise, which direct the assembly process.

It is for this reason that natural selection has generally ensured that these two processes occur together.

The system that Theraulaz and Bonabeau (1995) choose in order to test their qualitative stigmergic model is the growth of the combs in wasp nests, by a building process described as *lattice swarms*. These are models in which 'individuals' or *agents* are allowed to move randomly in a three-dimensional space or lattice. In this they can build by depositing standard bricks, according to rules specified by the particular model, but each based upon configurations of local stimuli (i.e. defined as within an agent's range of perception), not upon any plan or blueprint. There is also no communication between agents. These swarms resemble social insect builders because the model does not depend upon individuals following a set sequence of behaviour, and permits building to occur simultaneously in different parts of the structure.

The rules incorporated into these lattice swarm models are entirely artificial, but designed to correspond to sensory and behavioural capacities known in wasps. So, although the models do not test how wasps build, they are a test of whether such a qualitative stigmergic algorithm could in principle create complex, dispersed architecture as seen in eusocial wasps. The results give strong support to the conclusion that it could (Fig. 4.10). A semblance of pattern or organisation is evident in architecture generated by *coordinated*

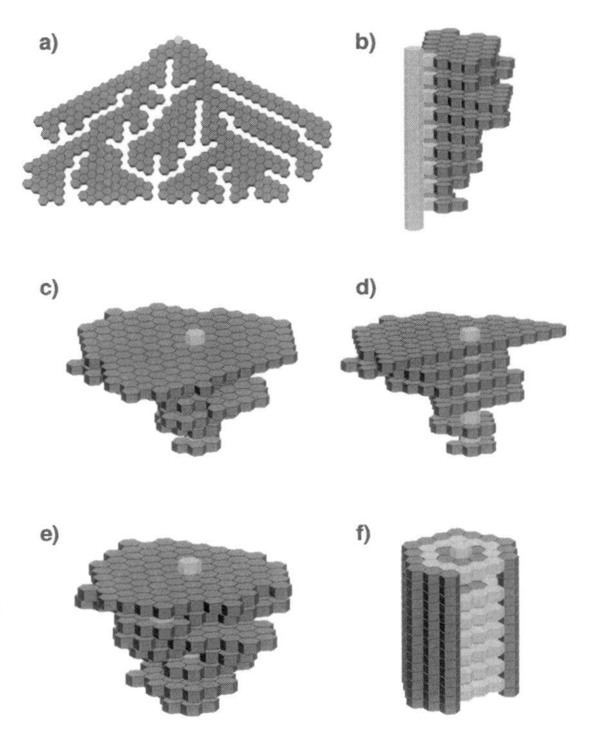

Fig. 4.10. Self-assembly by *lattice swarms*. Examples (a)–(f) are generated by coordinated algorithms using hexagonal building units. These resemble examples of actual wasp nest comb architecture, as indicated below. (a) *Parapolybia*, (b) *Parachartergus*, (c) *Vespa*, (d) *Vespa* (with section cut away), (e) *Stelopolybia*, (f) *Chartergus* (envelope partially removed to show internal structure). (Permission of Guy Theraulaz.)

algorithms; that is by algorithms where the local stimulating configurations created at a particular stage, have non-overlapping properties with those of a previous or subsequent stage, so avoiding conflicting cues that could disorganise the building process.

Coordinated algorithms have the property that if any of them is rerun, they produce a similar architecture to that of the previous run (Fig. 4.10), whereas rerunning non-coordinated algorithms does not necessarily produce the same result. Coordinated algorithms also prove to have the property of *modularity*, a repetition of architectural elements in the arrangement of combs, a particular feature of eusocial wasp nest designs (Wenzel 1991), which in some cases approximate that shown by a particular wasp genus. Non-coordinated algorithms on the other hand do not generate architecture with known biological parallels.

Two other characteristics of the coordinated algorithms revealed by this lattice swarm model are noteworthy; their rarity and their robustness. Considering the total of all algorithms that can be generated in relation to a behavioural space, each characterised by its own set of stimulating conditions, coordinated algorithms turn out to be quite rare, occupying only a small proportion of the behavioural space.

Applying hierarchical cluster analysis to these, demonstrates that any two architectures that closely resemble one another also have similar algorithms. It also turns out that coordinated algorithms are very robust to the addition of random behavioural rules. For the most part these have little or no effect because the architecture is so constrained, that random rules can hardly ever be applied. This demonstrates that the growing structure not only offers new stimuli, but also imposes constraints, which direct the assembly process towards a functional outcome. It is in fact built by a system that combines self-assembly with self-organisation.

4.5 Instinct learning and cognition

So far this chapter has supported the view that the building, not only of individual shelters, but also of relatively enormous and architecturally rich structures, could be achieved following a set of simple decision rules applied to locally available stimuli. In fact, a leaf rolling beetle like *Chonostropheus chujoi*, by making an assessment of leaf size on the basis of information gained during a leaf survey (Sakurai 1988*b*), may be making a more complex assessment than a termite assisting in the construction of a mound, the dimensions of which are orders of magnitude greater than its own body length. The behavioural organisation of both, however, does seem within the capacity of genes to codify, so does any building behaviour remain that cannot be explained in this way and, if so, how is it organised?

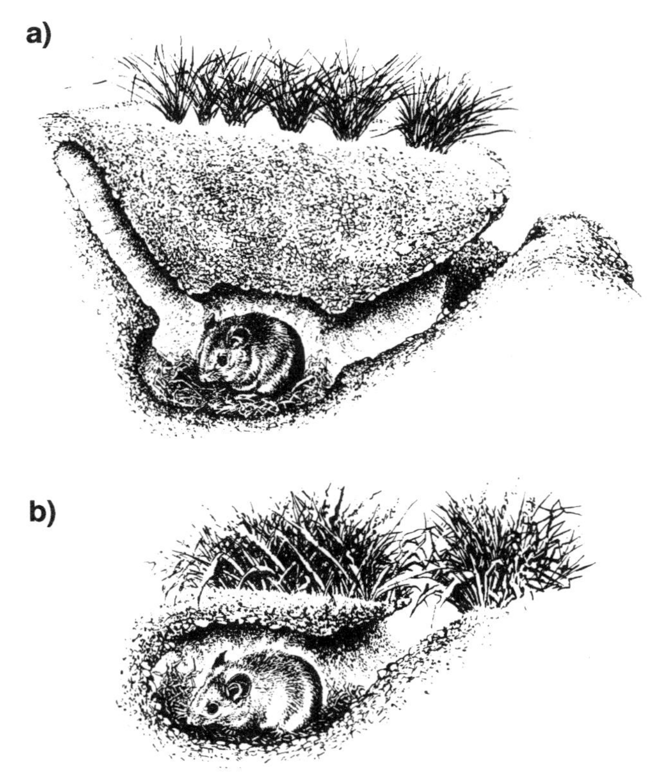

Fig. 4.11. Genetic differences in burrow architecture of rodents. (a) Burrows of oldfield mice (*P. polionotus*) have more than one exit tunnel from the nest chamber, with a plug blocking all but one. (b) Burrows of deermice (*P. maniculatus*) have a single short tunnel from the nest chamber to the surface. (Adapted from Hoffmann 1994 with permission of Cambridge University Press, after Dawson *et al.* 1988.)

4.5.1 Developmental determinants of building behaviour

Very little evidence is actually available on the relative contributions of genetic and environmental influences on the development of building behaviour and the manner of their interaction.

Oldfield mice, *Peromyscus polionotus*, construct a burrow system where more than one exit tunnel connects to the nest chamber, a plug blocking all but one exit. Deermice, *Peromyscus maniculatus*, dig a simple chamber at the end of a short burrow (Fig. 4.11) (Dawson *et al.* 1988, Hoffmann 1994). Both species, after 20 generations reared in laboratory cages without an opportunity to dig, excavate a species-typical burrow. Hybrids between them dig burrows characteristic in all details of *P. polionotus*. F1 hybrids are fertile and, if backcrossed to the recessive *P. maniculatus*, yield offspring showing varying mixtures of burrow characters of the two species (Dawson *et al.* 1988). Burrow architecture it seems is determined by a number of loci at which *P. polionotus* alleles are dominant.

Evidence of learned building behaviour is poorly documented and while this could be due to a preponderance of genetic determination in the control of building behaviour, it does also seem to have been due to neglect. The weaving with flexible strips of vegetation by weaver birds could entail the learning of fine motor control. Collias and Collias (1964*a*) compared the nest building behaviour of young male village weavers, *P. cucullatus*, with or without experience of flexible green vegetation used in nest building. A preference for flexible over rigid was more common in experienced over naive birds. Naive yearlings were unable to weave a single strip in their first week of being offered flexible reed grass, whereas experienced controls were quite proficient. Even after 3 weeks of practice, birds of the deprived group were only capable of weaving 26% of strips, compared with 62% by controls (Collias and Collias 1964*a*, 1973). Orb web spiders alter their web designs in response to a variety of environmental experiences (Section 6.3.3). Whether any of these are examples of associative learning is unknown but they indicate the possible occurrence of it among invertebrate as well as vertebrate builders.

Clearly some kinds of building do involve learning. This might enhance building performance by, for example, improved conceptual appreciation of the nature of the building programme or, at a more humble level, improved motor coordination. The study of animal tool behaviour provides a body of research where a range of learning capacities can be studied.

4.5.2 Tool manufacture and use

Tool use has sporadic occurrence in the animal kingdom; tool manufacture is rarer still (Beck 1980). The manufacture of a tool is by definition a construction behaviour and even the use of tools involves manipulation skills. It is therefore appropriate that tools should be considered alongside animal constructions as a whole, against which at first glance they seem unimpressive in scale and complexity. This chapter has presented evidence that the construction of complex architecture could be accomplished with a simple nervous system making a series of simple decisions based on immediate rather than recalled information. It should therefore strike us as surprising that it is for the manufacture of such simple objects as tools that we find claims for animals of advanced cognitive processes.

Because we are the pre-eminent tool-making species and are very curious about the evolutionary origins of our remarkable abilities, we need to be very careful not to over-interpret superficially similar behaviour in other species. But if tool manufacture and use do require long periods of learning, and provide evidence of insight and inventiveness, then the conclusion must be that a manufactured object, tool, nest, etc., by itself gives misleading evidence of the mental process involved in creating it. Tool use by insects is likely to be overwhelmingly genetically determined and unlikely to involve insight (Alcock 1972). It therefore ranks among the lowliest of examples of environmental modification by non-human animals, of which the construction of large termite mounds is arguably supreme. But there is another kind of environmental manipulation in a world where learning, thinking behavioural innovation are rewarded. Here, non-human tool use is a lowly expression of abilities that do illuminate the path that leads to the flowering of human technology.

This distinction is not a new one. Ingold (1986), in an essay which seeks to clarify the different senses in which the term tool is applied in human and animal biology, makes a distinction which incidentally but rightly includes all construction behaviour. The distinction is between constructions that conform to the Dawkins (1982) concept of extended phenotype (a product of the animal's genotype, externally expressed), and those that result from a mental plan. The latter only in Ingold's terminology are *artefacts*, that is, the product of technology; *technology* being a body of knowledge on which the creation of the artefact depends. This distinction illustrates again that constructed objects themselves can be a poor guide to the complexity of the underlying neural processes, which only experiment can expose.

Beck's (1980) review of the subject of tool use showed that to define a tool is difficult and in some respects arbitrary. His collected examples are miscellaneous and, while some do support the view that tool manufacture and even use, may require manipulative skill and insightful problem-solving, others strongly support the reverse (Hansell 1987*b*). Examining tool behaviour here in the context of building behaviour serves as a counterweight to judging it against human capabilities. It also raises the questions of whether and how advanced mental faculties might provide any advantage in the evolution of building.

Chimpanzees were reported by Lawick-Goodall (1968) to use leaves as sponges and to select and prepare fine flexible plant stems to insert into termite mounds, feeding on the termites pulled out hanging on to the stem in defence of their nest. Subsequent studies have described other chimpanzee tools to feed on termites, for example, stiff sticks that with one end can be used to dig into *Macrotermes* nests, while the other end is brush-like, apparently to pick up the termites (Sugiyama 1985). At another site, chimpanzees have been found to use a *tool set*, with a straight and sturdy digging stick for breaking into the *Macrotermes* mound after which a finer, brush-ended probe is used to pick up the termites (Suzuki *et al.* 1995). Tool use is evidently a feature of all chimpanzees communities, and among non-human primates, only chimpanzees have been reported with tool sets and tool kits (different tools for different tasks) (McGrew 1993).

But, in spite of the habitual use of tools by chimpanzee and the variety of tools used, the construction abilities displayed in their preparation seem modest; stems are broken off plants and associated leaves removed. The specifications for the preparation of sticks may be quite particular in terms of length or thickness (McGrew 1974, 1992), but, the importance attached to them depends upon the level of understanding or cognitive ability displayed by the chimpanzee, and therefore the extent to which it can be said to have a technology in the sense of Ingold (1986).

Evidence of learned skills and problem-solving is probably best documented in the use by a few chimpanzee groups of the combination of a hammer and anvil to crack open hard-shelled nuts. For this to be effected the nut must be placed on a flat anvil and struck with a hand-held hammer. Matsuzawa (1994) found that no chimpanzee under 3 years old could accomplish this. Infants handled nuts and stones from an early age, but not until 2 years of age did they place a nut on a stone. By 3 years a nut was placed on a stone and hit with the hand, shortly after which the full combination of actions was attained, but refined levels of accomplishment took almost 10 years to achieve.

Use of a *'meta-tool'* (a tool that supports the action of a tool) has been claimed in this nut cracking behaviour (Matsuzawa 1994). An experienced nut-cracker was anecdotally recorded as placing a stone under an anvil to make the top surface more level and so prevent nuts rolling off. This is very simple in terms of construction behaviour, but in this context, was taken as evidence of insight and problem-solving.

Social learning is apparently also involved in the development of nut-cracking skills. Infant chimpanzees take an interest in and interact with adult nut-crackers from an early age suggesting observational learning. A few recorded incidents involving a mother leaving nuts or a hammer near to an anvil, offering a youngster a better hammer or adjusting position of a nut on the anvil, hint at an active educational process (Boesch 1991, 1993). This tool using behaviour does therefore seem to be dependent on cultural transmission, although genetic influences could also be involved. Confirmation of a culture requires that the transmission of the behavioural characteristic is not constrained ecologically in other habitats but never the less does not occur. Under this definition McGrew *et al.* (1997) found an absence of nut cracking among chimpanzees at a site in Gabon in spite of the presence of suitable nuts and potential implements to crack them; Whiten *et al.* (1999) documented 39 different behaviours that are characteristic of a particular group of chimpanzees but absent in others in spite of suitable conditions; some of these concern tool use.

Use of tools ranging from probes and hammers, to more marginal examples such as bait fishing has been found to occur in 104 bird species, a higher number than has previously been recognised (Lefebvre *et al.* 2002). This survey also found a larger residual brain size in true tool users compared with borderline tool users, and a similar difference between them in the size of the neostriatum, an area of the bird brain regarded as comparable with the mammalian neocortex. Their interpretation is this: brain development equips birds for the 'complex cognitive processes involved in tool use'. Much more scarce, however, is evidence of such processes.

The woodpecker finch *Cactospiza pallida*, which is one of the Galapagos finches, not only uses a cactus spine or sharp stick to extract insect prey from crevices, but can also modify the tool to make it more effective. Experiments on the development of this behaviour using captive birds have shown that captured adult finches that were not tool users, remained unable to use tools after being housed with tool users. Naive juveniles housed with experienced tool users did however, all develop tool use, but so did all members of a control group of juveniles without adult company and in a comparable length of time (Tebbich *et al.* 2001). Juveniles can develop the skill without being reinforced with prey, but show a strong inclination to manipulate tool-like objects. The authors conclude, with commendable caution that trial-and-error learning appears to be linked with genetic predisposition in the development of this tool behaviour. Social learning is not confirmed and certainly not necessary.

More elaborate looking tools are made by the New Caledonian crow *Corvus moneduloides*. Like woodpecker finches, they make tools to extract insects from crevices in trees. They may use stick probes but they also make two different tools that depend for their effectiveness on the principle of the hook: a hook-ended twig, and a stepped-cut ribbon of *Pandanus* leaf. The creation of the hooked twig depends upon the crow making a cut with its beak at the base of a fine twig that includes a recurved spur from the main stem. The hooks on the *Pandanus* leaf are the serrated edge of the leaf itself, but the tool is made in a rather precise routine. The bird first makes a small, angled cut into the leaf before tearing towards the leaf tip, following the direction of the parallel leaf veins. It then makes a second angled cut away from the leaf edge, and a second tear, before cutting directly in from the leaf edge to meet the tear at its base to separate a stepped tool, with hooks along its edge pointing towards the wider end. Alternatively, the cycle of cuts and tears may be repeated to create a tool with up to five steps (Fig. 4.12) (Hunt 2000*a*).

Hunt (1996, 2000*a*,*b*) is keen to stress the relevance of the behaviour of this crow to the study of evolution of humans. In making the case, he points out the presence of local and regional differences in the manufacture and use of tools and upon the population-level occurrence of 'handedness' in the cutting of the tools.

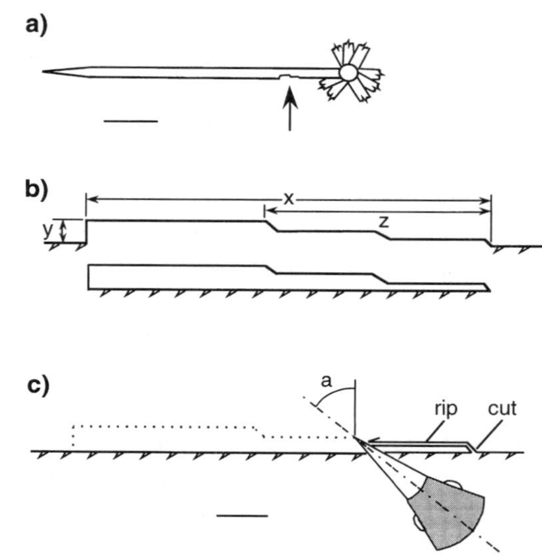

Fig. 4.12. Tool construction by the New Caledonian crow (*C. moneduloides*). (a) Stylised cross-section of the leaf crown of a pandanus tree viewed from above (scale bar = 35 cm). (b) A stepped cut tool with three steps, cut from a pandanus leaf. The directional spines of the edge of the leaf are shown along its border. Above it is shown the outline of the edge of the leaf from which the tool has been cut (x = tool length, y = tool width, z = length of tapered section of tool), (c) To make the tool the crow first makes a cut to form the pointed tip, then rips a strip of leaf away before making the second cut. The angle a at which the cut is made shows some regional variation, but the final cut is typically made at right angles to the edge of the leaf (scale bar = 20 mm). (Adapted from Hunt 2000*a*.)

The distinctive local differences Hunt refers to as 'traditions', although without evidence of their persistence over time. This, it is made clear, is not intended to imply cultural transmission, which at present remains unstudied. On handedness, Hunt (2000*a*) contrasts the lateral brain specialisation in the absence of symbolic language observed in the crows, with its presence in humans. This, says Hunt, provides us with a better model for looking at the neurophysiology and behaviour of a non-human tool-maker than chimpanzees, and argues that the complexity of the construction routine, the precision of the manipulation and the consistency of the output indicates a level of tool manufacture in these crows comparable to that shown by humans no more than 100,000 years ago.

Hunt and Gray (2003) argue for a single origin of the leaf cut tool, with cultural transmission and innovation leading to regional variation and greater tool complexity. This, based as it is on the outlines left in the leaves after tool removal and without any direct behaviour evidence, is a degree of interpretation that would be considered surprising in the context of bird nest building.

Great caution is needed before embracing Hunt's interpretation of crow tool complexity, because animal construction behaviour shows that there is a well trodden alternative path towards standardisation of building units, that of the extended phenotype (Dawkins 1982). Standardisation in the manufacture of building units, even complex shapes like the leaf panels of the caddis larva *Lepidostoma hirtum* (Hansell 1972) was taken in Chapter 2 as evidence of stereotypy. The Caledonian crows show manipulative skill, but so do nest building weaver birds. If we defined each of the fastenings used by the village weaver (Collias and Collias 1964*b*) (spiral coil, overhand knot, slip knot, etc.) as a tool, would we be more impressed? Evidence of the ontogeny of the pandanus leaf cutting routine is needed to determine its dependence upon learning and understanding. Certainly a detailed study of the ontogeny of weaver bird nest building should similarly yield valuable information.

Laboratory studies on the cognitive abilities of New Caledonian crows have begun. Two adult crows taken into a laboratory environment, were required to select a tool from a rack of sticks of varied length in order to extract a piece of meat from inside a horizontal transparent tube. The crows showed no hesitation in making use of the tools provided, generally choosing either a tool the length of which matched the distance of the meat from the tube entrance, or the longest stick provided. This, concludes Chappell and Kacelnik (2002), demonstrates an ability to respond flexibly to the problem in a manner comparable to primates, but conclude that it remains to be seen whether this reflects specialisation in tool behaviour or an overall highly developed cognitive ability. McGrew (personal communication) raises a pertinent question: Is the problem presented to the crows one of perception or of cognition?

Weir *et al.* (2002) presented the meat reward in a bucket at the bottom of a vertical transparent tube. Given a choice of a straight or hooked piece of wire, the two adult crows chose the hook and, more surprisingly, when one of the two birds was left only with the straight wire, it bent the end of it to create a hook to obtain the food. Woodpecker finches in a laboratory test have also shown an ability to modify twig tools to make them effective (Tebbich and Bshary 2004). These are very simple construction behaviours, but do seem to indicate cognitive abilities resembling those to tool-using primates.

Ingold (1986) makes a distinction in the context of tool manufacture between the making of *artefacts* and the making of other constructed objects; the former alone is the product of a *'design intention'*. If Inglod's (1986) dichotomy between making artefacts and simply manufacturing an object was applied to all animal building, then it seems possible that some, but a very few, living species would be credited with making the former, and that their qualification would be tool manufacture. Evidence taken from across all construction behaviour does show that much of invertebrate and vertebrate construction can be accounted for in terms of a stigmergic programme. This leaves species where the control of building imposes a range of increasing demands upon the central nervous system: short-term retention of and comparison between sensory information, learned changes in manipulative ability, and the understanding of design goals. The extent of these is uncertain at the moment because of insufficient detailed information on the ontogeny of construction movements, and of experimental studies on problem-solving or the repair of selected damage. The most advanced of these species may be building to a plan and have a degree of self-awareness, but they represent one end of a continuum that it is not helpful to subdivide, at least at the moment.

5 Mechanics, growth, and design

5.1 Introduction

Chapter 1 recognised two broad functional categories of animal architecture, houses and traps. The traps are generally nets; these have a short operational life compared to the life of the builder, are constructed quickly and are only operational when completed. A house on the other hand may endure the lifetime of the individual or colony that builds it, starting small and growing as the occupant(s) also grow. This chapter considers two problems relating to all these structures, whether trap or house. Why don't they fall to pieces, and how do they solve the problem of growth, while still remaining operational?

The purpose of a trap is to capture prey. It therefore has the quite circumscribed job of intercepting and restraining moving prey. It was, however, argued in Chapter 2 that this was a demanding job in engineering terms that has been solved in large measure by the evolution of sophisticated, self-secreted materials. Nets depend for their capture effectiveness on their ability to bear loads in tension. The orb web of spiders is a relatively simple structure that operates in this way. Structures that are entirely in tension are easier to study and understand than the rare ones that are wholly in compression, or the more common structures that experience a combination of tensile and compressive stresses. Orb webs, as simple structures made of rather standardised materials in tension have provided a convenient model system for understanding mechanical design in animal built structures. Significant progress has therefore been made through the study of them; this is in contrast with our understanding of any other such structure.

A house is a rather non-specific functional category. It provides shelter, but from what? The security it provides may be mechanical or thermal; it may have additional ancillary functions such as food storage. Houses may be on the ground, in it, or suspended above it. Houses as a category are mechanically varied, made of all possible kinds of materials, and often structurally complex. This combination of features makes them much less well understood than the best-studied nets.

5.2 Growth and design

5.2.1 Gnomons

The most economical of the net structures, the two-dimensional spider orb web being an example, are built in one session as complete structures according to a comprehensive design. Some houses are also built in this way; the pupal cocoons of lepidoptera, for example, are built as one continuous behavioural sequence. Many other species, however build houses in which they live for an extended period enlarging it as they grow. This includes species like caddis larvae, which live in individual houses, or colonial species such as vespine wasp, where an initial single comb of a few cells, ultimately becomes a stack of combs each with hundreds of brood cells attended by hundreds or thousands of adults.

One of the predictions upheld in Chapter 3 was that natural selection had favoured simple building rules and repetitive actions. If that hypothesis is applicable to architecture that grows with the builder, we would predict that building rules are conserved during growth. In this context, Thompson (1942) drew attention to the property of certain figures to remain unaltered in proportions with repeated incremental additions. These were referred to by Greek geometricians as *gnomons*. One such figure is a cone which, after repeated additions to the base, becomes a larger cone of the same proportions. In fact the additions could be to all sides but the base, or even to all of them including the base, and the shape of the cone would remain the same.

A caddis larval case approximates to a cone. Although not pointed at the fine end, it can be added to in small increments during larval life. This produces cone-like mineral cases in which the size of the sand grains does enlarge gradually from the narrower to the wider end. This appears to accord with the predictions of the conservation of building rules for particle size selection and case diameter determination during growth. However, the larval leg and head skeleton which are used to select and attach the sand grains, grow only at each moult and therefore in a step-wise manner, while case dimensions appear to grow gradually (Hansell 1984). If this is indeed so, then for the gradual enlargement of case diameter and particle size to occur, building rules cannot be conserved, but must be gradually adjusted during the course of each instar.

In some caddis species the larval case may undergo a radical change in design at a particular point in larval development. For *Lepidostoma hirtum* this is a change from a sand grain tube of round section to one of square section made of cut leaf panels, producing a hybrid architecture in the third instar (Hansell 1972). This sudden and marked change in case architecture, which is also seen in some other caddis species, contravenes the prediction of economy in construction behaviour programmes, and indicates some other priority, perhaps a change in habitat part way through larval development.

5.2.2 Spirals, helices, and helicospirals

The growth of a gnomon need not be in straight lines. The growth of the spiral shell of the Cephalopod *Nautilus pompilus*, for example, is gnomonic; it retains the same proportions at any stage of its growth. A spiral is a curve, which continually increases its radius of curvature as it moves away from its point of origin (Thompson 1942). The shell of a nautilus is an equiangular or logarithmic spiral, the same type of spiral as that formed by the scaffolding spiral in the orb webs of araneoid spiders (Fig. 4.6). The merit for the nautilus of this growth pattern is that the distance between neighbouring whorls of the spiral become progressively larger, allowing for the increasing size of the organism.

A helix is a curve which does not alter its radius of curvature, but each turn is displaced along the axis of rotation, as in a DNA molecule. This allows extension of the helix at either or both ends. One further growth form combines the features of helix and spiral. The shell of many gastropod molluscs is a cone coiled along an axis of rotation such that the curve in space generated by any given point makes a constant angle to the axis of the enveloping cone. This form is described as a *helicospiral* (Thompson 1942).

These alternative growth patterns can be described by three axes of variation: expansion rate, translation rate, and radial displacement. If this is represented as a three-dimensional space, it can be seen that the shells of

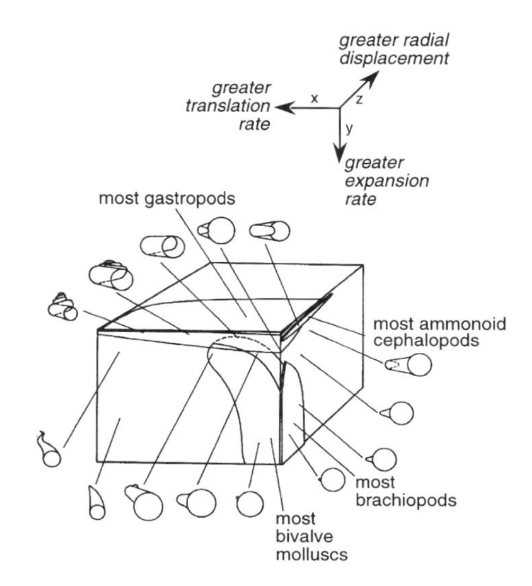

Fig. 5.1. Potential spiral growth forms shown in a three-dimensional space with axes: x = translation rate, y = expansion rate, z = extent of radial displacement. The parts of the space occupied by the major shelled invertebrate groups are shown. The boundary delimiting the gastropod molluscs, which the houses of some animal builders resemble, envelops only a small minority of the available space. (Adapted from Currey 1970.)

gastropods, and indeed of other shelled invertebrates such as cephalopods and brachiopods, only occupy a minority of the available space (Fig. 5.1), perhaps the result of functional adaptation or of evolutionary constraint (Raup 1966; Currey 1970).

The possible advantages of the gastropod shell growth form are various, but two fundamental ones are that the contact between neighbouring whorls increases shell strength, a second is that overlap of the walls saves building material. Heath (1985), testing a prediction of maximum efficiency in the use of shell material, showed that there was a U-shaped relationship between the ratio of external shell area and the volume it enclosed when plotted against the extent of whorl overlap. This curve gives an optimum measure of overlap, where the measure of new shell surface over enclosed volume is at its lowest. However, examination of a sample of terrestrial gastropods showed that the degree of overlap was greater than that predicted by the economic hypothesis, indicating some other selection pressure possibly for greater shell strength.

Caddis larval cases show some curvature in a large proportion of species for unexplained reasons. In the Helicopsychidae, which contains around one hundred described species, the growth form is that of a helicospiral (Fig. 5.2) (Wiggins 1977). The case growth pattern of at least one species of bagworm, *Apterona helix* (Psychidae), is also in this form (Wheeler and Hoebeke 1988). Williams *et al.* (1983) demonstrated that the cases of *Helicopsyche borealis* were more resistant to crushing than straight sand grain cases built by other caddis species of comparable or larger size occupying the same habitat. They concluded that this was an adaptation to allow the larvae to move through the interstices of the gravel without being crushed. However, the strength of the *Helicopsyche* case will depend upon how massively it is built so, without this comparison with straight-cased caddis species, it is not possible to conclude that its comparative strength derives principally from its spiral form. This leaves no convincing evidence of the functional significances of spiral designs exhibited by case building caddis larvae and Psychidae. A useful first step would be a more detailed description of their form, which allowed them to be located in the three-dimensional space alongside the Gastropoda (Fig. 5.1) (Raup 1966), and tested against the hypothesis of Heath (1985) of optimal whorl overlap.

Good mechanical protection with economic use of material is the suggested function of the helicospiral retreat built by the spider *Achaearanea globispira* (Theridiidae). This is as a silken tube enclosed in a spheroidal envelope of sand grains (Fig. 5.2) Henschel and Jocqué (1994). It is inferred that the enlargement of this tube along its length represents successive extensions as the spider grows to maturity. This is supported by the evidence of double retreats resulting apparently from the attachment to a female retreat of one built by a mature male. The retreat of the male, which coils in the opposite direction to that of the

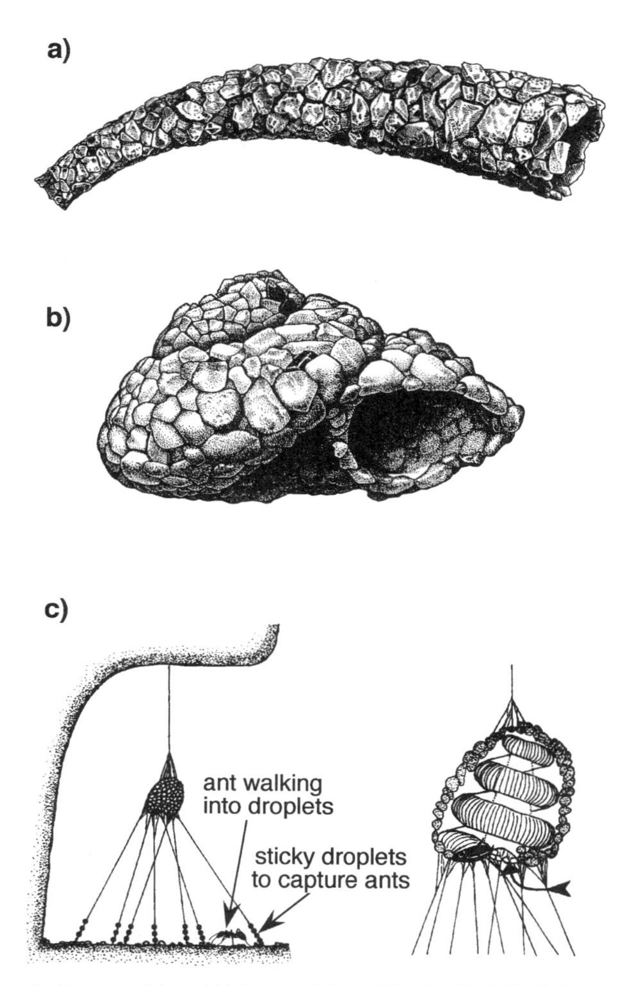

Fig. 5.2. Helicospiral houses: (a) and b) Adapted from Wiggins (1977), (c) from Henschel and Jocqué (1994). (a) The case of *Pseudogoera singularis* shows the curvature and widening towards the anterior typical of many caddis larvae. (b) The case of the caddis larva *Helicopsyche borealis* shows a helicospiral growth form. (c) The silken retreat of a female bauble spider *Achaearanea globispira* is covered with sand grains and suspended above capture threads bearing sticky droplets for catching ants. A section through the retreat shows a helicospiral tunnel and central chamber with a lightly sealed emergency exit (arrow) (Sticky droplets and ant enlarged for clarity).

female, is a simple helix, apparently because it is all built when the male is already adult.

The nests of bees and wasps, which are enlarged as the colony grows, are not constrained, as are the cases of caddis larvae, by the increasing dimensions of a single occupant. These social insect nests do grow in a number of different ways including a spiral. A well-established colony of the wasp *Agelaia areata* has a comb in the form of a loosely rolled sheet of paper with the spacing

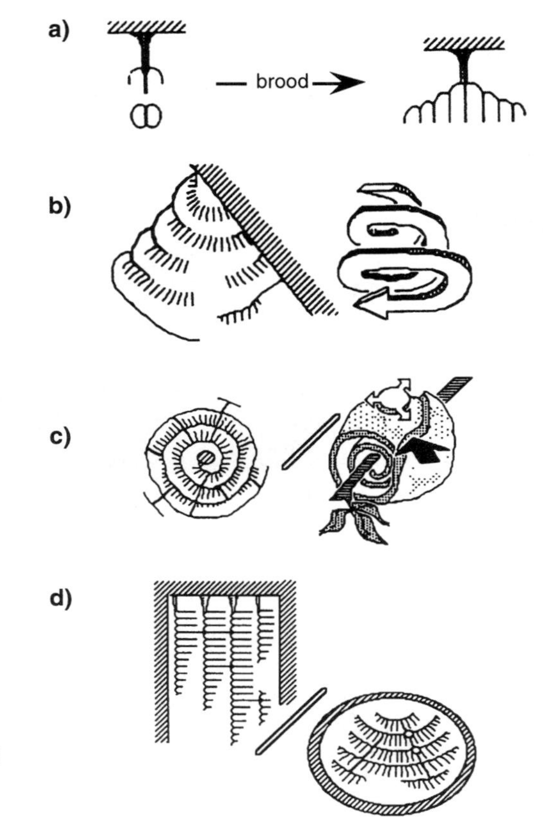

Fig. 5.3. A selection of the architecture and growth patterns of nests of eusocial Vespidae with names of associated genera or species: (a) *Polistes* and *Mischocyttarus*. Single comb suspended from petiole. (b) *P. raphigastra*. Helicospiral comb growth form. (c) *Agelaia areata*. Spiral extension of comb on the axis shown. (d) *A. testacea*. Concentric, vertically oriented combs, with cells cantilevered out from the axis of suspension. (Adapted from Wenzel 1991.)

between successive layers of a comb remaining constant to form an arithmetic spiral. In the polistine wasp *Polybioides raphigastra* however, comb growth is in the form of a helicospiral enclosed in a nest envelope of the form of that of a gastropod mollusc (Fig. 5.3).

Six genera of Polistinae have independently evolved an expanding spiral nest form, twice as a growing sphere, four times as a helix (Fig. 5.3) (Wenzel 1991). A helical pattern of comb growth is also shown by some meliponine bees of the genus *Trigona* (Michener 1974). In all these, the comb design offers the possibility of indefinite comb growth with minimal or no change to the building rules.

Among burrowing species, helical construction, while not common, is shown by a variety of taxa; helicospirals are, however, unknown. There is no obvious structural advantage for a helical burrow and the possible disadvantage that successive passes of the helix, one under the other, could undermine the system. The vertical separation of successive turns of the helical burrow depends upon the angle of descent and the radius of curvature but, a general lack of detailed descriptions makes assessment of rival hypotheses of structural, constructional, and operational advantage of these hard to make. From a constructional point of view, a helical burrow can be regarded as having two components, a ramp and a curve. A possible constructional advantage for a ramp is that, compared with a vertical shaft, spoil removal up a ramp may be easier (Dworschak and Rodrigues 1997). This argument is unaffected by the radius of curvature of the ramp, as the length of burrow required to reach a given depth is the same for a given angle of descent. The spoil removal explanation may have validity for some species, but is unconvincing for the helical ramps in the burrow system of the thalassinidean shrimp *Biffarus arenosus*, for example; it has vertical shafts in the upper part of the burrow system while the helical burrows are lower down (Bird and Poore 1999).

A possible constructional advantage for burrow curvature could be anatomical asymmetry which allows easier digging on one side than the other. Male crabs of *Ocepode ceratophthalma*, for example, construct a burrow of anticlockwise descent if their left chela is longer and vice versa for an enlarged right chela. The chelae of females are symmetrical and their burrows U-shaped (Farrow 1971). However, in thalassinidean mud burrowing shrimps, including *B. arenosus*, *Axianassa*, and *Callianassa*, individual burrow systems are frequently found containing helices of both sense (Dworschak and Rodrigues 1997; Bird and Poore 1999). These are thought to be built by a single individual, but this is difficult to confirm. Burrows of the polychaete worm *Notomastus latericeus* (Powell 1977) and the extinct Miocene beaver *Palaeocastor* (Martin and Bennett 1997) both exhibit dextral and sinistral helices, although there is no obvious asymmetry in the body plan of either species.

As with constructed helices, like *Trigona* nest combs, a significant constructional advantage for a helical burrow may be that it can be extended indefinitely using the same behavioural rules. This explanation fits well with elegant design of the ichno-fossil burrow system of what is apparently an extinct species of solitary bee of the late Cretaceous-early Tertiary, found in Uruguay. A reconstruction of this system shows a helical ramp of 16° off which, at half-circuit intervals, branches a more curved passage of 20° incline that ends in a club-shaped brood chamber. The repetition of these three elements, ramp, side tunnel, and brood chamber, creates a stack of brood cells neatly arranged above one another inside a helical burrow (Fig. 5.4) (Genise and Hazeldine 1998).

Alternative operational hypotheses of helical burrows have also been proposed but equally remain untested. The differentiation of the burrows of *B. arenosus* into upper, vertical shafts and lower helical passages, suggests to Bird and Poore (1999) that the helical sections are to better exploit the food supply found at the lower levels. The advantage of the helical burrows of the Miocene beaver *Palaeocastor* (Martin and Bennett 1977) may have the advantage over

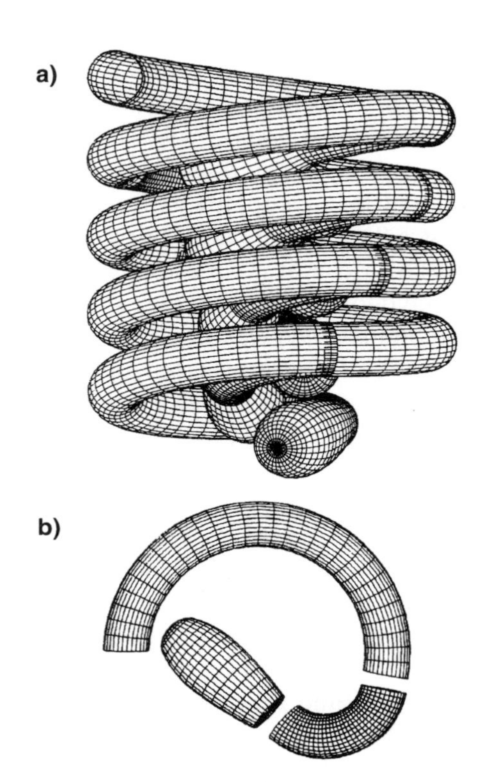

Fig. 5.4. (a) Helical fossil nest burrow, possibly of a solitary bee, shows repetition of the three elements (b) helical ramp, side tunnel, and brood chamber. (Adapted from Genise and Hazeldine 1998.)

a straight ramp of making the burrow more easily defensible or the owner less detectable to a pursuing enemy. It could however, have served to minimise competition between neighbouring burrows. These beavers apparently lived in colonies with burrows no more than 2 m from one another. Their tight helical burrows would have ensured that a spacing established above ground would be maintained below it.

5.2.3 Modules

Many social insect colonies are founded by a single queen or royal pair. Almost all of these species make fixed nests in which to live, and need a structure that remains constantly in operation yet expands to fit the growing colony. Given free space to grow after nest attachment, selection should favour a simple programme of repeated addition. However, Wenzel (1998) describes the layout in mature nests of some species of *Protopolybia* and *Brachygastra* as chaotic. In these, new patches of comb and envelope seem to have been added in no orderly way in relation to the rest of the nest structure. This probably reflects selection for behavioural flexibility in response to variability in their typical nests sites among the branches of trees.

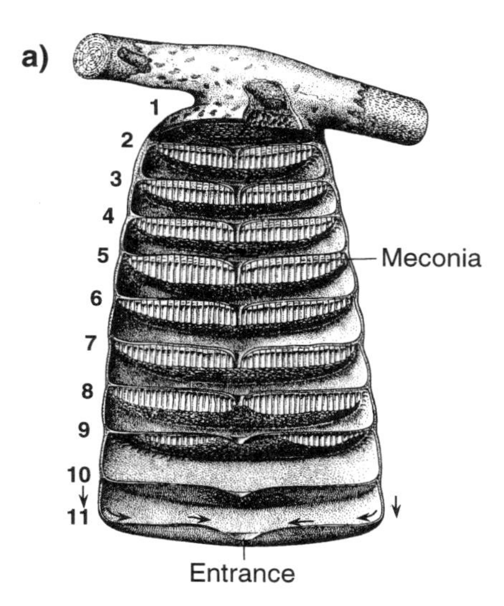

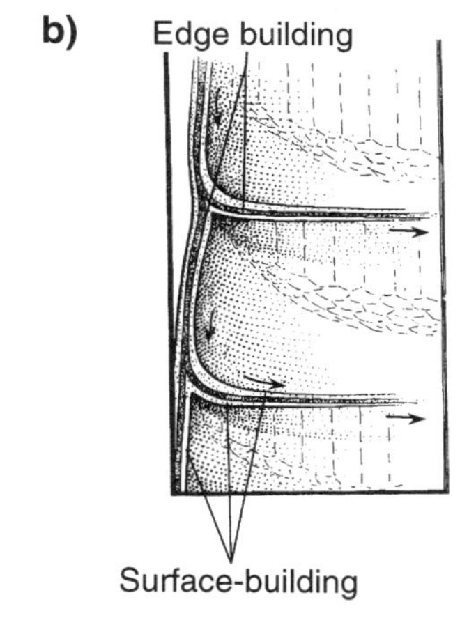

Fig. 5.5. (a) Enlargement of the nest of *Chartergus chartarius* is modular, with the addition of new envelope at the bottom, and construction of a comb on the previous external surface. (b) The new envelope is created by adding material to the edge of the structure but as the envelope at any point needs to bear more combs added below it, it is strengthened by the plastering of material to its inner and outer surfaces. (Adapted from Jeanne 1975 with permission from University of Chicago Press.)

Like helical growth, modular enlargement permits continuing growth while conserving building rules. Modular nest growth is shown by species of *Polybia*, *Chartergus* (Fig. 5.5), and *Epipona* (Jeanne 1975; Wenzel 1991, 1998), where the lower, external surface of the current envelop serves as the base of the next comb to be added after a new envelope has been built over it. In Section 4.4.4 it was shown that the *Epipona*-like design was one of a family of virtual nest designs generated by *coordinated algorhythms* (Theraulaz and Bonabeau 1995). These have the property that the construction rules reliably build the same architecture on repeated occasions. They also have the property of generating modular architecture typical of social wasp nest growth (Wenzel 1991).

The hexagonal cells of which wasp combs are typically composed are, of course, themselves modules. They represent the primary nest architecture, where the comb arrangement is the secondary architecture. Hexagonal patterns where junctions are formed by three arms separated by an angle of 120° are found repeatedly in biology, and represent one of only three regular patterns of polygons (triangle, square, hexagon) that can divide up a plane surface and leave no spaces between. Of these, the hexagon encloses the most area for a given perimeter and so, in terms of material used to manufacture it, is the most efficient (Pestrong 1991).

Bee combs are two layers of cells back to back so that each has six neighbours to the side, and a further three at the back. Were this pattern to be repeated over the cell opening, the enclosed space would be a *rhombic dodecahedron*, a figure that, if composed of similar plane surfaces, encloses the most volume for a given surface area of any equivalent form. (Thompson 1942). Bee combs exhibit both modular growth and economy in use of materials.

The rolled up comb pattern of *Agelaia areata* (Fig. 5.3) results in part from comb cells being wider at the opening than at the base. Curvature of combs, although not to this extreme, is quite a common feature of wasp nest design (Wenzel 1991). However, it has been known since the eighteenth century that a curved surface cannot be created with a regular array of hexagons (Galloway 1988). The spherical molecule of carbon, C_{60}, Buckminster-fullerene (alternatively a modern association football) is composed of 20 hexagons and 12 pentagons. Insertion of pentagons in the hexagonal array is also used by wasps to create curvature in their combs (Wenzel 1989).

5.2.4 Constraints on architectural form

Architectural forms are constrained not only by the properties of materials but by the limitations of the equipment used in the building process. Chapter 3 as a whole demonstrated the general lack among builders of specialised constructional anatomy. The relationship between building unit mass and body mass for building species has not been compiled for animal builders. To do this would be instructive in revealing, for example, the advantage of building in an aquatic rather than a terrestrial environment, or whether species are collecting bricks to maximise total energetic efficiency or to minimise numbers of journeys. However, virtually without exception, the unit size that can be handled is constrained by the size of an individual builder, and in the majority of cases the largest building unit that can be manipulated is much smaller or lighter than the builder itself. Even in advanced eusocial species, individuals do not cooperate to manipulate much larger building units. Inevitably an exception can be found. Weaver ants (*Oecophylla*) could not draw together the leaves which form the walls of their purse nests, without the ability to link their bodies to form chains (Höllolobler and Wilson 1990).

Another exceptional constructional behaviour for non-humans, is the use of a scaffolding, while in our species it is invariably involved in the construction of anything much taller than a human individual. A rare animal example is the scaffolding or auxiliary spiral of the orb web of spiders. This is a device that helps spiders to span the gaps between radii when laying down the capture spiral. Humans also use buildings to produce buildings. The most ancient of these may be the use of arch templates or 'centering' to support the arch stones (voussoirs) before the insertion of the keystone that allows the arch to support itself. The absence of this constructional device does not appear to limit the absolute size of animal-built structures. Termite mounds of *Macrotermes* and *Amitermes* are large enough to hold millions of builders, but even if desirable, they probably could not construct a soil roof to span a 'great hall' within the mound.

Another feature almost completely absent from animal architecture is moving parts. This is unsurprising because highly effective moving parts can be supplied by the builder itself for locomotion, prey restraint, or whatever. However, the limitations of construction equipment also rule out all but the simplest kinetic elements in animal architecture. Nevertheless, moving parts are present in some traps and some houses. Sliding junctions that allow a capture thread on an orb web to move through a neighbouring radius are found in the orb webs of araneoid species (Craig 1987*a*) probably to assist in the absorption of the kinetic energy of moving prey.

The most obvious need in a house for a moving part is for a door. A number of species of invertebrates and vertebrates use an armoured part of their own body to block the doorway, the pronotum of the thorax, for example, in the caddis larva *Goeracea genota* (Wiggins 1977). An equivalent architectural device is apparently present in the nests of the termite *Prohabitermes mirabilis*. In this species the doors take the form of spheres of mud and faecal cement that are found, two to four in number, loose in some nest chambers. The spheres have a diameter slightly greater than that of openings between chambers and have been found cemented into doorways, apparently to block access (Tho 1981). The problem of a hinged door has been solved by trap door spiders such as *Ummidia* (Ctenizidae) by the use of silk and a gravity closing mechanism (Bond and Coyle 1995). The door itself is a silk-covered plug of earth that fits into the bevelled entrance of the home burrows. The creation of this clearly requires some behavioural complexity as well as the appropriate materials.

5.3 Mechanical design

5.3.1 Lack of data

A 6 m mound of *Amitermes* can be considered as a structure in compression, largely due to the dead weight of its own material bearing downwards.

Artificially increasing the load on the mound would eventually lead to *shear* failure as cracks penetrated the mound approximately diagonal to the direction of stress (Gordon 1978). However, when subjected to the lateral force of the wind, the mound acts as a beam anchored at one end. It will resist the bending stress of the wind by opposing tensile stresses on the windward side and compressive stresses on the leeward. Under extremes of such stress, the mound might fail by the propagation of a crack through the mound in the direction of the wind.

I am not aware of any data on the strengths of termite mounds under such stresses, or indeed any accounts of natural failure. This could be a lack of scientific interest, or the total absence of natural failure in a highly over-engineered structure. More probably the structure is engineered to cope with quite different sorts of problem, for example, mechanical damage by termite predators or erosion by rain. The absence of relevant data in this case renders even speculation sterile. But this does reflect a general lack of information on natural mechanical failure of any animal built structures, or experimental testing of them. This necessitates rather broad hypothesising on their mechanical design features and similarly general predictions about properties. However, it also provides an opportunity simply to point out some aspects of structures that show interesting complexity awaiting more detailed study. An exception to this pattern are spider webs which, for reasons of simplicity of materials and structure given earlier in this chapter, have proved very convenient to study, although it is evident that even here there are aspects of their functional design that are still little known.

5.3.2 Beams

A beam with its weight supported at both ends, is in compression in the upper part, while the lower part is in tension. Consequently there is a line, which in fact runs through the centre of gravity of the beam, which is an axis of neutrality with respect to stress. The distribution of stress in a cantilevered beam is essentially the same except that, since the beam is attached at only one end, it sags down at the other, putting the upper beam surface in tension and the lower in compression. A grass stem is a vertical cantilevered beam, which is bent by the wind. As the stresses it experiences are greatest at its edges and zero along its axis, the material is distributed accordingly to create a hollow tube of round section because the direction of the wind may be from any quarter. Stresses due to bending are resisted not simply by the area of cross-section of the beam but by its shape.

The stresses in a bending beam are not simply due to the magnitude of the bending force, but rather to the bending moment. This, for a cantilevered beam subject to a downward force at its free end is the bending force × the length of the beam, for a beam supported at both ends, subject to a downward force in the middle, it is the force × length/2.

A number of bird species make nests as complex arrays of stiff beams of round cross-section. These take the form of solid or hollow rods, in practice sticks or grass stems. These may be made into simple platforms as in the nests of some dove species (Columbidae), deep cups as in the grass stem nests of *Sylvia* warbler species, or domed cups, as in the nest of the black-billed magpie (*Pica pica*).

Both twigs and grass stems used in bird nests are generally dead when collected. In life both are designed as very long cantilevered beams and may retain some of the virtues of this in death. In bird nests, the unsupported lengths of any sections of beam of such materials are short, although the bending forces they experience are unknown. We know little of any aspect of the mechanical properties of these structures, although it is striking that completed nests of these materials can exhibit considerable robustness. One possible reason for this is that the nest possesses some of the properties of a three-dimensional roof beam structure known as the *reciprocal frame* (Chilton *et al*. 1994; Hansell 2000). This has the properties that the completed array can span a distance greater than that of an individual beam, that beams lock each other in position, and that stresses from localised loads are spread through the structure.

One consequence of the creation of a hollow beam is a dwelling opportunity within it. This is a house design adopted by many species of caddis larvae, the bagworm family (Psychidae) of Lepidoptera and some other insects. Describing these homes as 'beams' could however be very misleading, suggesting as it does that these are designed wholly or particularly to resist bending. Predation studies on caddis larvae show that for some species, resistance to bending may be important in defence against, for example, fish predators (Otto 1987), but not against predation by larvae of the water beetle *Dytiscus* (Johansson and Nilsson 1992). The caddis case crusher devised by Otto and Svensson (1980) to measure caddis case strengths, supports the case at two points and applies a load mid-way between them. Failure occurs through the bending of the case till it buckles. This simple but sensitive device has been used in subsequent studies (e.g. Johansson 1991), but its relationship with the forces that, for example, a sand grain caddis case (Fig. 5.2) is designed to resist, is unclear. We need better-designed test procedures.

Where the bending moment acting upon a beam is in a consistent direction, the beam section can exhibit a bilateral rather than a radial symmetry, with the material again distributed away from the axis of symmetry to the location of greatest stress. The engineering I-section beam is a popular human solution, but beams supporting the combs of some wasp species face the same demands. The combs of nests of many paper wasps (Polistinae) are connected to an overhead substrate by means of a petiole of nest material. As the comb is enlarged, so the cross-sectional area of the petiole is too (Downing and Jeanne 1986). The attachment of the petiole to the comb may be in the middle or towards the edge (Figs. 5.3(a) and 8.3(a)). Where an asymmetrically supported comb is not hanging vertically, it exerts a bending moment on the petiole. Selection favouring economy of design predicts that material in the petiole beam should be distributed so as to resist bending with the minimum use of material. As the direction of the bending force is known, a section resembling the engineering I-beam would be the most efficient.

Polistes and *Mischocyttarus* nests have no envelope, so the petiole has the important role of limiting invasion of the comb by predatory ants by minimising the probability that a foraging ant will encounter it. This should select against a petiole with larger circumference in proportion to its cross-sectional area. This hypothesis predicts that, for symmetrically supported combs, the petiole should be circular in section and that for cantilevered combs in these genera, a dissected profile should be more pronounced where predation pressure from ants is lowest. Jeanne (personal communication), from examining nests of species from tropical (greater ant threat) and temperate (lesser ant threat) regions, has the impression that these predictions might be upheld by these two genera, and that the Old World genus *Belonogaster* might provide additional evidence.

The most complex animal built beam structures are the paper and wax combs of social wasps and bees. These are almost invariably hanging cantilevered structures. They may be horizontal combs attached at one end, or as a balanced cantilever attached in the middle, or as vertical combs where the length of the cells themselves represents the length of the unsupported projection (Fig. 5.3) (Wenzel 1991). This variety represents the strength and versatility of the comb structure.

The strength of wasp or bee comb is additionally complex because the materials of which they are composed changes over time. Vespine wasp comb cells at the beginning of the season, for example, contain young brood only and are made entirely of paper. On pupation, however, these larvae line their cells with cocoon silk, so enhancing the strength of the cell walls but by an unmeasured amount. The same alteration occurs in the brood comb of honeybees. This may be regarded as a complexity at the level of the materials, but this boundary is blurred. The possible mechanical role of the meconium in wasp nest combs is one on a similar scale. Polistine and vespine larvae prior to pupation, press their voided gut contents into the roof of the cell. This may have a role in making the comb more rigid (Cole 1998). In some Old World genera such as *Ropalidia*, the meconium is removed from the back of the comb by the adults. This provides an opportunity for testing the possible mechanical role of the meconium by comparing the rigidity of *Ropalidia* combs with those of the species of *Polistes*, for example, where the meconium remains.

5.3.3 Thin sheets

A horizontal sheet of paper supported at each end may not even have the strength to support its own weight. However, if the sheet is bent into the form of an arch with its ends fixed, it becomes stiffer and stronger. This can be dramatically enhanced again if the curvature of the roof is present in all vertical sections, and the roof is therefore in the form of a dome. A domed roof is in effect, a circle of arches that are linked together laterally to form a monolithic structure. The consequence of this in practical engineering terms is illustrated by comparing the thicknesses of concrete roofs built in the form of an arch or of a dome. In the former the thickness will be of the order $\frac{1}{20}$–$\frac{1}{30}$ of its radius and for the latter $\frac{1}{200}$–$\frac{1}{300}$ its radius (Salvadori 1980).

Paper domes are common in social wasp nest designs, covering combs located on trunks, branches or leaves. Using such a light material makes the building of large enclosed spaces possible. These are created initially by the adding of individual pulp loads to the edge of the advancing sheet. These, because of slight differences in individual pulp collection sites, are evident as a series of small stripes on the envelope exterior. Junctions between these loads represent potential lines of weakness. Cole *et al.* (2001) demonstrated a marked anisotropy in the envelope material of *Dolichovespula norwegica*. The strength in tension of the paper was almost eight times greater when stressed along the direction of application of the pulp loads than when stressed perpendicular to the direction of application of pulp loads.

Wenzel (1991) noted that the envelope of *Synoeca* exhibited corrugations and that these were aligned perpendicular to the direction of application of pulp loads. This orientation may enable them to resist bending along the axis of potential weakness in the envelope (Fig. 5.6). However, there is more to the structure than this; inspection of the inner surface of the envelope does not reveal the lines of separation between material loads because they have been subsequently plastered over with additional loads, which almost certainly strengthen the material in tension and possibly also in compression.

5.3.4 Attachments in tension, adhesive and behavioural

Any structure that bears a load in tension has two components a tension strut and an end fitting to secure each end. In the tension strut, which is typically narrow in relation to length, stress acts in tension along the axis of the strut, with shear stress acting at 45° to it (Gordon 1978). A tension member of specified material supports a load in proportion to its cross-sectional area. However the volume, and hence broadly speaking weight, of an end fitting increases as the power of $\frac{3}{2}$ of the load (Gordon 1978). It is therefore more economical to

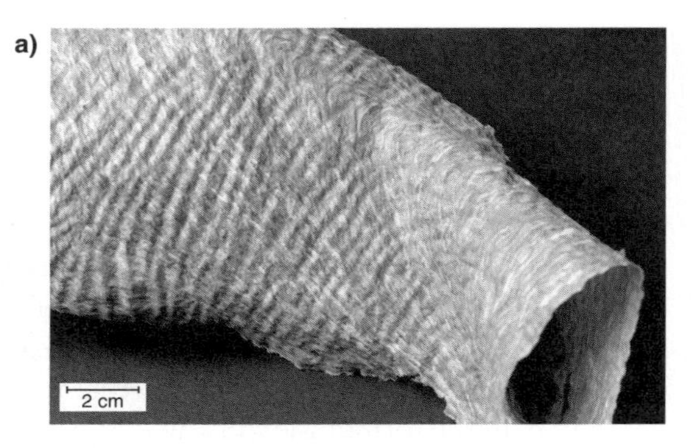

Fig. 5.6. (a) The paper envelope of a nest of a *Synoeca* species, with entrance spout on the right. The direction of the orientation of the fibres of the building material (streaks, representing the application of building material loads) is perpendicular to that of the axes of corrugations that run round the envelope. (b) Envelope detail, external view: the junction between two building loads of fibrous plant material running across the middle of the picture is a potential source of weakness in the envelope. (c) Envelope detail, internal view: the junctions between original building loads is concealed by later 'plastering' with more fibrous material. (Photos: (a) Glasgow University Media Services), (b) and (c) Margaret Mullen.)

divide a tension force between several attachment points, rather than concentrate it into one.

Attachment devices in animal built structures have received very little attention beyond simple description. Studies on spider webs, for example, have concentrated overwhelmingly on the threads and very little upon the thread attachments. Spider web attachments are very small, and therefore probably cheap in proportion to the quantity of silk in the threads themselves. In addition, web failure occurs typically in the threads rather than at the attachment points. Taken together, these features may have contributed to a lack of

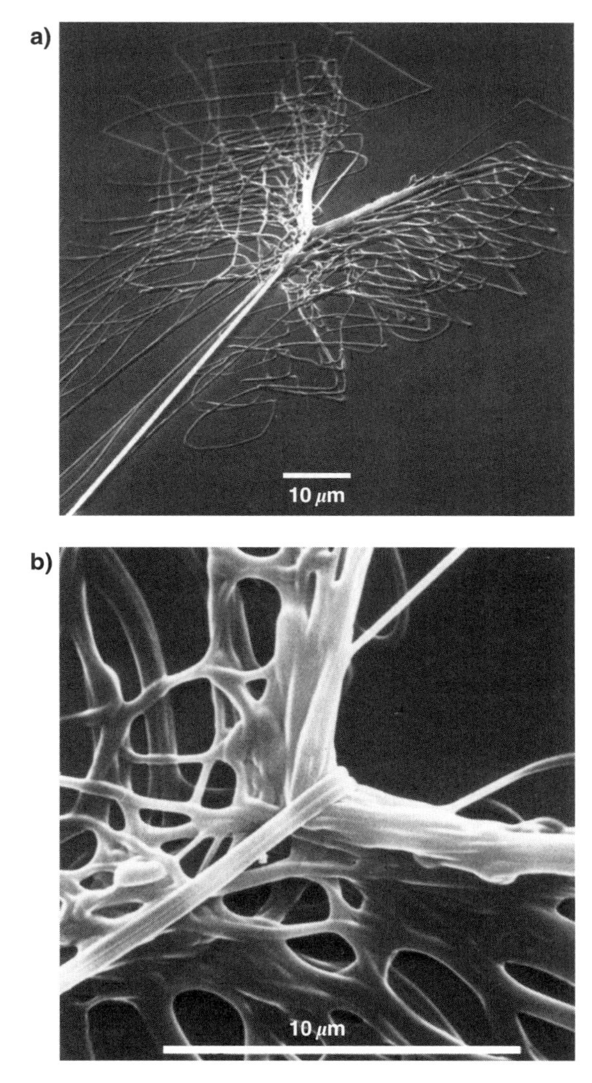

Fig. 5.7. Silk thread attachment device of the spider *D. socialis*. (a) General view, (b) Detail showing design features worthy of investigation, in particular the primary insertion of the tension thread at the back of the attachment, which will result in the highest stresses being experienced in the posterior rather than the anterior part of the diagonally spreading attachment surface. (Permission Karin Schuett.)

scientific interest in spider web attachment devices. At the descriptive level however, they can be seen to have strong design features. Those of *Drapetisca socialis* have a pad of threads which spread out as two wings which make contact either side of the double-stranded tension thread that emerges from the posterior of this attachment complex (Fig. 5.7) (Schütt 1996).

The method of adhesion of this device to the substrate could either be molecular forces between the attachment material and the surface material, or by a simple process of interlocking resulting from the initially plastic secretion of the spider invading the irregularities of the attachment surface before solidifying. It is unknown which is true.

Adhesion is only one of the two principles of attachment found in animal structures. In the other, the builder uses its behaviour to manufacture some form of fastening. This is principally exhibited in the nest building of birds and involves the use of collected materials, either vegetation strands or silk (Hansell 2000).

Collected silk is rather easy to use. It forms the looped component of a Velcro fastening, in which some other material (usually plant) forms the hooks (Section 3.3.1). For the nest of the origma *Origma solitaria*, the attachment is of spider web silk looped over irregular incrustations on the roof of a cave (Hansell 2000). Spider silk as a structural material is very largely confined to nests of under 30 g dry weight, but this is for the nest structure as a whole, not just for the attachment (Hansell 1996*b*). Constructing end fittings out of strands of vegetation requires more complex behaviour (Section 3.3.4). Weaverbirds (Ploceinae) and oropendolas must use these to suspend their nests, but their load bearing capacity and durability is unstudied.

5.3.7 Struts in tension

The nest of the wasp *Polistes fuscatus* is an open comb attached to the overhanging surface by a nest petiole (Fig. 5.3(a)). The growth of the comb is initially radially symmetrical about the single central support. As it grows the petiole is strengthened by the addition of paper pulp and oral secretion, however adding weight to the comb does not stimulate petiole reinforcement unless it causes the comb to tilt (Downing and Jeanne 1990). Tilting the comb by the addition of asymmetric weight may stimulate the construction of secondary comb supports but these, rather than being placed above the additional weight, are attached to the opposite side of the comb so preventing that part of the comb moving closer to the overhead support surface. Such a petiole should, of course, be regarded not as one structure but two, a narrow strut and an attachment disk. Damage to either of these does apparently provide a stimulus to repair them, but only a weak one (Downing and Jeanne 1990). These responses of *P. fuscatus* do not show strong engineering priorities in the wasp's behaviour. More obvious is a response preventing too much contact between the comb and the attachment surface, probably to limit the known threat of predation by ants.

The nests of wasps of the genus *Dolichovespula* grow by the suspension of additional combs below the previous one, each comb radially symmetrical and expanding by the addition of cells to the margin. As cells are added, so eggs are laid in them and brood raised ensuring radial symmetry in the weight distribution of any comb. In a mature nest the upper comb may support as many as three below it suspended from ribbon-like struts, raising the question of how the supports are strengthened to bear the increasing weight as the nest grows.

In *D. norwegica* and *Dolichovespula sylvestris*, the density of nest supports below the first comb is greater than that beneath all other combs. This indicates an adaptive response to the increased stress. How this is mediated is unclear since the support provided to any comb is correlated not only with the area of comb supported directly below it, but also with the amount of brood in that comb, and to the total comb suspended. Cole (1998) speculates that the wasps could judge where to add additional supports either by the amount of separation between combs, which might change under increasing load, or by differences in resonance frequency of vibration of the nest in different locations.

Differences are evident between these two *Dolichovespula* species in the number and distribution of comb supports. The mechanical significance of this is unknown, but it does demonstrate again that more detailed descriptions of the nature of built structures would itself be valuable in directing research. Certainly the broader comparison between the nests of *Dolichovespula* and *Vespula* do allow useful mechanical predictions to be tested.

At the level of the building materials, the comb and envelope paper of *D. norwegia* have longer plant fibres and a greater tensile strength than those of *Vespula vulgaris* (Cole *et al.* 2001). At the structural level, the comb supports of *Dolichovespula* are ribbon-like, while those of *Vespula* are thin vertical struts (Fig. 5.8.), and the comb expansion in *Dolichovespula* conserves the radial symmetry while those of *Vespula* may become quite asymmetric. These differences are consistent with the differences in the likely mechanical demands imposed on the nests of these species in their respective nest sites. *D. norwegica* nests are typically located in exposed sites such as bushes or trees, while those of *V. vulgaris* are found in protected sites such as underground cavities or roof spaces (Spradbery 1973). The tougher paper, balanced cantilever comb design, and ribbon-shaped comb supports of *Dolichovespula* conform to an expectation of more severe mechanical stresses, for example, from wind. *V. vulgaris* nests, in their protected sites can function successfully with less tough engineering and with flexibility in the pattern of comb growth to exploit the unpredictable dimensions of a nest cavity (Cole 1998; Cole *et al.* 2001).

In the modular nest design characterised by *Polybia* and *Chartergus* wasp species (Figs 5.5 and 8.3(c)) it is the nest envelope that provides all the support for the combs. As additional combs are added to the bottom of the nest, the upper part of the envelope bears more and more weight. Accommodation to this enhanced stress is achieved by repeated addition to the initial envelope of new layers of material plastered on to the outer and inner surfaces (Jeanne 1975).

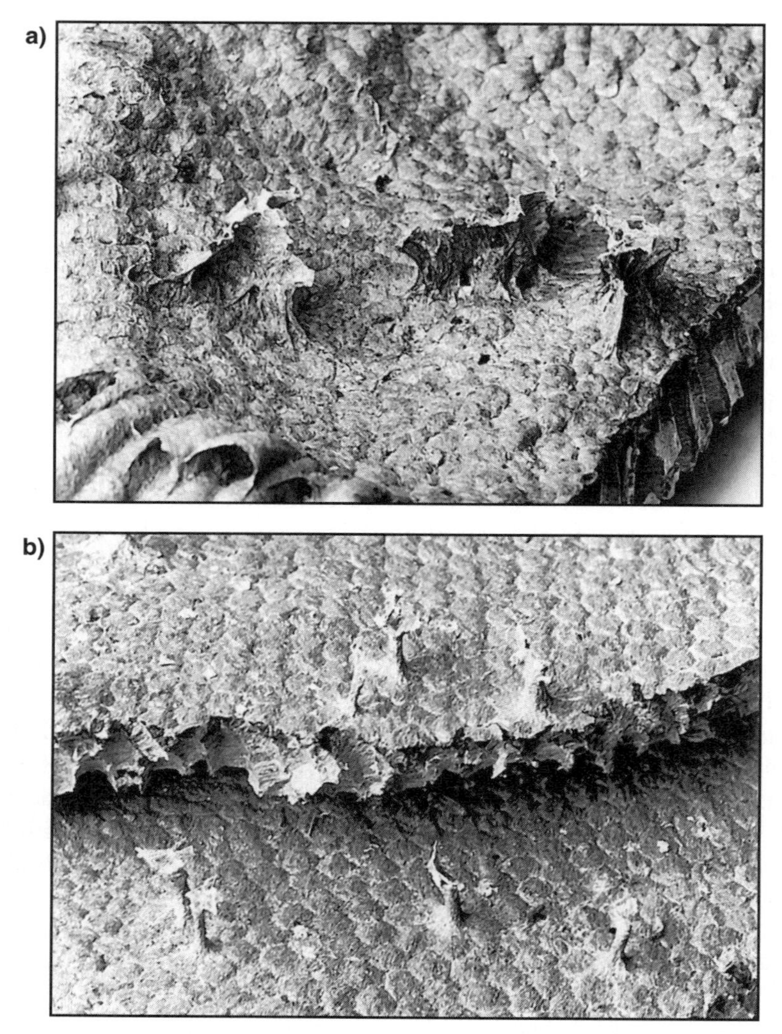

Fig. 5.8. Comb supports in the nests of vespine wasps. (a) Comb supports in *Dolichovespula* nests are sinuous and ribbon-like. (b) Comb supports in *Vespula* nests are simple vertical struts. (Photo Glasgow University Media Services.)

With age therefore, the nest takes on the qualities of a stout, round-sectioned beam reinforced with transverse bulkheads.

5.4 The spider orb web: engineering in tension

5.4.1 Introduction

The orb web design of spiders conveys an impression of elegant simplicity. It provides a large surface for the capture of prey yet is economical in the use of materials, simple in design, and has all elements in tension. Yet the weight of these spiders differs across species by more than two orders of magnitude (Craig 1987*a*); they also differ in ecology of their websites, the orientation of their webs and the time of day that they forage. Inevitably this is accompanied by differences in the prey that are caught. In consequence orb webs, although superficially resembling one another, are engineered to achieve different objectives dependent upon spider species.

5.4.2 Maxwell's lemma: a minimum volume of material design

One functional explanation of the orb web is that it is selected to minimise the use of silk in prey capture. For a structure that bears loads entirely in tension, this can be assessed by comparison with the *minimum volume design* also known as *Maxwell's lemma* after the nineteenth-century physicist, James Clerk Maxwell.

For a structure built of one material, all of which is in tension, Maxwell's lemma states that, if the stress in all members is equal and equal to the breaking stress of the material, then the structure is built with the minimum volume of material to resist the forces acting upon it (Denny 1976). Alternatively, Maxwell's lemma for minimum volume design states that: 'if (1) every member of a structure is built of the same material and under uniform tension and (2) if stress (force/area) is the same in all members and equal to the breaking stress of the material, then a minimum amount of material has been invested in the structure to withstand a specified set of loads' (Craig 1987*a*).

The cost of silk as a building material, which has influenced various aspects of web building (Section 6.3.3), could have selected for orb web designs that approach minimum volume status. Failure to meet the criteria would nevertheless be of interest since it might indicate conflicting functions for the web or constraints on the ability to minimise silk use.

In *Araneus* species the cross-sectional area of the main structural threads depends upon their location. In webs of *Araneus sericatus*, radii typically contain two silk strands, frame threads 4–8, and mooring threads 8–10 strands (Denny 1976). This reflects the relative tension forces to which they are subjected. The ratio of these forces in the webs of *Araneus diadematus* for radii, frame, and mooring threads are 1 : 7: 10, but the ratio of stresses (force/cross-sectional area) between radii and mooring threads was measured by Wirth and Barth (1992) as only 1.0: 1.5, a distribution approximating that expected from a minimum volume design.

Bearing in mind an *A. diadematus* rests at the hub of the web, it might be predicted in accordance with a minimum volume design, that there should be more radii in the upper than the lower half of the web, yet the reverse proves to be the case; there are two to three times as many radii in the lower compared to the upper half of the web (Wirth and Barth 1992). The functional reason for this is apparently that spiders can run more rapidly down the web to reach prey than up, giving the portion of the web below the hub greater prey capture value (Rhisiart and Vollrath 1994). As a result of this, the tensions in the radii in the upper half of the web are two to three times that in the lower, mirroring the differences in the proportion of radii. In addition, tensions are concentrated into a small number of radii that have tensions of up to 100% greater than the rest. These high tension radii are characteristically those that were built before the completion of the web frame—the primary web structure. Yet, in spite of their higher tensions, Wirth and Barth (1992) found they had no greater diameter nor were composed of more threads than were secondary radii.

In spite of some unevenness in the stress distribution among radii, there is evidence that radial tensions are adjusted during construction. Many orb web species including *Araneus*, add radii by first laying a thread from the hub to the frame, but then returning directly, drawing up the thread and spinning a replacement. By observing the deflection of frame threads towards the hub brought about by the tension of the initial radius and its replacement, Eberhand (1981*a*) showed that the second thread is inserted at a lower tension. Adjusting the tension during construction appears also to have the aim of pre-stressing the web to a required level that assists prey capture. Wirth and Barth (1992) demonstrated this in the web building of the two araneoid species *A. diadematus* and *Nephila clavipes*, observing that similar tensions occurred in web anchor threads, whether they were built within a rigid frame or to fine glass rods that bent in the direction of the web hub under the forces exerted by the web.

Further evidence of the evening out of stresses in the web is found in the changing diameter of radii from hub to frame. As each turn of the auxiliary spiral is added under tension across the radii, it reduces the tension in the radial threads nearest the hub. The result is that, when the auxiliary spiral is complete, there is a range of tensions along each radius, the lowest at the hub end, the greatest at the frame end (Wirth and Barth 1992). The capture spiral, which replaces the auxiliary spiral, is also applied in tension, ensuring that in the completed web the gradient of tensions in the radii persists, although it is less steep than initially, due to the greater extensibility of the viscid capture thread compared with the auxiliary thread.

The minimum volume design hypothesis predicts uniformity of stresses through the structure, and consequently that a section of radius towards the periphery of the web, will have a cross sectional area that is greater compared with the end nearer the hub. This is supported in the design of the web of the araneoid orb spinner *Zilia diodia* where radii are found to be composed of a double thread of dragline silk at the periphery of the web, but reduced to a single thread towards the hub (Zschokke 2000) (Fig. 5.9). The proportion of the length of the radii that is double is, as predicted, also greater in the upper half

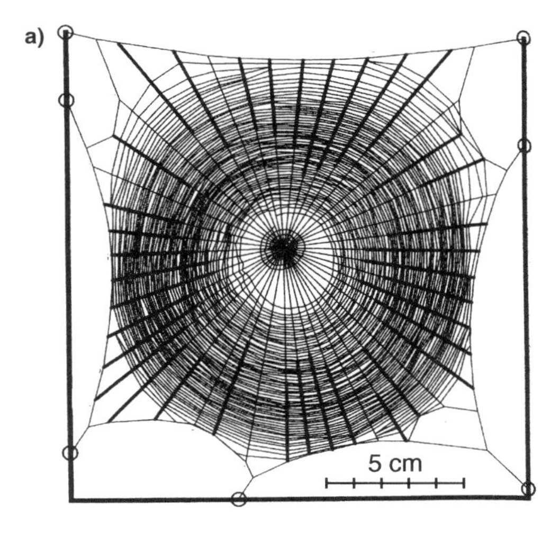

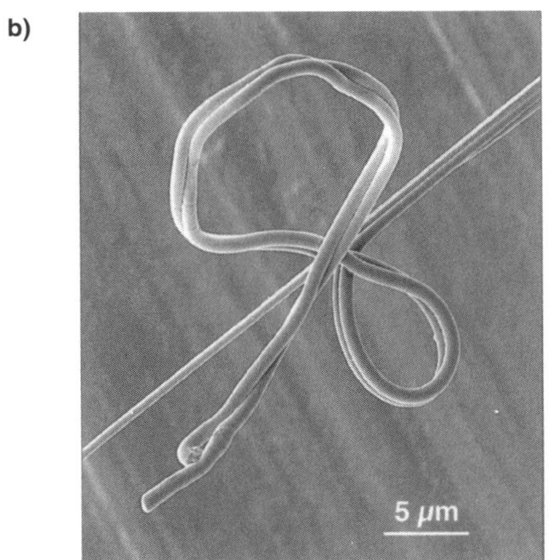

Fig. 5.9. *Zilia diodia* orb webs have radii of single threads in the central part of the orb, but have double threads in the outer sector of the orb where stresses are greater. (a) Bold lines along the radii indicate the extent of the doubled thread. (radii without any bold part were not analysed) adapted from Zschokke 2000. (b) Scanning electron micrograph of a radius at the point of transition from double-stranded to single-stranded thread. (Photo S. Zschokke 2000.)

of the web compared to the bottom half. This serves to compensate for the weight of the spider at the hub and the fewer radii in the upper half of the web.

5.4.3 Orb web structure: ecribellate species

The ecribellate orb web builders of the Araneoidea were derived from the orb web building Deinopoidea, through the innovation of a new capture principle, the viscid spiral thread. The majority of research on the functional design of orb webs has been conducted on Araneoidea, so it is convenient to consider them first. Ecribellate orbs in general show adaptations to place orb webs with a vertical orientation in open sites in order to capture fast moving prey, although inevitably there are exceptions. This has resulted in web designs that must absorb and dissipate the energy of impacting prey. This is facilitated at the level of the silk material through the extension of dragline and capture thread silk as described in Section 2.4.3; Fig. 2.11. However, the structure of the web also contributes to the absorption of the kinetic energy of the prey in additional ways at a number of levels.

Kohler and Vollrath (1995) noted that, during extension, an isolated single radius of the web of the ecribellate orb spinner, *A. diadematus* exhibits a low Young's modulus (low stiffness) value over the first 5% of extension. Although this could be an attribute of the silk, it may in part be due to twisted or buckled sections of the thread. These could be regarded simply as poor workmanship, but are nevertheless an attribute of the thread above the level of its biochemical composition.

Decelerating prey over a longer distance during thread extension diminishes the maximum force acting on the web. But a section of capture thread extending under the impact of insect prey is not anchored to fixed points but to flexible radii, in turn attached to flexible frame threads. These, as has been noted, are pre-stressed and are stiffer than the viscid spiral (Lin *et al.* 1995). Their extension allows energy to be dispersed spatially through the web.

The consequence of this is that, on impact of prey, a wave of disturbance travels through the web following the lines of greatest stiffness, which are the radii (Lin *et al.* 1995). Only a proportion of the absorption of the energy of impact can be explained by the hysteresis of the threads undergoing extension, the remainder is explained by the effects of *aerodynamic damping*. With thread diameters of around 1 μm, and velocities of movement of about 1 m s^{-1}, inertial forces of web movement are small enough to be ignored, while viscous forces of the web being driven through the air are significant. The drag force per unit length of thread is proportional to its velocity but independent of its diameter. Lin *et al.* (1995) incorporated these effects into a model predicting the displacement of a radius within a web after the impact of an object resembling in weight and speed a flying insect (Fig. 5.10). The model showed a strong damping effect, reducing the web displacement to about half in four cycles of oscillation. Removing the effect of aerodynamic damping but leaving the hysteresis properties of the radial thread silk, produced a markedly reduced damping effect.

Measurements in which real webs had prey-like pellets fired at them, confirmed the damping effects of the model, while isolation of a single radius by cutting away all capture spiral connections halved the damping effect, showing the contribution made by spreading the displacement through the web. The frequency of web vibration, however, remained the same whether or not the spiral connections were intact, consistent with the hypothesis that major forces restoring web shape are provided by the prestressed radial threads.

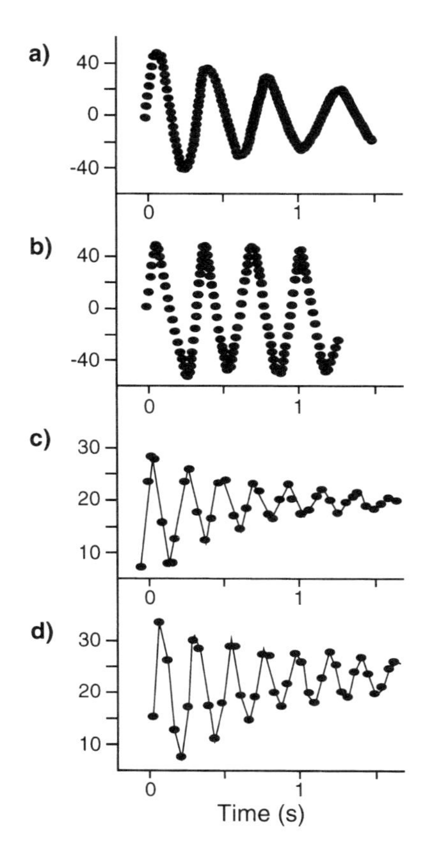

Fig. 5.10. Actual and simulated aerodynamic damping action of the web structure on thread oscillation. (a) Output of computer model showing out of plane displacement of a radial thread at point of impact of a moving object. (b) The same simulation but with the effect of aerodynamic damping removed. (c) Laboratory measurement of out of plane displacement in a test web struck by a moving prey-like object. (d) Measurement of displacement in the same radial thread after all contacts with the capture spiral was severed. This shows marked reduction of the damping effect. (Adapted from Lin *et al.* 1995.)

5.4.4 Orb web variation between ecribellate species: adaptive engineering

Comparative studies on orb web builders have revealed the extent to which webs with a superficially similar design can embrace rather different mechanical principles in the capture of special prey.

Craig (1987*a*) compared the insect prey caught in the webs of five ecribellate orb web builders (Araneoidea) from the same site in Panama and tested their webs for the maximum energy of prey that could be restrained, by determining the force of a falling piece of metal foil just sufficient to break a web held in a horizontal orientation. The species varied greatly in size with adult females of the largest species *(Micrathena schreibersi)* weighing 146 mg, while at the other extreme, those of *Leucauge globosa* weighed only 2.7 mg. Considerable differences were found between species, with webs of *M. schreibersi* proving to be able to absorb kinetic energies at least three orders of magnitude higher than those of *Cyclosa caroli* and four orders of magnitude greater than webs of *L. globosa*. Considering this evidence, and that of the prey caught in webs, Craig (1987*a*) ranked the spiders from high to low energy webs in the order: *M. schreibesi, Mangora pia, C. caroli, L. globosa,* and *Epilineutes globosus*. Broadly, therefore bigger spiders construct high energy absorbing webs. Comparisons of these webs at the level of the materials through to the whole orb structure reveals that it is a combination of characters that is responsible for this rank order, but that the functional role of high energy and low energy absorbing webs is rather different.

At the level of the materials, Craig (1987*a*) found that there was no trend in tensile strength of thread in relation to energy absorbing rank order for either frame or viscid spiral threads. Nevertheless, as bigger spiders were found to produce larger diameter threads, there were considerable differences in the energy absorbing capacity of their threads, with the breaking energy of viscid threads varying by two orders of magnitude, and frame threads by three orders of magnitude across the range of species.

The extensibility of frame threads showed no trend across species, although those of *M. schreibersi* showed greater extensibility than the rest. For viscid spiral threads there was a trend, with greater thread extensibilities in those of high energy compared with low energy webs. At the level of the web, the differences between high and low energy absorbing webs were that the former had threads that were more highly pre-stressed, and they tended to have a higher ratio of radii to spiral turns although for this feature *M. schreibersi* did not fit the general trend. The effect of these characters is that high energy absorbing webs are rather stiff but with highly extensible capture spirals. In consequence, the energy of local impact can be more readily dissipated through the web. This makes the orb web design important for the capture of large, energetic prey. The low energy webs are by contrast, of low stiffness and the orb web design is incidental to their capture of small, slow flying insect prey. This, speculates Craig (1987*a*), has liberated small araneoid species to evolve alternative web designs to exploit new habitats.

This interpretation is supported by measurement of the energy absorbing capabilities of whole webs, showing whole webs of *M. schreibersi* at an order of magnitude higher than the energy absorbing properties of their frame silks and two orders of magnitude higher than the energy absorbing properties of their viscid silk. In contrast, in *L. globosa* (a low stress web with a radius to spiral ratio of less than one) the energy absorbing properties of frame silk is higher than the energy absorbing properties of the web as a whole (Craig 1987*a*).

The web of *M. schreibersi* does however show that, while broad trends may be discernible, there may be alternative solutions to particular prey capture problems. This species, in spite of being ranked as having the highest energy absorbing web, has a relatively low number of radii and the highest frame thread extensibility. These features, coupled with the high tensile strength of the threads, means that it relies more on thread extension and less on web stiffness than the other high energy absorbing web species.

The function of some other observed differences between these webs remain unclear. Sliding junctions between capture threads and radii occur in all the species studied but, for example, their occurrence was least in the webs of *C. caroli*, whereas in *M. pia* all junctions showed a moderate amount of sliding. *M. schreibersi* web junctions showed the highest degree of sliding, but they represented less than 20% of all junctions (Craig 1987*a*).

5.4.5 The webs of cribellate species

Ecribellate spiders evolved from orb building cribellate ancestors (Craig 2003). The radii and frame threads in both groups are essentially the same material, and it is the nature of the capture principle of the spiral thread that distinguishes them (Section 2.4.3). These differences, however, have contributed to differences between them in the functional design of the orb web.

The cribellate spiders of the Deinopoidea (Uloboridae) are comparable in size to smaller species of the ecribellate, Araneoidea, and Opell (1997), comparing the horizontal orb webs of four cribellate species with the vertical orb webs of four araneiods, found some similarities between them. In both groups there was a positive correlation between spider weight and web area, indicative of a relationship between capture potential and metabolic need, and between spider weight and number of spirals per web area.

Two important differences are apparent between cribellate and ecribellate species in relation to their capture threads. First, for spiders of comparable weight, the cross-sectional area of capture spiral axial threads is greater in araneoids than in deinopoids; second, there is a positive correlation between capture thread cross-sectional area and spider weight in araneoids (already noted in Section 5.4.4), which is not present in deinopoids. This according to Opell (1997) shows that the evolution of the new, adhesive capture principle and accompanying change in web orientation from horizontal to vertical, was accompanied by an increase in importance of web extension in the dissipation of energy of the faster moving prey that could now be intercepted.

The ratio of radii to spiral turns was also greater in the cribellate orb webs than in those of araneoids. This difference cannot be attributed simply to differences in web orientation, because the horizontally oriented web of the ecribellate *L. globosa* has a ratio of radii to spiral turns of less than one (i.e. smaller than that shown by comparable vertical orb araneoids). The high density of radii in the webs of the deinopoids, plus the high Young's modulus of cribellar capture threads makes these webs quite stiff (Opell and Bond 2001). This suits deinopoid webs to the capture of low energy prey by absorbing energy of impact through local thread hysteresis mechanisms. This shows a minor emphasis on absorbing the energy of impact of prey by aerodynamic damping compared with webs of araneoid species (Lin *et al.* 1995).

As with ecribellate orb webs, such a general assessment of mechanical design is not sufficient to explain differences in detail between or even within cribellate orb building species. The horizontal orb web of the deinopoid spider *Octonoba sybotides* occurs in two forms distinguished by their stabilimenta, one linear through the hub, the other spiral outside the hub. Individuals apparently spin one or other type of web in accordance with their energetic state (Watanabe 1999*b*). Satiated spiders tend to create linear stabilimenta, while hungry spiders built spiral ones. The latter spiders also respond more rapidly to small flies in their webs than do those with linear stabilimenta. Watanabe (1999*b*) reasoned that it might be that webs with a spiral stabilimentum vibrated more to struggling small prey than did webs with linear stabilimenta. Small prey should produce vibrations of smaller amplitude and higher frequency than larger prey. Radii under greater tension, should respond better to these aspects of small prey and this is achieved more effectively by spiral rather than linear stabilimenta.

By controlling the level of hunger, Watanabe (2000) showed that while spiders responded equally rapidly to large prey regardless of web type, spiders on webs with spiral stabilimenta responded more rapidly to small prey than did those on webs with linear stabilimenta. What advantage in that case is the linear stabilimentum? Watanabe (2000) proposes that the weight of a well-fed spider raises thread tensions and that the linear stabilimentum by lowering tension in the radii may in some way help the spiders to target larger prey.

5.5 Design and aesthetic sense

An essential aspect of human design is that the object, whether artefact or architecture, should not simply be functionally effective but also look 'right'. The consequences of this sensitivity to the visual world that we have created have been profound, but additionally it raises questions on its origins that we may be able to answer through the study of other living species. Non-human primates appear to be disappointing models for such a quest, while bowerbirds (Ptilonorhynchidae) by contrast appear to be the best candidates available (Hansell 2000). To propose that an animal has an aesthetic sense is to claim that it obtains some satisfaction or pleasure from an experience. Darwin (1871) made this clear, when expressing his view that bowerbirds did assess conspecific displays in this way.

Good correlational and empirical evidence now supports the view that bowers are the product of sexual selection through female choice (Borgia *et al.*

1985; Madden 2003*a,b*). It is also apparent that differences between species in bower design can evolve rapidly and with limited effects of phylogenetic constraint (Kusmierski *et al.* 1997; Uy and Borgia 2000). Female choice is influenced by the presence of particular structures or ornaments; in the satin bowerbird (*Ptilonorhynchus violaceus*), for example, females have been shown to favour males having bowers decorated with relatively high numbers of blue objects, and with a twig avenue that is firm and symmetrical (Borgia 1985*a,b*). Males of the spotted bowerbird (*Chlamydera maculata*) have been observed to prominently advertise those ornaments that females find most attractive, especially in their case green *Solanum* berries (Madden 2003*b*).

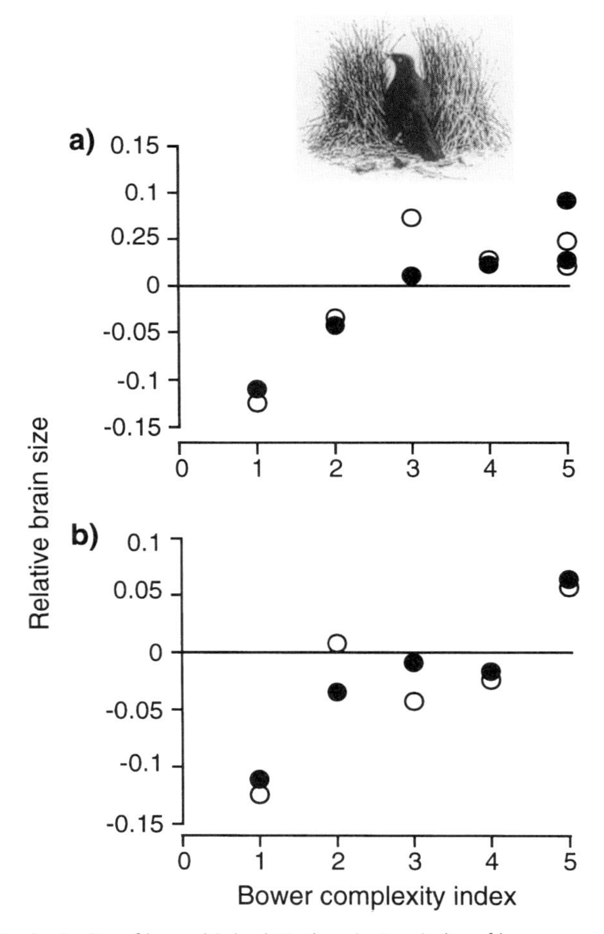

Fig. 5.11. Relative brain size of bowerbirds plotted against an index of bower complexity. (a) Avenue building species and the catbird; (b) maypole building species and the catbird (• = males; o = females). (Adapted from Madden 2001.)

Bowers differ from the sexually selected plumage traits, such as that conspicuously seen in the peacock, in that the male does not simply grow the display but constructs the display through his own behaviour. Consequently female choice selects for males that possess certain kinds of brains. Bowerbirds have larger brain sizes when compared with other medium-sized passerines of comparable ecology, and within the bowerbird species, there is a strong correlation between relative brain size and bower complexity. Significantly, the relationship holds not only for males but also, although less strongly, for females (Madden 2001) (Fig. 5.11). The enhanced size of the female brains is of particular interest because, whereas the size of male brains can be related directly to the building of the bower itself, for the female it cannot. This allows two possible explanations, either that the brain size of females is a secondary consequence of selection for large brain size in males, or that where males need enlarged brains to create bowers of high quality, females need enlarged brains to judge the quality of those bowers.

None of this tells us anything about aesthetic sense in bowerbirds, although the possibility has been raised by previous authors. Diamond's (1982, 1988) view is in favour, largely on the grounds of the extreme complexity of the displays of some species. In contrast, Borgia *et al.* (1987) maintain that the demonstrated preferences for certain bower ornaments and other simple criteria such as bower symmetry, are consistent with rigid genetically determined preferences. This, they say, argues for a lack of general aesthetic criteria. The law of parsimony would also appear to resist the inclusion of a pleasure mechanism, as it is simpler to hypothesise that the birds have some more direct way of assessing the design qualities of bowers. Miller (2000), in the context of the evolution of human aesthetic sense, turns this argument on its head by reasoning that a pleasure system, by bringing together the assessment of a variety of disparate criteria, could make mate choice simpler.

Bowerbird displays exceed the complexity of the song or plumage displays of most bird species, not least because they are multimedia presentations. The structure of the bower may itself have many elements of space and colour, but it also provides a stage on which the male may exhibit vigorous movement, display dramatic plumage and produce complex vocalisations. Vocal mimicry is a common feature of bowerbird displays. It has been recorded, for example, in the mat-displaying, Archbold's bowerbird (*Archboldia papuensis*) (Frith *et al.* 1996), and the maypole-building, golden bowerbird (*Prionodura newtoniana*), a species which includes elements of the calls or songs of at least 22 other species (Frith and Frith 2000*a*).

In view of this complexity, the demonstration that a particular kind of ornament affects female choice, while important, neglects the possibilities for interaction between the display elements; that is, that there is not simply ornament but design. Attention is now beginning to turn to this. Patricelli *et al.* (2003)

demonstrate that female satin bowerbirds are more tolerant of intense and vigorous displays of males that have bowers of better quality and more ornaments.

I believe there is the potential to go further and test predictions of an assessment mechanism which uses an aesthetic judgement. Embracing Griffin's (1998) argument that animals through their communication provide insights into what animals are *'thinking and feeling'*, we should be examining in greater detail at what both male and female bowerbirds are doing. In Hansell (2000), I argued that existing knowledge of bowerbird displays was consistent with an *'art school'* hypothesis, in which males are seen as having an extended period of time to develop an appreciation of the effect their displays may have upon females.

Anecdotal evidence in support of the hypothesis is provided by the following behaviour in immature males: they may take 6 or 7 years to reach a level of performance that attracts females (e.g. satin bowerbird—Borgia 1985*b*, 1993; golden bowerbird—Frith and Frith 2000*a*), they exhibit practice building behaviour, either alone or in student groups (satin bowerbird—Vallenga 1970), and they observe mature males at work and the work that they have completed (satin bowerbird—Vellenga 1970; golden bowerbird—Frith and Frith 2000*b*). These indicate the great potential for future developmental studies, involving the manipulation of such experiences.

Some observations on the behaviour of mature males also encourage the art school view: males may exhibit care in the placement of ornaments (spotted bowerbird—Madden 2003*b*), and 'artistic mannerisms' in the arrangement and rearrangement of ornamants (*Amblyornis inornatus*—Diamond 1988). Males may also show individuality or innovation in the use of ornaments: (*A. inornatus*—Diamond 1987), or exhibit local or regional cultural differences in bower construction and ornamentation which are matched by female preferences (spotted bowerbird—Madden 2003*a*; *A. inornatus*—Uy and Borgia 2000).

The art school hypothesis also predicts features of the behavioural mechanism concerned with bower construction for which we currently have no evidence, but which are testable. In particular this includes physiological evidence of specialised brain mechanisms, including a 'pleasure' mechanism.

But these predictions only relate to male courtship behaviour, thus ignoring half of the question, the nature of female preference. If males are being sexually selected by females to exhibit very refined skills and judgements, then the art school hypothesis makes further predictions about females. In particular that females will need a period of learning in order to assess effectively all the information provided in a male's display. Indeed the presence of an extended learning period in females, rather than largely genetically determined responses, would be important support for the hypothesis. If males are training to be artists, females should be training to be 'art critics'. There is now evidence of changes in female preference in relation to age. Female satin bowerbirds ranging in age from 1 year to more than 3 years, having visited a bower when the male was absent, were found to be more likely to return if it was experimentally supplemented with blue feathers. However, where the females had observed males displaying at the bower, first and second year females were more likely to return to decoration enhanced bowers compared with controls, but the three-plus age class of females were not (Coleman *et al.* 2004).

6 Building costs, optimal solutions, and trade-offs

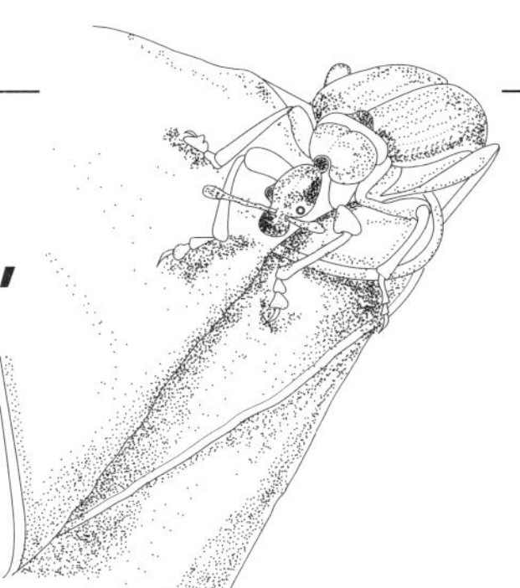

6.1 Introduction

Natural selection will favour building behaviour that allows the builders to breed early, have more offspring, and survive better, provided that these attributes do not interfere with one another. If, on the other hand, building affects life history traits in such a way that they impact upon one another, for example, that the cost of building delays reproduction, then there is a situation of *trade-off* between life history variables.

In the absence of trade-offs, the optimality of a design can simply be judged according to the effectiveness of the design in relation to its cost. However, in a situation where there are trade-offs, it is first necessary to discover how building affects life history and, from that, how it impacts upon fitness. The optimal design in this case is the one that maximises fitness, subject to the constraints imposed by the trade-off.

In Section 1.1, a distinction was made between home builders, trap builders, and burrowing foragers; this chapter considers the implications of the costs of these activities. Trapping food and burrowing for it, both concern foraging success, and so are considered together as variant foraging strategies. Foraging success can often be directly observed as feeding rate, calculated as energy intake, and thereby compared in the same currency with energetic costs. Chapter 1 shows that trap building is not widespread, and is mainly expressed by the net building caddis larvae and web building spiders; burrowing foragers are largely represented by the rodents. These three groups of organisms form the basis in this chapter of three case studies on the theme of foraging costs.

Homes are for protection, so the benefit they provide may be measurable as some consequence of the failure to invest in protection. Costs may still readily be measured in terms of energy consumption or additionally, time, but appropriate measures of benefit may vary. For homes that are largely concerned with raising young, an appropriate measure of benefit may be reproductive success, for individual homes it could be survivorship. This chapter uses bird nest building as the basis for a study on the costs of home building to raise young, and caddis larval case building to examine costs of building for personal security.

6.2 Building homes

6.2.1 Birds and nest building

6.2.1.1 Estimations of nest building costs

The sources of costs in nest building may be as simple as the collection of the material followed by its assembly, but could include costs of processing the material to make it suitable for construction or even the costs of material synthesis (Fig. 2.1). The proportions of these costs vary between species, allowing the consequences for species with different cost distributions to be dissected and compared. For a nest of collected materials and simple assembly behaviour it has been assumed that costs are fairly accurately estimated as the total cost of flights out to collect material plus the cost of the return flights carrying the load.

Withers (1977) calculated the cost of constructing a 600 g nest by the cliff swallow (*Hirundo pyrrhonota*) as 122 kJ, based on an estimation of about 1,400 collecting trips, each of which includes 20 s of flight, 10 s of mud collecting time, and 30 s of assembly time at the nest. A similar calculation by Collias (1986) for the village weaver bird (*Ploceus cucullatus*) collecting leaf strips, gave an estimated total flight distance for the 40 g male of 325 km, at a total cost of 158 kJ.

The black-billed magpie (*Pica hudsonia*) provides an example of a bird that builds a relatively large nest. Sticks are brought to the nest singly, their number in the nest therefore providing a minimum value for the stick collection journeys. Mud for the cup lining is brought as pellets, the recovery of which can be used to determine the number of journeys to collect the observed weight of this material.

For a particularly large nest with a weight of 76 times the weight of an adult magpie, the number of journeys was estimated at 2,564 collection trips requiring a flight distance of 276.2 km. Incorporating the horizontal and vertical transport costs of the weight of the two materials for the return half of the journey gave a total nest building cost of 209.1 kJ (Stanley 2002). Nevertheless, this worked out at no more than an average of 2.61 kJ adult^{-1} day^{-1}, or 1% of estimated metabolisable energy intake of an adult magpie per day.

We still have too few estimates of this kind, but the cost of nest building seems likely to vary greatly between species. The weight of a completed nest compared with that of the individual or pair that builds it certainly shows large differences between species. The nest of the woodpigeon (*Columba palumbus*) at about 150 g weights only about $\frac{1}{3}$ of that of the bird, while the massive stick nest of the hamerkop (*Scopus umbretta*) is said to weight hundreds of kilos although built by a pair of birds that together weight less that a kilo (Hansell 2000).

The scarcity of a nest material may also be an important contributor to nest costs and is one requiring more detailed examination. The nest of the long-tailed tit, at 26 g, is only about three times the weight of an adult bird but is composed of possibly 3,700 pieces made largely of four materials, moss, spider cocoon silk, lichen, and feathers (Hansell 2000). Of these, the spider cocoons and the feathers appear superficially to be the most difficult items to find and therefore costly in time and energy to locate. Hansell and Ruxton (2002) tested how readily available feathers were for woodland passerines to locate, by dispersing 1,000 feathers between 20 sites each week for a 16 week period, in a 3 year study. The marked feathers found in 14 recovered long-tailed tit nests showed that the birds travelled an average of only 45 m to collect a marked feather and, as only about 3% of feathers per nest were marked, naturally occurring feathers were, contrary to our perception, quite readily available.

Birds typically build a new nest each building season, but in a minority of species a proportion of nests can be refurbished and used again the following season. Such species have provided an opportunity to assess the cost of building by comparing the reproductive performance of the *rebuilders* with *refurbishers*.

The eastern phoebe (*Sayornis phoebe*) is such a species. It builds a nest of similar design and materials to the barn swallow (*Hirundo rustica*). Only the female builds, allowing a direct link to be made between nest building costs and subsequent reproductive effort. Initial evidence that nest building caused clutch reduction compared with the lesser effort of refurbishment (Weeks 1978), was contradicted by Conrad and Robertson (1993). A third study (Hauber 2002) in which the reproductive performance of females with their nests experimentally removed was compared with those provided with one, found no significant clutch size reduction in the first brood of rebuilders compared with females provided with nests. However, birds of the nests-removed group completed their first clutches later and showed a lower probability of attempting a second clutch. This shows that, while nest building is necessary for reproduction, its costs may limit reproductive success; a situation of trade-off.

6.2.1.2 The trade-off between nest reuse and nest parasite burden

It appears from the evidence above that the optimal solution to minimising nest building costs is, where possible, the refurbishing of a nest from last year. But, reuse of nests carries with it a potential additional cost; the risk of exposing the clutch to haematophagous arthropods surviving in the old nest material (Section 1.3.1).

The breeding success of great tits (*Parus major*) in nest boxes harbouring the flea *Ceratophyllus gallinae*, was compared with that of pairs breeding in sterilised boxes by Richner *et al.* (1993). Chicks in the treated boxes grew better regardless of clutch size, whereas over the study period of 14–18 days of age, the weight of chicks of small or large clutches from infested nest boxes either did not rise or actually fell. The nutritional condition of chicks measured as body mass/tarsus length was also found to be sharply reduced in the parasitised group at 17 days, as the chicks were nearing fledging.

Great tits are *secondary cavity nesters*, that is to say that they do not create cavities themselves, but depend on locating existing cavities, some of which are produced by the efforts of *primary cavity nesters* such as woodpeckers. Between these are weak cavity nesters, such as in the black-capped chickadee (*Poecile atricapillus*). So, primary cavity nesters represent species that create expensive nest sites but may have the option to reuse them, while secondary cavity nesters can only choose existing cavities although they may have the option of rejecting low quality ones.

Aitken *et al.* (2002) determined the rates of nest cavity reuse for 20 species in three guilds of cavity nesting bird and mammal species in British Colombia. The proportion of nest cavities reused in a subsequent year was highest for secondary cavity nesters (48%), it was 28% for primary cavity nesters, and 17% for weak cavity nesters. This shows that cavity reuse by excavators such as woodpeckers, species that could be paying high energetic and time costs for nest cavity excavation, is in fact low, while as expected, species that do not create their own cavities are most likely to be found in a cavity that has been used before.

Species looking for a nesting cavity might nevertheless be able to minimise the potential cost of ectoparasite infestations, if there was a surplus of cavities and if they could detect the presence of an expectant ectoparasite population. Aitken *et al.* (2002) predicted that, as parasite populations in an empty nest cavity were likely to decline over time since the last nesting attempt, a cavity would be more attractive for reuse if it had not been used last season, than if it had. However, this *lag hypothesis* was rejected since in the species studied, the difference proved to be in the reverse direction.

Reasons have been advanced at to why the detection of nest cavity infestation might not always result as predicted above. Great tits will choose a flea-free rather than a flea-infested nest box (Oppliger *et al.* 1994), and European pied flycatchers (*Ficedula hypoleuca*) do discriminate against nest boxes containing last year's nest material in favour of clean ones (Merino and Potti 1995). However, this discrimination was not found in a Swedish population of pied flycatchers (Olsson and Allander 1995) possibly because the shorter breeding season in the more northerly location alters the priority to one of accepting a cavity that has proved successful in the past.

For some secondary cavity nest building species, there is an alternative to tolerating an infested nest cavity; that is to fumigate it. A number of bird species include green plant material in the nest, among them cavity nesters such as the European starling (*Sturnus vulgaris*) and members of the tit family (Paridae). Starlings have been shown to have a preference for certain aromatic herbs characterised by volatiles with properties of inhibiting or retarding growth of microbial and arthropod populations. Herbs selected by starlings have been shown to inhibit development of haematophagous arthropods (Clark and Mason 1985, 1988; Clark 1990). This *nest protection* hypothesis is not without challenge, and evidence that green plant materials in starling nests reduce ectoparasite numbers in starling nests disputed (Brouwer and Komdeur 2004). Alternative hypotheses have been proposed for the use of green plant materials in bird nests, for example, the *mate attraction* and *drug* hypotheses.

It is now apparent that it is only the male starling that carries green plant material to the nest and that he can be induced to carry more material by the presence of a female, so supporting the mate attraction hypothesis (Brouwer and Komdeur 2004). The Corsican population of blue tits (*Parus caeruleus*) appears to continue to introduce aromatic plants into the nest cavity during the chick rearing phase (Petit *et al.* 2002). These volatiles, it is suggested, may be effective in protecting chicks against night biting mosquitoes (Lafuma *et al.* 2001).

More information is needed in these studies on the costs of material collection. These would seem to be of two possible kinds. The first is the energy and time costs for the parent(s), the second is the possible harmful effects any plant volatiles have on chick development and survival. However, the claims of the drug hypothesis (Gwinner *et al.* 2000) is that the volatile compounds actually benefit the chicks, in this case by stimulating their immune system, so protecting them from infection. This study found that starling chicks in nests with added herbs had higher body weight, higher haematocrit levels, and survived better till the following year compared with no herb controls.

6.2.1.3 The costs and benefits of nest insulation

Where a nest is sited in a tree or similar location, some nest materials must serve the function of maintaining its structural integrity during incubation and chick rearing phases. This structural nest wall may also serve the additional function of reducing the heat loss from the eggs during their incubation. Nest building costs will be expected to increase linearly with the collection of more materials, while insulation should increase steeply initially, but with diminishing effect. The advantage of greater amounts of material was shown by Redman *et al.* (1999) for a mammal, the short-tailed field vole (*Microtus agrestis*). The thicker the nest wall the greater the insulation. Even at 20 g of nest material, the most observed, the insulative capacity of the nest had not reached its asymptote. The benefit of this investment in insulation will reduce parental expenditure on energy to heat young or eggs. Consequently, if nest materials were only to serve the function of insulation, the optimum amount of material with specified insulating properties could be calculated.

The negative relationship between amount of nest material and costs of incubation, have been confirmed experimentally. Removal of the feathers from nests of barn swallows increases the rate of heat loss from the eggs and the duration of the nestling period and, although the number of chicks fledging is unaltered, there is evidence of greater parental effort to compensate (Møller 1991). Tree swallows (*Tachycineta bicolor*) readily build nests in nest boxes, which are in the form of a grass base lined with feathers. Experimental removal of feathers from these nests not only results in an extended incubation period but also, in this case, a significantly lower fledging success (Lombardo *et al.* 1995).

In species which are intermittent incubators, one bird only is responsible for the incubation so, while it forages, the eggs cool. The rate of cooling when the bird is absent, and its rate of energy expenditure on incubation when present, will depend on the quality of nest insulation and the ambient temperature. Reid *et al.* (1999) tested two hypotheses of optimal incubation pattern, using starlings as the test species: the *egg temperature* and *parental energy* models.

The former model predicts that incubation bouts will be shorter when clutch cooling rate is slower, because the bird terminates the incubation bout when the clutch reaches a certain threshold temperature. The latter predicts the reverse, that is, that incubation bouts will be longer when clutch cooling rates are slower, because here, the decision to leave is determined by the energy level of the incubator, which will decline less rapidly when insulation is better. These were experimentally tested by the insertion of small heating mats into nests of the experimental group. Incubation and foraging bout lengths were then compared with matched controls.

Results support the parental energy model. Overall, incubation sessions by the starlings were significantly longer in experimental nests than in controls, as were the lengths of daytime incubation bouts. Across all of the 33 study nests, the mean daytime incubation bout length was negatively correlated with nest cooling rate. Duration of foraging bouts, by contrast, did not differ significantly between experimental and control nests. The implications of this for nest quality is that a bird in a better insulated nest is, by making energy savings during the incubation phase, enabled to spend longer in the nest before its energy level falls to the threshold precipitating the next bout of foraging.

However, Reid *et al.* (1999) point out that the majority of studies on the duration of incubation bout lengths, like that described by Møller (1991) for the barn swallow, have found support for the egg temperature hypothesis. Confirmation of both models can however, be reconciled if it is argued that foraging bouts will be unaffected by ambient temperature provided that food supply is abundant and a bird can gain enough energy per bout to maintain both clutch temperature and own body condition. In a less favourable environment, however, this may not be the case, resulting in incubation efficiency being constrained. In such circumstances, the optimum incubation pattern would be to spend less time incubating when the clutch cools slowly. Both models demonstrate the benefit of a better insulated nest in reducing incubation costs, but the optimum amount of nest material will depend upon the costs of its collection and, as it turns out, other possible costs.

Some evidence supports the view that sexual selection might operate on male nest builders to build larger nests (Soler *et al.* 1998). On the other hand, Møller (1990) demonstrated a positive correlation between the size of blackbird (*Turdus merula*) nests and predation of artificial eggs placed in them. The association of greater nest volume with higher predation is a possible explanation for the selection of materials of low thermal conductance to line nests, because they may ensure maximum heat conservation for minimum material volume, so limiting nest size (Hansell 2000). An alternative however is that less material ensures lower building costs, the outer nest layer being the minimum to meet the mechanical demands placed upon the nest.

The absence of nest linings in a high proportion of nests (Hansell 2000) shows that in a substantial minority of species such insulation as is required can be met by the structural layer alone. However, a trade-off may be involved here between nest insulation and rate of nest drying. Slagsvold (1989 *a,b*) found that smaller nests dried out more quickly than larger nests, and suggested that this might lead to nests built in more exposed sites being constructed with thinner walls and less absorbent materials than those in more sheltered sites. In this way, a lower level of insulation when the material was dry, would be compensated by the prevention of serious loss of heat from a nest that stayed wet longer. This needs to be tested.

Reid *et al.* (2002), in order to determine whether a nest was effectively designed to minimise heat loss, chose the pectoral sandpiper (*Calidris melanotus*), a species that constructs a simple scrape in the ground lined with plant material. This has the advantage that there is no need for materials to perform a structural role in the nest. This species breeds in the cold environment of the Arctic tundra, and only the female incubates, leaving the eggs exposed during frequent foraging excursions. This permits the thermal insulation provided by the nest to be studied on its own. The task of this nest is to limit convective loss of heat from the eggs to the air, and conductive loss into the ground.

The cooling coefficient of a warmed model egg equipped with a temperature sensor was found to be 9% lower if placed in a bare nest scrape than placed on the ground immediately adjacent to the scrape. The addition of a natural nest lining to the scrape reduced the egg cooling coefficient by 25% compared with an unlined scrape.

The effect of the thickness of the lining in reducing the rate of heat loss was investigated using artificial scrapes lined with different amounts of material, while ensuring that the position of the model egg was always 30 mm below ground level. When the scrape was screened from the wind, the egg cooling coefficient was positively correlated with scrape depth (Fig. 6.1). However, when exposed to the wind as well, the egg cooling coefficient was increased markedly for shallow scrapes, but was not significantly affected in scrapes deeper than 7 cm.

The curve relating the depth of the scrape lining with egg cooling coefficient (Fig. 6.2) shows that the thickness of the material in natural scrapes approximates closely to the minimum required to minimise heat loss.

It seems that pectoral sandpipers breeding in these cold conditions experience conflicting selection pressures to minimise heat loss: deeper scrapes diminish the convected heat loss from the eggs due to the wind, but increase the conductive heat loss from the eggs to the ground. However, it can be seen from Fig. 6.1 that the mean clutch depth of 30 mm, is one close to the maximum depth at which the ground temperature remains relatively high.

The composition of the lining materials by females was also shown to differ in its proportions from those available around the nest. In particular it was found that the most represented material in the nest lining compared to its

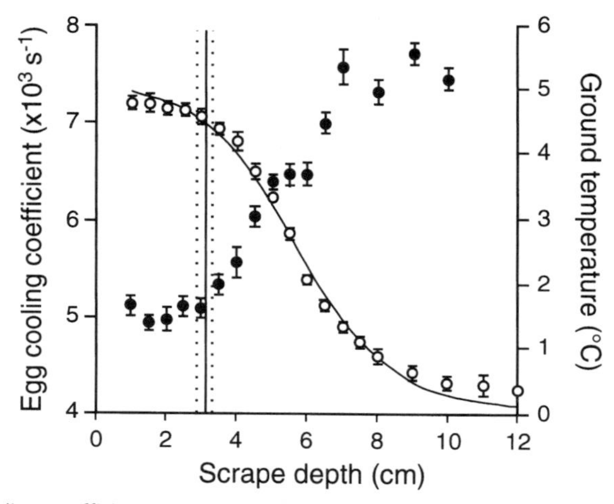

Fig. 6.1. Egg cooling coefficient and ground temperature in the nest of the pectoral sandpiper (*C. melanotus*) in relation to scrape depth. With wind excluded, egg cooling coefficient was correlated with scrape depth (•). Ground temperature decreased non-linearly with depth (o). Vertical solid and dashed lines represent mean ±1 SE observed clutch depth. (Adapted from Reid *et al.* 2002.)

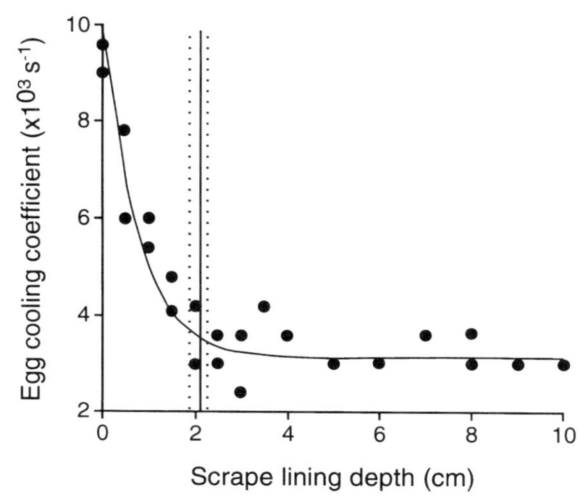

Fig. 6.2. Egg cooling coefficient in relation to the depth of lining in experimental nest scrapes. Vertical solid and dashed lines represent mean ±1 SE of the depth of lining observed in actual nests. (Adapted from Reid *et al.* 2002.)

availability was a species of lichen (*Dactylina arctica*). This was not the most effective insulator when dry, but was among the most effective when damp, an almost invariable condition prevailing in their habitat, and showed the least difference in its effectiveness between dry and damp conditions.

6.2.2 Caddis larval cases: silk expenditure and a trade-off between larval security and adult fecundity

Caddis flies do little feeding as adults, consequently resources acquired during the larval phase must be invested in growth, larval case construction, and adult gonad production. The energetic cost of constructing the larval case includes searching for and manipulating building materials, but these must also be fastened together by silk. Expenditure of silk for case construction during the larval phase therefore will affect protein available for growth and reproduction, a potential trade-off between larval security and adult reproductive performance.

Caddis cases show both within-species and between-species differences, and a variety of hypotheses have been proposed to account for them, but the influence of predator pressure on case design is supported by a number of studies. Observations made of predation attempts by brown trout (*Salmo trutta*) and bullhead (*Cottus gobio*) on four species of larval caddis, three of which construct organic cases (*Limnephilus rhombicus*, *Limnephilus pantodapus*, *Glyphotaelius pellucidus*) and one a mineral case (*Potamophylax cingulatus*), have demonstrated that the mineral case gives better protection than the leaf case examples. Bullheads are unable to ingest it and it provides the highest survival rate against predation attempts by trout; however, in brown trout, the handling time to consume *L. pantodapus* larvae, which have a weak cylindrical leaf case, is significantly longer than that for *P. cingulatus* larvae in their mineral cases. The protection it offers is apparently afforded by its length relative to the size of the larva, rendering it more difficult to handle (Johansson 1991).

Variability in case construction was shown by a larval population of *Limnephilus frijole* occupying a site largely free of predators. When these were offered to fish and dragonfly nymphs in an experiment where crypsis was excluded, larvae in cases that were stronger and wider were more likely to survive either predator, whereas case length had no clear effect (Nislow and Molles 1993).

Potamophylax cingulatus is one of those species that in earlier instars builds a vegetation case, which it later changes to one of sand. The vegetation cases are however, longer relative to the larva than are mineral cases. These vegetation cases, when experimentally shortened to more closely match the length of the larvae, render the larvae at greater risk of trout predation than unaltered ones. Larvae in mineral cases suffer less predation than those in the vegetation cases. Mineral cases can withstand greater compressive stress before collapse than can those of plant materials, but, relatively more silk is required for their construction (Otto and Svensson 1980). The cases of the caddis larva *Silo pallipes*, and some other species occupying swift flowing streams, have massive stones attached to their sides, normally two on each side. Removing the stones to test between the rival hypotheses that the lateral stones either provided ballast to prevent cases being washed away, or that they provided protection against predation, Otto and Johansson (1995) found convincing support only for the latter.

The weight of evidence therefore does seem to support the view that caddis larval cases are designed to protect against predation, although this does not exclude other possible functions. The greater amount of silk and greater strength of sand grain cases compared with vegetation cases (Otto and Svensson 1980) does provide some support for the view that better protection is more costly. This should be an advantage to larvae that invest in protection, conditional upon predation risk. Some evidence at least supports this view.

Cases of *Hesperophylax magnus*, another limnephilid species, were found to have a higher mineral content in habitats containing greater numbers of fish and invertebrate predators, which is positively correlated with resistance to crushing (Fig. 6.3(a)). This is shown experimentally to be a conditional response. Larvae from populations that built weak cases in nature, in the laboratory built stronger cases in the presence of predaceous stonefly larvae (Fig. 6.3(b)) (Molles and Nislow 1990).

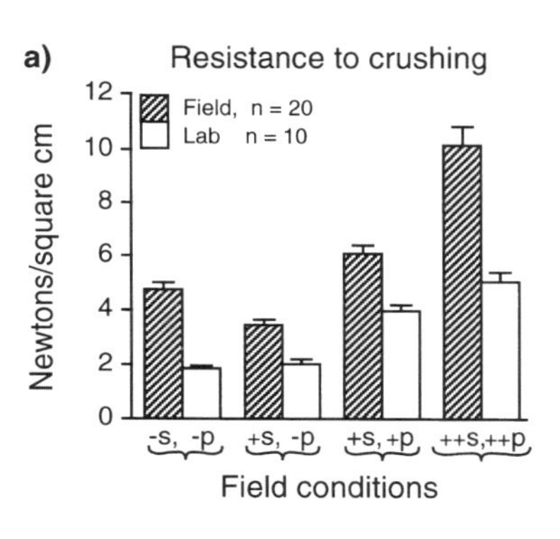

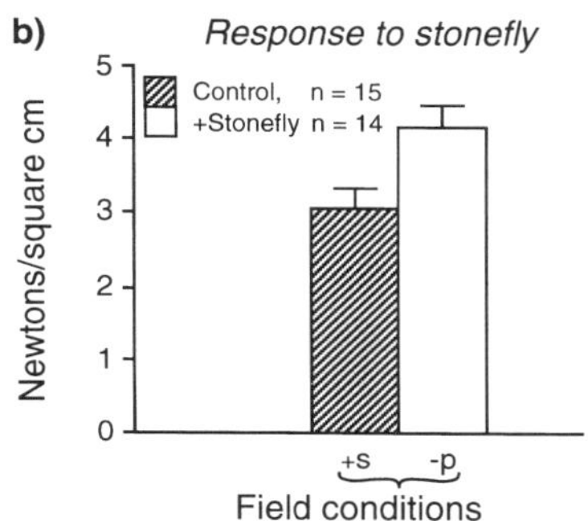

Fig. 6.3. Resistance to crushing by cases of the caddis larva *H. magnus*. +/−s is an indication of the presence or absence of specialised case building by larvae in that habitat. +/−p indicates the presence or absence of predators in that habitat. Bars indicate one SE of the mean. (a) Resistance to crushing of cases built in the field and in the laboratory by *H. magnus*. The strength of cases of larvae is correlated with the degree of predation risk and likelihood of occurrence of specialised protective cases in that habitat. (b) Cases of *H. magnus* larvae were significantly stronger when built in the presence of the predaceous stonefly nymph *Hesperoperia pacifica*. (Adapted from Molles and Nislow 1990.)

Bearing in mind this correlation between building cost and level of protection, how much should a larva invest? The problem, given that essentially all feeding is confined to the larval phase, is that there must be a trade-off between larval security and adult reproductive effort. This trade-off was investigated by Stevens *et al.* (1999) for the mineral case builder *Odontocerum albicorne*.

Final instar larvae of this species were required to rebuild their cases, and aspects of their later development and reproductive potential compared with controls. This demonstrated that experimentals rebuilt their cases so that at pupation these matched the weight of those of controls, although they had a lower silk content. Nonetheless, as a consequence of the rebuilding, the total silk expenditure of experimentals exceeded that of controls in both sexes (Fig. 6.4). Experimentals did not undertake any compensatory feeding before pupation, or delay their pupation date, but did have a shorter pupation period, possibly because pupation duration is mass dependent.

The effects of the differences between larval treatments were evident in only some of the measured adult features; the forewings were shorter and the thoraxes lighter in experimentals of both sexes compared with controls, however, no differences were found between the abdomens of

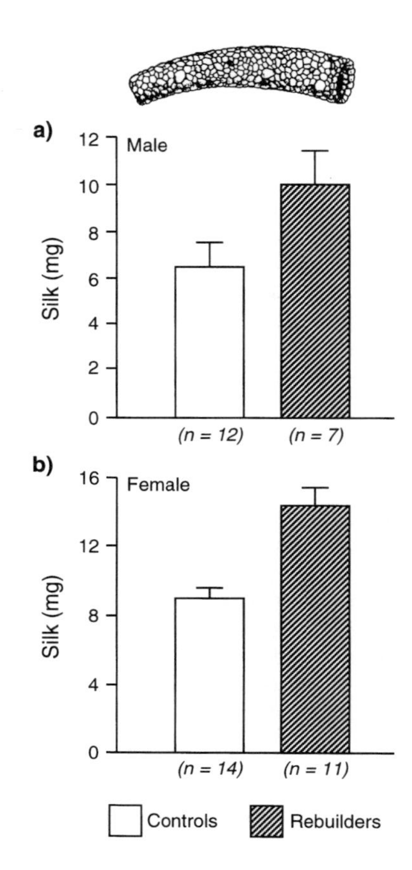

Fig. 6.4. Cost of case reconstruction in fifth instar caddis larvae *O. albicorne*. Total silk produced by larvae forced to rebuild compared with controls. Data presented as means ±1 SE. n = sample size. (a) Males, (b) females. (Adapted from Stevens *et al.* 1999.)

experimentals and controls in respect of either their masses or their nitrogen content. It seems therefore that the consequence for adult *O. albicorne* of additional expenditure on larval security may be a diminution of flight capability as an adult. The reason for this could however, be either one of developmental constraint or of strategic preservation of adult reproductive potential. A case rebuilding experiment on *G. pellucidus* allowed these rival hypotheses to be tested.

Odontocerum albicorne is a species with a short adult lifespan, the imago eclosing with the gametes already formed. By contrast *G. pellucidus* lives several months as an adult, during which time the gonads reach maturity. Stevens *et al.* (2000) predicted that in this species the hypothesis of strategic allocation of resources for adult life should conserve the allocation to the thorax at the expense of the abdomen, while the developmental constraint hypothesis predicts the same outcome as in the experiment on *O. albicorne*, since resource allocation is seen to be constrained by some other resource partitioning imperative.

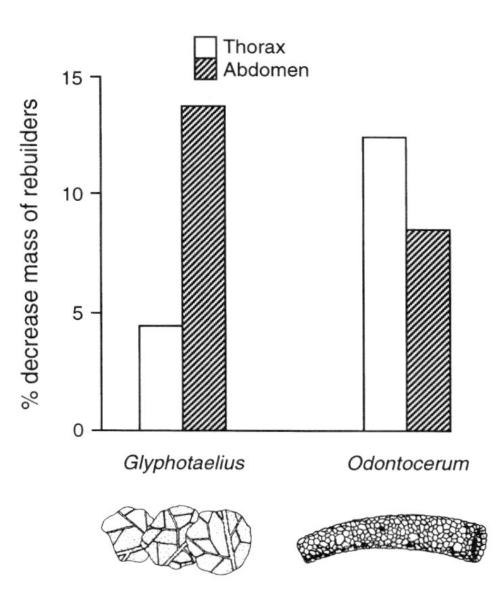

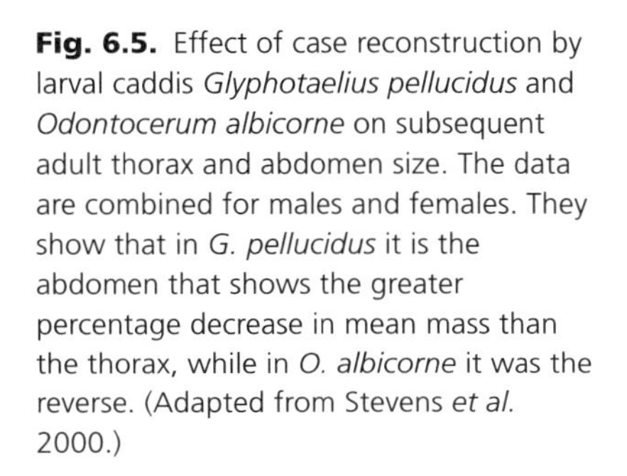

Fig. 6.5. Effect of case reconstruction by larval caddis *Glyphotaelius pellucidus* and *Odontocerum albicorne* on subsequent adult thorax and abdomen size. The data are combined for males and females. They show that in *G. pellucidus* it is the abdomen that shows the greater percentage decrease in mean mass than the thorax, while in *O. albicorne* it was the reverse. (Adapted from Stevens *et al.* 2000.)

The case rebuilding experiment on *G. pellucidus* showed that, on emergence as adults, the allocation of resources was indeed different from that in *O. albicorne*. Thorax masses did not differ significantly between experimental and control groups, although those of females were larger than those of males, while the masses of the abdomens of the rebuilding group were relatively smaller (Fig. 6.5).

The reduced abdominal allocation in experimental *G. pellucidus* probably results in reduced reproductive output when breeding does occur, however, with a long adult life before reproduction, the priority of this species may be to conserve thorax mass as flight muscle. It remains possible that some of this mass could later be relocated to develop reproductive tissues.

6.3 Foraging with burrows and webs

Fossorial rodents share with net spinning insects the need to construct in order to obtain food; in the case of the former it is feeding burrows, in the case of the latter it is nets. Each of the two groups have special circumstances relating to their foraging costs. For fossorial foragers it is the high cost of travel to obtain food items. This cost has been estimated at between 360 and 3,400 times greater for subterranean compared with above-ground travel (Vleck 1979). This is the subject of Section 6.3.1.

Trap construction is very largely confined to the arthropods (Table 1.1) and it was argued in Section 2.5, generally makes use of self-secreted materials in the form of nets or adhesive threads. Self-secreted building materials of any kind are rare among the vertebrates and, although self-secreted trap materials are used by Appendicularia (Chordata), they are absent from the true vertebrates. The relative absence of self-secreted materials in vertebrates may be due to problems of scale. The costs for a larger organism to secrete a useful amount of building material may be too high to make it economical. There is insufficient empirical evidence to judge this, but little swifts (*Apus affinis*) make their nests largely from their own salivary mucus and, although the energetic cost of this is unknown, it takes a mature pair 1.8 months to complete a nest and a yearling pair as much as 4.6 months. Usurpation of nests is common and accompanied by conspecific infanticide, even when only one member of the pair is replaced (Hotta 1994), evidence in this case of a high cost in time expended in nest building quite apart from any energetic cost.

Nearly all net spinners among the caddis larvae and web spinning spiders are sit-and-wait predators and, as such, they can potentially enhance their capture rate in one of three possible ways: First, the net could be placed across the predominant direction of prey movement, which should select for careful sitting of the net. Second, the predator might cause the medium in which the prey is suspended to move to the net. This could be achieved by pumping or fanning, which economically is much more likely in an aquatic medium rather than in air. Third, the predator might attract the prey to it by means of bait or a lure. This is more likely to be successful against aerial prey such as flying insects, which are able to move actively under their own power, than in water.

The special circumstance of foraging costs for net or web spinning forager is that all the costs needed must be paid in advance of obtaining benefit from it. The need to invest in a net before foraging can begin, should select for adaptive responses based on current or recent foraging experience. This could permit a minimising of costs in the following ways: first the careful siteing of the net, second the structure of the web, third, the giving-up time in a particular foraging location. In addition, different species of net spinner should exhibit differences in conditional net building behaviour related to their particular ecology.

6.3.1 Mammal burrows

As a consequence of the relatively much higher costs of subterranean compared to above-ground foraging, predictions of optimal foraging that have been supported by studies of above-ground foragers, need to re-evaluated and tested for fossorial species. The high cost of burrow digging should also lead to economic design of the burrow system wherever possible.

6.3.1.1 Optimal burrow segment length

The pocket gopher (*Thomomys bottae*), is a fossorial rodent that extends its burrow in segments, each of which consists of a horizontal length of feeding

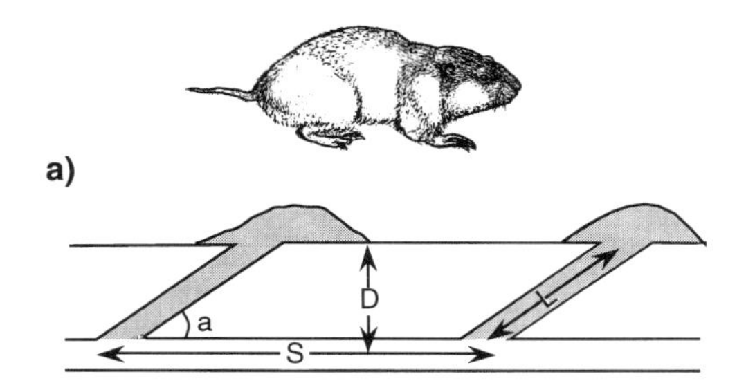

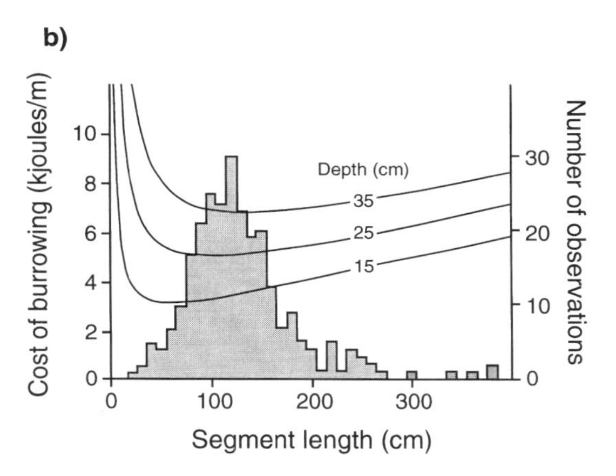

Fig. 6.6. Optimum length of burrow segments dug by the pocket gopher (*T. bottae*). (a) Diagrammatic side view of burrow design: *S* = feeding tunnel segment length, *D* = depth of the tunnel, *L* = the length of the lateral tunnel to the surface, and *a* = the angle it makes with the horizontal. (b) Cost of burrowing in relation to segment length. The three curves describe the energy cost of burrowing in relation to segment length for feeding tunnels of stated depth. The histogram shows the frequency distribution curve of segment lengths in a study area. (Adapted from Vleck 1981.)

tunnel and a branch that separates from it at an upward incline until it reaches the surface. The inclined burrow has a role in allowing the spoil excavated at the face of the feeding tunnel to be disposed of above ground. The cost (length) of digging a spoil-removal tunnel is fixed, and determined by its slope and the depth of the feeding burrow below the soil surface. The cost of digging the feeding burrow is, however, not fixed since, as the work-force moves forward, so the distance the spoil has to be moved back to the nearest inclined branch to the surface gets greater, raising the cost of burrowing. Based on these costs, which provide no evidence of trade-off, it is possible to calculate the segment length that will minimise burrowing costs.

Vleck (1981), using estimates of the cost of burrowing by *T. bottae*, and that of pushing the loose soil out of the burrow, calculated the optimal segment length to be 1.22 m (Fig. 6.6). The observed length of the segment was 1.33 m, longer than predicted, although a moderately close fit. However, this model does ignore the possible benefit of locating food by concentrating on the feeding burrow. The plateau zokor (*Myospalax baileyi*) extends its burrow in the same manner as the pocket gopher, and Jianping (1992) found that segment length in this species conformed to the predictions of Vleck's (1981) model. However, he also found that it was equally predicted by an alternative model of optimum energy return per burrowing cost. This incorporates the gain of food intake, which is only provided when burrowing horizontally.

Burrowing should be relatively cheaper for larger than smaller individuals, because the cost of digging a burrow is directly proportional to its cross-sectional area. Selection should therefore favour large body size in fossorial species, except that this is conditional upon an increase in food density required to support them, a condition that is rarely met. Nevertheless, the prediction is supported by the occurrence of the largest fossorial rodents occurring in the most productive habitats both within and between species (Vleck 1981).

The same size advantage will also apply to non-fossorial, burrow dwelling mammals, yet they too are generally small. The badger (*Meles meles*), at 10–20 kg in weight, is among the larger species. However, without being able to quantify the benefit, it is hard to predict what might be the size limit for this group of burrowers. The burrow might be defensive, thermoregulatory, or both, and its value in relation to investment unknown. There might also be a mechanical constraint, since there is likely to be a correlation between burrow diameter and the likelihood of its collapse.

6.3.1.2 Burrowing costs and optimal prey choice

An important prediction relating to optimal food-type choice, generally supported by above-ground foragers, is that the food species will be collected preferentially not relative to their abundance, but according to their profitability, or net value (Pyke 1984). Heth *et al.* (1989) tested this prediction against the behaviour of a fossorial forager, the mole rat *Spalax ehrenbergi*, by recording the plant species hoarded in its burrow systems. The results showed that the frequencies of plant species in mole rat stores did in fact correspond to their relative abundance within a foraging territory. These fossorial rodents therefore exhibit neither definite preference, nor direct their search to any particular species.

Heth *et al.* (1989) explain this departure from the pattern for above-ground foragers as due to the high cost of burrowing compared with that of moving above ground, and to the inability of burrowers to detect food items except at very close range. For above-ground foragers, differences between food types in terms of handling time or food value, as well as search time, may be significant. For subterranean species, however, foraging costs will be overwhelmingly a function of the distance between food items, and largely independent of the type of food item (Fig. 6.7). The consequences of this for food discrimination is that a subterranean forager that discriminates between food types will substantially increase search expenditure and therefore lower

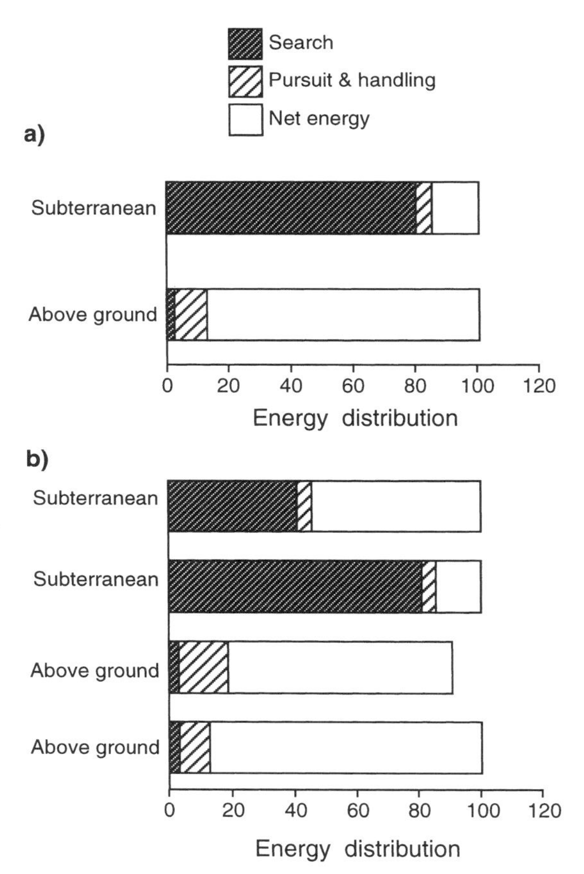

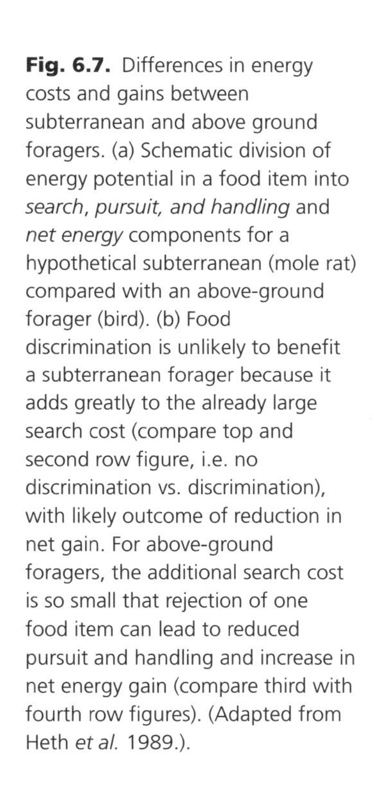
Fig. 6.7. Differences in energy costs and gains between subterranean and above ground foragers. (a) Schematic division of energy potential in a food item into *search*, *pursuit, and handling* and *net energy* components for a hypothetical subterranean (mole rat) compared with an above-ground forager (bird). (b) Food discrimination is unlikely to benefit a subterranean forager because it adds greatly to the already large search cost (compare top and second row figure, i.e. no discrimination vs. discrimination), with likely outcome of reduction in net gain. For above-ground foragers, the additional search cost is so small that rejection of one food item can lead to reduced pursuit and handling and increase in net energy gain (compare third with fourth row figures). (Adapted from Heth *et al.* 1989.).

net energy gain, whereas for the above-ground forager, where travel costs are slight, rejection of food types with lower energy value or higher handling time increases net benefit.

Because builders by their activity change their habitats, it should be anticipated that the consequences for building costs may sometimes be complex. So, whereas fossorial rodents, while moving, experience much higher costs than above-ground foragers, in a resting state, they experience some energetic savings. The naked mole rat (*Heterocephalus glaber*) has a resting metabolic rate only 40% of that predicted for a rodent of its size, and the lowest body temperature recorded for a fossorial rodent (32°C). This has been interpreted variously as saving on the cost of thermoregulation, minimising problems of dissipating metabolic heat (assisted by hairlessness in this species), or a response to depressed O_2 and heightened CO_2 concentrations (McNab 1966, 1979).

6.3.1.3 Burrowing as a cost of parental care

Fitness gain has been used as a measure of benefit for burrowing of male European ground squirrels (*Spermophilus citellus*). This was measured as mating success, and compared with paternal effort in terms of digging a litter burrow, which a male only does for a litter of which he is the father (Huber *et al.* 2002). Male digging behaviour reduced his foraging time, leading to weight loss, but benefits to the offspring were evident in their increased mass at natal emergence. There was, however, found to be a trade-off between mating effort and paternal effort. Males with an average of only one mate or more than two, showed burrow digging at only one litter burrow, whereas digging at two litter burrows was shown by males with two mates. Most parenting is therefore shown by males of intermediate mating investment, with males of higher mating success showing less parental effort.

6.3.2 Aquatic net spinners: caddis larvae and others

The sinking rate of an object suspended in a fluid medium is determined by its *terminal velocity*, that is, the speed at which drag equals weight (Denny 1993). Drag is determined by the 'frontal' area of the object, its velocity of descent, and the density and viscosity of the fluid. Water has a density about 830 times greater than air, the exact value depending upon its temperature, and a viscosity 70 times greater. The consequence of these differences is that suspended particles will fall out of the air stream more rapidly than out of the water current. Filter or suspension feeding is therefore more likely to evolve in an aquatic medium than in air.

Aerial plankton does occur as pollen grains, fungal spores, or bacteria, but larger potential food items such as insects need to use their own power of flight to stay suspended. A spider web, therefore, does not so much filter the passing air as intercept food items travelling in a direction of their own choosing through it.

In the aquatic environment by contrast, suspended items travel predominantly in the direction of the medium and consequently can be filtered by a net placed across the current. This suspended material includes organic debris and weakly mobile organisms from bacteria to invertebrate larvae, in the size range from 1 to 1000 μm, referred to collectively as *seston* (Wildish and Kristmanson 1997).

Some species have evolved a means of enhancing flux through the capture net, or making flow more constant by their own pumping action. In this way, they are not only able to enhance capture rate, but they also have to commit additional costs. In the tranquil environment of the open ocean, plankton is not subjected to severe stresses that water imposes on attached organisms in streams and rivers. Freed from this physical constraint, Appendicularia have developed capture nets much larger than themselves, powered by their own pumping action. The explanation of this advanced by Acuña (2001) is that they and other pelagic tunicates such as *Doliolum*, which has an internal filter, have been able to accomplish this in a quite nutrient poor environment, because of the low cost of the filter apparatus relative to the live body mass of the organism. This is achieved in the *Doliolum* by having the body form of a large empty barrel, with the filter mechanism intercepting particles from the water current driven through it. The body is, however, mostly inert, gelatinous, and largely composed of water. Their tunicate (urochordate) relatives, the Appendicularia, have achieved the same result by building an inert filter and dwelling, through which the diminutive organism drives the current with its tail (Fig. 2.9).

Caddis larvae of the Hydropsychidae are the dominant fresh water organisms that construct capture nets to feed on the seston. It is now apparent that these nets do not, as was formerly assumed, act like sieves. Instead, particles adhere to the threads of the net, allowing particles much smaller than the mesh size to be captured (Alstad 1987*a*). Particles strike the threads for essentially one of two reasons: either they are in the fluid stream bound for direct interception by the thread, or if they are denser than water, their inertia causes them to attempt to continue in a straight line deflecting them from the 'streamline' passing around the net fibre and into a path of interception (Wildish and Kristmanson 1997) (Fig. 6.8). This inertial kind of impact will be enhanced by the speed and density of the particle, and is more likely in a fluid of low viscosity.

The cost of spinning a net is closely related to the amount of silk in it The energetic costs of net production were determined for *Polycentropus*

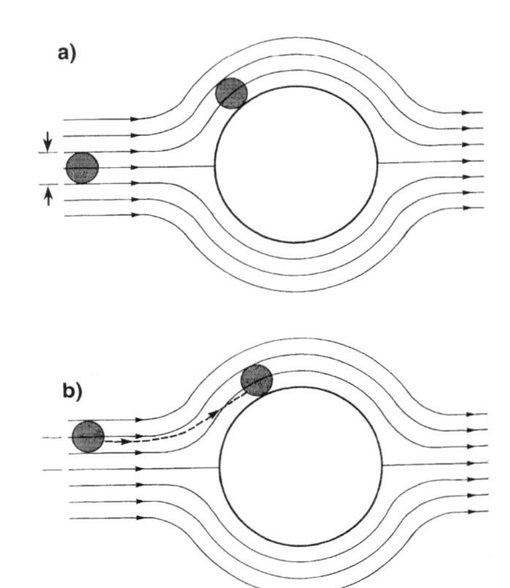

Fig. 6.8. Schematic representation of the capture of sestonic particles by the silk thread of a capture web: (a) by direct interception and (b) by inertial impaction. (Adapted from Wildish and Kristmanson 1997.)

flavomaculatus from the respirometry of larvae at rest during net spinning, plus bomb calorimetry of the silk net (Dudgeon 1987). This showed that metabolic rates of the larvae were raised 17% during net spinning and that nearly 76% of the total cost was in the silk itself, which in caddis cannot be reclaimed through ingestion. A larva can use more silk in order to either reduce the mesh size or enlarge the net area. Both of these will influence the particle capture rate, as will also the seston density and the water flux through the net. The optimal net design for a given habitat is, however, subject to the trade-off that, as increased current speed imposes greater stress on the net, a greater proportion of the current is deflected round rather than through it. This may be alleviated by an increase in mesh size, but this risks more seston particles passing through the net rather than being retained.

Mesh sizes in the Hydropsychidae, which vary from as much as 500 μm to only a few microns across (Wallace and Malas 1976), are quite species-typical and, in many species, of rather regular mesh size throughout an individual net. This regularity of mesh size within species, and difference between species has been the focus of debate because is reflects either particle size specialisation or alternatively the outcome of trade-off. Meshes of different size are associated with different habitats. Larger mesh sizes are typical of faster flowing waters with mesh of decreasing size found in progressively slower moving water. Two functional hypotheses have been invoked to explain this distribution: the *capture rate hypothesisi* and the *particle size hypothesis*.

The capture rate hypothesis maintains that the functional architecture of the nets in different habitats results from a trade-off between water filtration rate (flux) and capture efficiency (Alstad 1987*b*). It is based on the assumption that, as nets capture particles much smaller than their mesh size, they are not highly selective in the size of particle they collect, a view supported by observations of Alstad (1987*a*). Therefore, according to this hypothesis, larger mesh sizes in faster flowing water is an adaptation to lower particle abundance in faster flowing streams than in the slower flowing eutrophic habitats, necessitating a larger flux through the net to raise capture rate to that of nets in the slower richer waters. The particle size hypothesis, however, argues that meshes are specialised to catch particular sized prey, small particles in the slower water and larger ones in the faster water. Brown (2004) shows in a field experiment that the nets of *Hydropsyche siltalai* (mesh 50 μm for later instars) do catch proportionally more larger particles from those available. She also confirms that capture is largely by direct interception not by sieving, and that capture efficiency, at a maximum of 2.5%, is an order of magnitude greater than previously estimated for a hydropsychid.

Loudon and Alstad (1990) turned to mathematical model based on fluid mechanics to simulate the performance of nets of different design operating in different water velocities containing particles of different sizes and

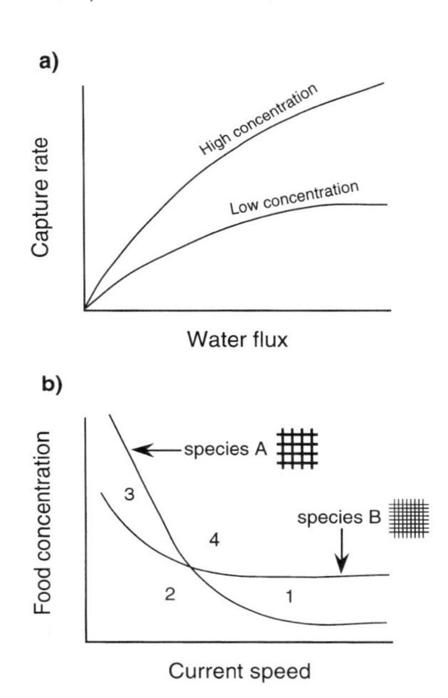

Fig. 6.9. Capture effectiveness of caddis larval nets. (a) These hypothetical curves represent the assumption that capture rate is related to seston concentration, but that capture efficiency declines with increasing current speed. (b) Model comparing the capture success of a larger more robust mesh of species A with the finer, more delicate mesh of species B. In quadrant 4 food concentration exceeds the marginal value for both, and in zone 2 is below that for both net builders. However, the species with the fine meshes (B) requires a lower resource concentration at slow current speeds (zone 3), because its capture efficiency is higher. Conversely, the robust mesh (A) can survive and function in fast currents, giving it greater effectiveness at low food concentrations (zone 1). (Adapted from Alstad 1987*b*.)

densities. Their results show that as mesh size increases for a given fibre diameter, so the velocity of the water through the net increases to approach that of ambient. However, the efficiency of particle capture does decrease with increasing mesh size, since the likelihood of a particle striking a thread becomes less.

The product of these contrary influences is a mesh size that theoretically maximises particle capture rate for given conditions of fibre diameter, net area, current velocity, and particle size (Loudon and Alstad 1990). The mesh size that maximises capture rate is shown to increase with particle size, supporting the particle size hypotheses; however, this is weakened by the demonstration that regardless of mesh size, the size of particles captured strongly reflects the available size distribution.

The simulation also compares capture success of a large area of net of large mesh size with a smaller area of smaller mesh size. This shows that the minimum level of food concentration for survival is higher for large mesh sizes at all current velocities. This contradicts the descriptive model of the capture rate hypothesis (Alstad 1987*b*) (Fig. 6.9) in which the marginal resource requirement is lower for small meshes at low current speed, but lower for larger meshes at high current speeds.

Larvae of the net spinning caddis *P. flavomaculatus*, forced to rebuild a net each day understandably lose weight, while larvae with the same net over the same 10 day period do not, but the rebuilders reduce the size of their nets although the weight of the nets remains fairly constant as a proportion of declining larval body weight (Dudgeon 1987).

A manipulated feeding regime was used by Dudgeon (1987) to test the prediction that in *P. flavomaculatus* larvae would minimise their cost/benefit ratio. The result was that starved larvae built the smallest nets relative to their body weights, compared with a group fed every other day, while nets spun by an *ad libitum* fed group were relatively even larger. This was interpreted as starved larvae minimising cost/benefit ratio by reducing costs, while the fed groups, although increasing costs, were reducing cost/benefit ratio by enhancing net gain.

Petersen *et al.* (1984) compared the net dimensions of larvae of *Neureclepsis bimaculata* in three different locations, one eutrophic and two oligotrophic. They found that larvae in the high quality (eutrophic) seston site spun nets of smaller size in relation to body mass than those in two oligotrophic sites. In all three sites, the area of the entrance to the funnel-shaped net of this species reduces with current velocity, but this is more pronounced in the nutrient poor sites, where their funnel entrances are substantially larger at lower current speeds. This can be interpreted as larvae in the nutrient poor sites trying, but unable, to compensate fully for the lower food concentration. This is evidenced by the smaller size at fifth instar of oligotrophic larvae relative to those from the eutrophic location. This failure could be due to one of two possible constraints: either the larvae are limited in the amount of silk they can invest, or the structural limitations of the silk to build a larger structure (Petersen *et al.* 1984).

6.3.3 Spider webs

Spider web designs vary greatly, and even the orb webs of different species, in spite of their superficial similarity, may differ in their capture and engineering principles (Section 5.4). The expectation is that between species, the webs will differ in their costs and in their energetic return on investment. Evidence presented below, demonstrates the validity of both these.

The expectation that prey attraction devices will evolve to enhance capture rate is confirmed by the function of stabilimenta in the webs of some day foraging orb web builders, and by the moth mimicking pheromones produced by some night foraging species (Section 1.4). Additional ways of enhancing prey capture are examined here, for example, the selection of prey-rich sites and web investments based on monitoring of current or recent foraging success.

6.3.3.1 Variation in cost-benefit relationship between webs of different spider species

The effectiveness of webs between species differs markedly. One reason is differences in the adhesive principle used in the capture thread (Section 2.4.3).

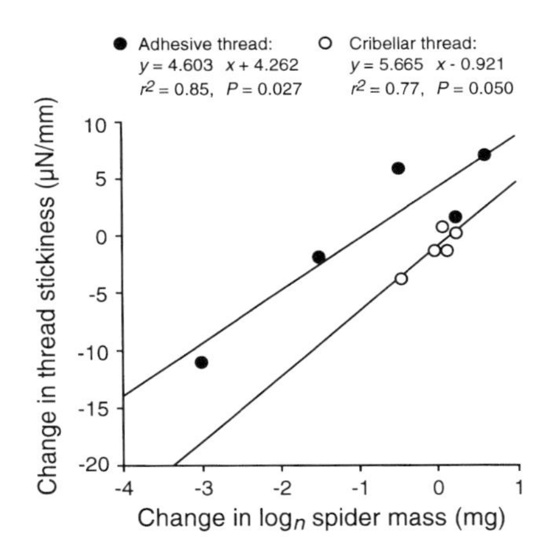

Fig. 6.10. The regression slopes of changes in spider thread stickiness in relation to change in spider mass, do not differ between the adhesive thread of ecribellate spiders (•) and cribellar capture thread (o). However, the line for ecribellate species is significantly elevated showing that, relative to spider mass, the adhesive threads are stickier than cribellar threads. (Adapted from Opell 1998.)

Opell (1998) calculated the costs of webs of four species of orb spinning Araneidae (Araneoidea) and eight Uloboridae (Deinopoidea) (four of them orb web builders) based on silk volume plus additional behavioural costs. He then compared their effectiveness in terms of stickiness of the web for equivalent web costs. Relative to spider mass, the ecribellate Araneidae invest less material in their capture threads than the cribellate Uloboridae and relative to the volume of material used, adhesive threads of the Araneidae are 13 times stickier than the dry capture threads produced by the Uloboridae (Fig. 6.10). Opell (1998) suggests that one consequence for the Uloboridae for the low profitability of their webs is an evolutionary constraint on their size.

Differences in the size of spiders are also associated with distinctive behavioural and ecological specialisations. Unsurprisingly, from the point of view of metabolic costs, there is a general trend in web building spiders for larger species to spin larger webs with larger diameter fibres that capture larger, faster flying prey than do smaller spiders (Craig 1987*a*). However, in a comparison between orb weavers typical of the tropical forest understorey (in this case all Araneoidea), Craig (1989) demonstrated that the smaller, more fragile webs of the species *Epilineutes globosus* and *Leucauge globosa* intercepted on average respectively 40% and 6% of the spider's weight in prey biomass/web, while the larger, stronger webs of *Micrathena schreibersi* and *Mangora pia* caught on average respectively only 2% and 4%. This difference in prey capture per web is also associated with the larger spiders limiting construction costs by building only one web per feeding period. Smaller species, with their higher rate of food intake, tolerate higher rates of loss of their fragile webs and will build three to five webs per feeding period.

The relationship between silk expenditure and capture success in web building spiders is subject to a variety of trade-offs; Miyashita (1997) identifies four in particular: (1) Prey capture efficiency versus web durability; (2) prey retention versus web visibility; (3) silk density versus web area; (4) silk density versus silk stickiness.

The mesh size of spider webs can be increased by increasing the average distance between turns of the capture spiral. This gives a larger capture area for the same length of capture thread, although with some increase in the total web cost because of the necessary increase in the non-capture thread. The relationship between mesh height (the distance between capture spirals) and captured prey size has proved difficult to resolve (Herberstein and Heiling 1998). In general the mesh does not seem to act like a filter, but occasional specific exceptions appear to indicate cases where mesh size may be adapted to prey size.

Rypstra (1982) compared the efficiency of the webs of three groups of species, tangle, sheet, and orb web spinners, in their ability to capture *Drosophila* in the absence of the spider. This showed that for all three groups, the efficiency decreased with thread density, towards an asymptote at which the efficiency of orb web spinners was greatest, sheet webs least, and tangle web spinners intermediate. The levelling off of efficiency with increased thread density was interpreted as a trade-off between increased probability of prey retention when the web was struck, against increased web visibility and therefore prey avoidance.

Miyashita (1997) in a comparison of foraging success of webs of three species of *Cyclosa* which differ in their adhesiveness, mesh size, and thread diameter, found that two of the species had a much higher capture rate than the third, but achieved it in different ways, one by the strength and adhesiveness of its web, the other by its small mesh size and choice of prey-rich websites. The lower success of the third species was apparently due in part to the fragile web but also to the nature of its attack behaviour, but these disadvantages were associated with some advantages in other life-history traits such as high survival rate.

An interaction between capture rate, web rebuilding costs, and foraging duration is shown in the comparison between a day and a night feeding orb spinning araneid species. *Nephila plumipes* feeds during the day, capturing mainly hymenoptera in its relatively permanent web. The similarly sized sympatric orb spinner *Eriophora transmarina* builds a new web every night to capture lepidoptera.

The web of the moth-catching *E. transmarina* is vertically elongated compared with a more circular shape for that of *N. plumipes*. Its mesh size is also bigger and its web weight relative to that of the spider is less than that of *N. plumipes*. There is no difference in average prey size of the two species, but *Eriophora* has nearly twice the capture rate of *N. plumipes*, although for a shorter foraging duration. The reasons for the higher capture rate in *Eriophora* may be due to: its larger capture area or its more adhesive thread since larger adhesive droplets are possible at night without visual detection. The barrier web on either side of the capture web of *Nephila*, which apparently serves to reduce predation, may also impede prey capture. The differences in mesh size between the webs may reflect the difference between capturing fast, day flying prey, and the relatively slower nocturnal prey, principally moths (Herberstein and Elgar 1994). As has already been emphasised in Sections 1.4 and 5.4, the superficial similarity between orb webs of different species distracts from important differences of adaptive detail.

6.3.3.2 When to move web location

In making the decision to move website, a spider is trading the cost of moving against the possible benefit of a better foraging site. Janetos (1982) devised a model by which a spider might determine the optimum time to relocate its web. It predicts that frequent web relocation will be favoured when the cost of moving is low or when the difference between good and poor patches is high. Testing this against natural web relocation frequency between species of different web designs supports the web cost prediction. It showed, for example, that relocation was low for *Agelana limbata* that builds a relatively costly sheet web compared with species building much cheaper orb webs (Tanaka 1989).

The prediction that within-species, individuals in poor habitats will move more frequently than in rich habitats is upheld by the orb web builder *Nephila clavipes* which, in a poor habitat, is found to move more often than in a rich habitat. However, it will only move when foraging success is very low, the result being that it grows more slowly and reaches adulthood at a smaller size than individuals in a rich feeding habitat. This, concludes Vollrath (1985), indicates that rather than maximise their foraging success, selection favours avoidance of the high cost of moving in terms of risks and investment in new web.

Caraco and Gillespie (1986) found the opposite relocation tendency in the long-jawed, orb weaving spider (*Tetragnatha elongata*). It showed a higher rate of web relocation when prey capture was high than when it was low. They explained this in terms of a risk sensitivity model in which spiders were shown to relocate more frequently when experiencing high capture rate with high variability, than when experiencing low but predictable capture rates. However, Smallwood (1993) favours an alternative explanation based on the greater intraspecific competition in good quality patches, demonstrating empirically with the same species, that in a rich feeding habitat, relocation rates were higher when spider densities were higher.

The movement of websites by foraging spiders is very low compared with the frequency of patch relocation by say, bird foragers. The consequences of this were studied by Nakata and Ushimaru (1999), who reasoned that when a spider first arrives at a new location it will have little information on patch quality until it sets up a web and starts to monitor capture success. They predicted that stochastic variation typical of any capture rate would prevent a spider from making a quick assessment. Consequently, even after the first day, its assessment might still be inaccurate, but over a period of a few days would become less and less so. If this was the case, then spiders would draw upon previous experience at the current capture site in order to determine when to leave.

They tested this by withdrawing prey from the webs of *Cyclosa argenteoalba* spiders that were either newly arrived or had been at a site for several days. The predictions were first that prey removal would raise the rate of relocation in both groups compared with controls, and second that spiders of the newly arrived group would be more likely to relocate because they experience only failure, whereas the others have experienced some success before the start of the experiment. It was found that both groups did indeed have an elevated relocation rate, and also the former were more likely to move. These results support the view that these spiders make use of recent capture experience in making relocation decisions. However, the direction of the prediction does seem to depend on how long it takes newly arrived spiders to make an assessment, and what 'attitude' the resident group take to their previous success and recent change of fortune. Both these deserve further examination.

A further cost affecting web profitability might be the level of web damage not associated with prey capture. This was examined by Chmiel *et al.* (2000) who subjected juvenile female *Argiope keyserlingi* to one of four treatments in a laboratory environment: no web damage, and no food; no web damage, with food (one fly per day); web damage (30–50% on one side of the web) and no food; or web damage, with food.

This experiment showed that the cumulative movement of spiders over the 7 days of the experiment was influenced by prey capture and web damage. Greatest movement was shown by individuals experiencing web damage and no prey, while least movement was shown by spiders obtaining prey without web damage. It also showed that both relocation distance and direction were influenced by the web damage. Individuals experiencing web damage moved their websites further than spiders experiencing no damage, and the direction of relocation tended to be away from the side of the web to which damage had been inflicted.

6.3.3.3 Flexibility in expenditure on webs within species

If spiders can assess the quality of a website before web building, we should predict that the likelihood of web construction at least depends upon the

presence of potential prey. This expectation has been supported by laboratory experiments on the araneid orb spinner *Zygiella x-notata*, which was found to be stimulated into web construction more rapidly in the presence of flying insects than in their absence (Pasquet *et al.* 1994).
Spiders that commit to building could potentially vary their investment in terms of either web design or silk composition. Evidence of conditional variation in web architecture is now strong and varied, and at least one piece of evidence supports diet related composition of silk itself (Section 2.5).

Prey capture experience has also been shown to affect the top-bottom asymmetry of vertical orb webs. Enlargement of the lower half of the web compared to the upper has been explained as adaptive since a spider is able to travel more quickly down the web to intercept prey than up (Masters and Moffat 1983), however, Heiling and Herbestein (1999) have shown that the extent of the asymmetry in webs of *Larinoides scleroptarius* can be influenced by experience. Depriving spiders of web building experience by feeding them directly with prey till they reach adulthood still resulted in them building functional orb webs, so showing a strong genetic component to their construction behaviour; however, the experienced spiders built more asymmetric webs. *L. scleroptarius* and *A. keyserlingi* spiders could also be manipulated to build respectively more or less symmetrical webs depending upon whether they were raised with the experience of capturing more prey in the web above or below the hub.

Level of hunger, spider weight, and the conflicting demands of growth, reproduction, and silk production are now also known to influence web design within a species. Sherman (1994) tested whether the orb web spinner *Larinoides cornutus* could adjust web expenditure in accordance with the relative demands of feeding and egg production. He predicted that: the more food-deprived spiders would build larger webs, the sated spiders would dramatically reduce web size, and the smaller webs would be produced prior to egg production. Results confirmed all predictions: the spiders did build larger webs when hungry and smaller webs when sated. Spiders also decreased thread length and web area before producing eggs.

The araneid orb web builder *N. clavipes* is another species that increases the diameter of its web with decreasing foraging success (Higgins and Buskirk 1992). The changing web area in relation to recent capture experience could be a consequence of body condition change, however, it could also be influenced by the modification of behaviour through recent experience. This was tested on *Z. x-notata* by Venner *et al.* (2000). They showed that, although internal energetic changes may have an influence, recent experience alone may have similar short-term effects. They compared certain web parameters of spiders given four different exposures to prey compared with 'no prey' controls. The prey experience groups were: complete predation experience from capture in the web to consumption of prey; detection of prey in the web, but prey then removed; capture of prey in the web, but prey confiscated before the spider begins feeding; and spider fed prey directly while still in its retreat.

This revealed differences between the groups in their subsequent behaviour. The prey detection alone group built their next web at the same time and of the same size as the unfed controls. So the mere presence of prey in the web did not affect them. Spiders allowed to experience the full capture sequence and feed, reduced the area and amount of silk in their next web, however, this was also true of spiders experiencing prey capture but having the prey confiscated. This shows that the experience of the capture alone can cause web reduction.

Watanabe (1999*b*) found that the type of stabilimentum incorporated into the web of the uloborid *Octonoba sybotides* was influenced by feeding regime and egg development. The spiders may incorporate either a linear or a spiral stabilimentum into the web. Experimentally food deprived spiders tended to construct the spiral form, while prey supplemented spiders exhibited the linear form. After egg production, when energy reserves would have been low, the spiral stabilimentum was again favoured. These results show the influence of capture rate and internal reserves on web design, although the type of stabilimentum produced in these experiments was not predicted nor the relative costs of the alternatives known. The trade-off between prey and predator attraction has been found to influence the probability of inclusion of a stabilimentum in the web of *A. keyserlingi* (Fig. 1.9). Webs built in dense vegetation where the spiders are more at risk of predation from insects such as mantids, are less likely to include a stabilimentum (Bruce *et al.* 2001), olfactory cues indicating the presence of the predatory spider *Portia labiata* (Fig. 1.10) were sufficient to decrease the likelihood of stabilimentum construction by juvenile *Argiope versicolor* compared with the equivalent stimulus from a non-predatory spider (Li and Lee 2004).

Individuals of *Parawixia bistriata* can build webs of two different designs depending upon the circumstances. This flexibility is evidently adapted to exploit two different prey sources. This species builds one web type at sunset; it is of fine mesh and has a small area. The timing of this web building coincides with the peak availability of small diptera. These spiders are, however, also able to produce a web during the day that is on average 82% greater in diameter and with fewer spiral turns, with a resulting mesh size that is three times that of the sunset webs (Sandoval 1994) (Fig. 6.11). The construction of these webs during the daytime is closely synchronised with the emergence of termite swarms from their mounds shortly after light rain. Daytime webs were never observed except in the presence of termite swarms, even though not all rain showers lead to termite emergence.

The amount of silk thread in the two web designs is not significantly different, so in this case the contraction of web size in the evening does not

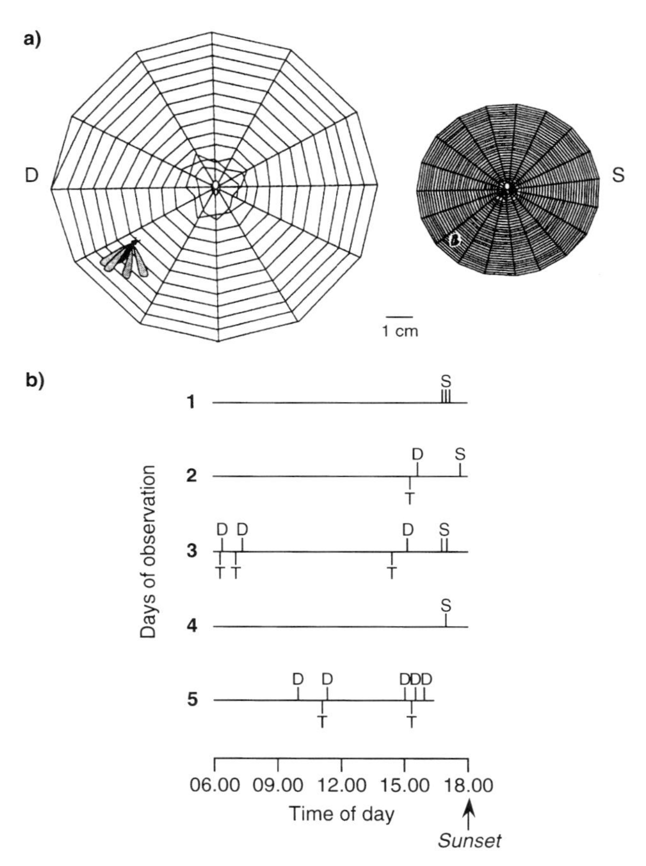

Fig 6.11. Conditional web design in the spider *P. bistriata*. (a) Use of different web designs for different prey types shown in schematic form. *D* = web of larger mesh and area spun during the daytime but only during termite swarming and *S* = the narrow mesh, smaller area web spun daily at sunset. (b) Timing of the building of the two web types in relation to time of day and the occurrence of termite swarms, shown for five separate days. *S* = sunset webs, *D* = daytime webs, and *T* = the occurrence of termite swarms. (Adapted from Sandoval 1994.)

support the hunger hypothesis for the change in web area. The explanation rather seems to be one of the designs appropriate to the capture of different prey. The larger mesh size of the daytime webs may render them weaker to localised impact of a flying insect, however, termite alates only undertake a single dispersal flight before shedding their wings and so are weak flyers. The adjustment of mesh size to prey size therefore supports the view that this is to maximise capture efficiency.

Similar foraging flexibility according to prey types has been demonstrated under laboratory conditions in which the parameters of orb webs of juvenile *Araneus diadematus* were measured when the prey provided were either the fruit fly *Drosophila* or mosquito *Aedes aegypti*. Both have similar mass, but the latter with its long legs, is a larger insect. On being fed one mosquito a day for 6 days, the spiders built webs of larger capture area and bigger mesh size. On being returned to a *Drosophila* diet, the web design was altered back to its original form. This experiment does not measure capture efficiency, but the responsiveness of the spider to available prey type and the concordance between mesh size and prey size supports that interpretatioin (Schneider and Vollrath 1998).

7 Animal architects as ecosystem engineers

7.1 Introduction

This is the only chapter of the book to deal with the subject of builders at the community level. Chapter 1, in examining functions, does touch upon interactions with other species closely associated with builders, for example, their predators or prey. But, in changing their own world, builders may incidentally change the world of a much wider range of organisms. This chapter explores how substantial and pervasive these influences are, and what are their nature and consequences.

7.2 Definitions

7.2.1 Ecosystem engineers

Ecosystem engineers have been defined as organisms that directly or indirectly modulate the availability of resources to other species by causing physical state changes in biotic and abiotic materials, so modifying, maintaining, or creating habitats (Jones *et al.* 1994; Lawton and Jones 1995). Within this definition these authors distinguish *autogenic* and *allogenic* engineers. The former are organisms that alter environments directly by their own structure as, for example, trees create woodland. Allogenic engineers on the other hand transform other materials, living or non-living, either mechanically, for example, by boring a hole in a branch or by some other means. Ecosystem engineering can therefore be viewed as a very broad concept; so broad in fact that concern has been expressed that, as almost any biological activity may change the physical environment, the concept will be diluted by including the most trivial effects (Reichman and Seabloom 2002). In part to clarify the concept, Jones *et al.* (1997) restate it, preferring the term *physical ecosystem engineer* to stress that trophic interactions and resource competition are excluded from their definition. The subject of this chapter is ecosystem engineering by animal builders and burrowers, a subject a little narrower than allogenic physical ecosystem engineers (Jones *et al.* 1994), which includes, for example, habitat modification brought about by the tree-felling activities of elephants.

7.2.2 Extended phenotype

Some aspects of phenotype are expressed beyond the limits of the bodies of the organisms that are responsible for them. To draw attention to this, Dawkins (1982) coined the term *extended phenotype*. Although some of these phenotypic effects may be large both in terms of their impact on the organism itself and on the surrounding environment, others may be very small. To avoid devaluing the whole concept Dawkins (1982), while making clear that any alteration of the environment beyond itself is an extended phenotypic effect, suggests that in practice the concept should be confined to those effects that can be shown to alter the fitness of the organism responsible for them. Beaver dams and gopher burrows may be obvious examples of extended phenotypes, however, Dawkins (1982) makes clear that his definition also embraces quite different phenotypic effects, for example, the manipulation by parasites of the physiology and behaviour of their hosts, topics that have no relevance to this book.

The burrow dug by an organism may protect it from predation or extreme temperatures, however as an extended phenotype, it may also exert an effect upon the environment in ways that alter the fitness of other organisms. Such examples of physical ecosystem engineering, need to be considered in this chapter. Jones *et al.* (1994) in fact, use the shorthand *extended phenotype engineering* to refer to them.

7.2.3 External physiology

The use of the term extended phenotype has not become widespread, although awareness of its importance has grown. Scott Turner (2000) explores Dawkins's (1982) concept in the context of changes to the 'physiological' conditions that animals effect in the environment immediately surrounding them by their behaviour. This he refers to as *external physiology*, which he shows by a variety of examples, mostly of animal ecosystem engineers, can alter the environment in very important ways for both the organism responsible and other species sharing the same habitat.

This concentration on the physiological effects of ecosystem engineering has certain very revealing aspects, the consequences of marine burrowers on sediment communities being a particularly powerful one. However, Scott Turner (2000) weakens the concept by then arguing that the extension of an animal's physiology beyond its conventional boundaries blurs the definition of individual in a way that necessitates consideration of whole communities as super-organisms. This Gaia-hypothesis interpretation is an unnecessary and misleading distraction from the ecological and evolutionary consequences of environmental modification (Hansell 2000). Laland *et al.* (2000) and Odling-Smee *et al.* (2003) have developed an evolutionary model, which incorporates selection pressures exerted by the inheritance of a habitat modified by ancestors; this is discussed in Section 8.4.

7.2.4 Keystone species

Species differ considerably in their influence over community structure. Identification of the most influential species in a particular habitat is therefore an important preoccupation of ecologists. Paine (1969) showed that the removal of one invertebrate predator, a starfish, from a rocky intertidal ecosystem, an association of moderate complexity, had marked effects on population density and species diversity. This demonstration challenges the view that the robustness or stability of a community is positively related to its food-web complexity (MacArthur 1955). Species exerting this disproportionate effect on a community have been termed *keystone species* (Mills *et al.* 1993). If such keystones are widespread across ecosystems`, then clearly, their identification is important not just for understanding how an ecosystem works, but also shaping management regimes for the conservation of endangered habitats. No species, whether primary producer or top predator, is precluded by the above definition from being a keystone. Indeed, builders and burrowers are categorised as *modifiers* by Mills *et al.* (1993), one of five categories of keystone species, the others being *keystone prey*, *keystone predators*, *keystone plant*, and *keystone link* (e.g. pollinators). So, if the concept has any validity, animal builders should feature strongly as keystone species.

7.3 Predictions and tests

This chapter examines the validity of the above definitions in the context of animal building. It also assesses whether animal builders are important ecosystem engineers, and whether they support the concept of keystone species. At a more detailed level the chapter examines whether animal builders, are net creators or destroyers of niches for other species, and whether they stabilise or destabilise ecosystems. Two further questions are raised for which we as yet have very little evidence: Do animal builders protect themselves against extinction, and how enduring is the legacy of their activity in an ecosystem?

7.3.1 Niche construction and habitat stability

Jones *et al.* (1994) express the view that not only may ecosystem engineers, by definition, alter environments, but also that they may exhibit a particular tendency to impact upon other taxa without reciprocal effects. This they refer to as an asymmetric species interaction. By giving new structure to the environment ecosystem engineers may also add new habitat niches, so acting to promote organismal diversity. Habitat modification by ecosystem engineers may create patchiness, an aspect also known to be important in promoting species diversity in ecosystems (White and Pickett 1985). Jones *et al.* (1997) point out that, by altering landscapes, ecosystem engineers also destroy habitats at least locally, with the consequence that the abundance of some species will be diminished although, on a larger scale local disturbance should increase habitat variety and with it species diversity. They therefore reassert the prediction of promoting diversity, while expressing the view that the evidence for this is currently weak.

Another important issue in the role of physical ecosystem engineers is their effect on ecosystem stability. Gurney and Lawton (1996) test the performance of a group of strategic models in which an organism has to modify the habitat in order to survive, either alone or in collaboration with conspecifics. The habitat can be in one of three states: *virgin* (uninhabitable unless modified), *usable* (but for a finite duration), *degraded* (but regenerating). The models do generate fluctuations in the occurrence of the engineers, and of degraded and virgin habitats, which depend upon the parameters set for rates of habitat change. However, under certain defined parameters, there is no effective habitat decay, resulting in a stable equilibrium. These effects provide some testable predictions for what the authors acknowledge is—in the light of insufficient empirical evidence—a very simple model.

From the examples they cite, it does however seem that Gurney and Lawton (1996) may not have drawn upon the full range of appropriate evidence. Typical of the examples given is habitat modification by beaver dams. These effects are characterised as being substantial but unsustainable, leading to degradation of the habitat and its abandonment after a few years (Johnston and Naiman 1990). This predicts that beaver populations spared from human hunting pressure, would show long-term cycles of abundance. But this may be untypical of the nature of habitat modification by the majority of animals. In the first place the modification of habitat by beavers is patchy so that, while at the level of the patch their population might show distinct fluctuations, at the ecosystem level numbers might be relatively stable. I have argued (Hansell 1987*a*) that nest builders and burrowers by their own activities should tend to stabilise their populations, bringing them close to the carrying capacity of the environment. This they do because, through their building activity they effectively create a more stable environment.

7.3.2 Resistance to extinction and the legacy of building

I also proposed (Hansell 1993) that the control builders exert over environments should tend to make their populations more resistant to extinction compared with equivalent non-builders. One way that they might bring this about is if, far from degrading the habitat and having to abandon it, they made it more suitable for themselves. Wilson and Agnew (1992) believe that, through environmental modification, an organism can create a positive feedback that switches the environment into a new state. An example from plant ecology might be where, through the growth of trees in an African savannah habitat, an understorey environment is created that favours termite colonies. These, through their enriching effect on soils, encourage further tree growth (Fig. 7.1) leading to increased soil moisture and a new community. The role of termites in this transition is integral and so supports the view that animal builders can be agents for ecological change.

This draws attention to another weakness in the assumptions of Gurney and Lawton's (1996) model for ecosystem engineers. Populations of ecosystem engineers may not as a rule be pioneers of virgin habitats; they may normally inherit a habitat already beneficially modified by their ancestors, a form of inheritance (see Section 8.4).

It also follows that, if ecosystem engineers such as builders and burrowers can exercise significant control over some environments and resist extinction, then their continued presence in some habitats may have a long history, and the physical alterations they have effected over that time be substantial (Hansell 1993). This may require an archaeological approach to understand more fully, but it is possible that site occupation times are

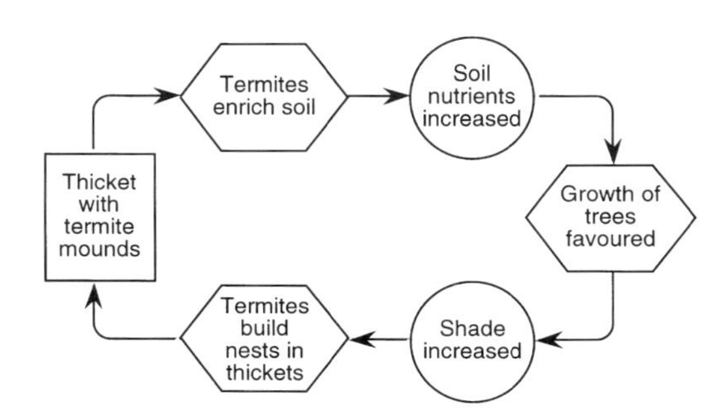

Fig. 7.1. A diagrammatic representation of how termites might, through their activity, alter the vegetation cover in a habitat and create a more favourable environment for themselves. (Adapted from Wilson and Agnew 1992.)

generally underestimated. It is speculated, for example, that some badger sets may have been in more or less continuous occupation for several hundred years (Neal 1986).

The remnants of nests built by some mud-dauber wasps (*Sceliphron* species) in Australian rock shelters occur with, and occasionally underlie prehistoric rock paintings up to 17,000 years old, demonstrating surprising durability (Roberts *et al.* 1997). In North America such nests are known to assist the establishment of mud nests by birds such as barn swallows and eastern phoebes on concrete walls under bridges (Weeks 1977). So, these wasp nests may be very small modifications to the environment, but illustrate that ecosystem engineers may also be influential through modest alterations acting, and possibly accumulating, over long periods of time. Even structures regarded as ephemeral deserve careful examination. Large webs of the social spider *Anelosimus eximius* may remain occupied and maintained for years (Nentwig and Heimer 1987). Similar webs built by other social spider species are known to form nest sites for certain birds (Hansell 2000).

7.4 Bioturbation of aquatic and terrestrial sediments

Perhaps the major and certainly most pervasive terrestrial environment is the ground itself, and one of the main marine habitat types is the benthic sediment layer. Chief among the ecosystem engineers in both these environments are burrow diggers. The question is, do they significantly modify them?

7.4.1 Aquatic sediments

Earthworms feed on the substratum in which they burrow, contributing to the recycling of nutrients and redistributing the soil layers. Darwin (1881) was probably the first biologist to draw attention to the importance of earthworms in conditioning soil and redistributing it on a large scale. This type of substratum disturbance is now referred to as *bioturbation* and substantial effects that it can have on the physics, chemistry, and biology of sediments are well recognised (Krantzberg 1985). Studies on bioturbation have looked at the burrowing patterns of individual species, tried to calculate overall rates of sediment displacement by burrowers, and looked at the consequences for nutrient circulation and community structure in this habitat

Marine sediments are generally soft and often rich in organic materials. Many burrowing organisms have exploited these properties to search for food or simply excavate refuges, in particular Annelida and Crustacea. Eight species of mud shrimps (Crustacea, Decapoda, Thalassinidea) found in British waters

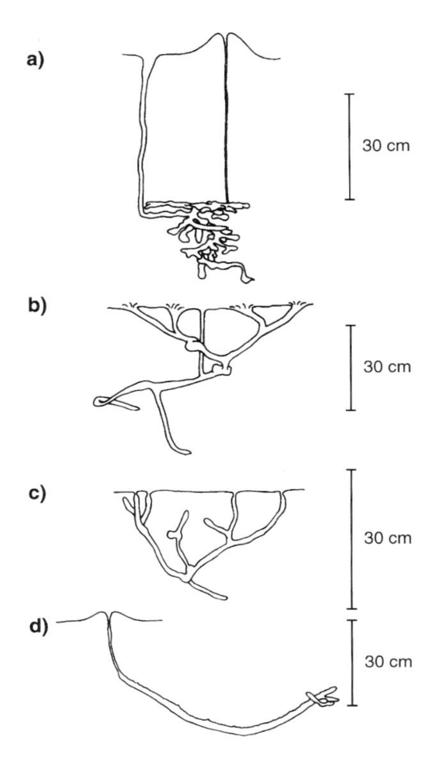

Fig. 7.2. The burrow morphology of three species of mud shrimp and an echiuran worm. The mud shrimps (a) *Callianassa subterranea*, (b) *Jaxea nocturna*, and (c) *Upogebia stellata*, and (d) an echiuran worm *Maxmuelleria lankesteri* (Adapted from Nickell *et al.* 1997.)

construct burrows of varying complexity (Fig. 7.2) (Hayward and Ryland 1990; Nickell *et al.* 1997). A survey of the Clyde Sea area on the west coast of Scotland reveals a mud-burrowing megafauna that includes, six species of Crustacea, including decapod, isopod, and amphipod species as well as three species of burrow dwelling fish. These occur at densities that Atkinson (1986) suggests must have an important effect on benthic community structure.

The effects of different burrows may depend upon the feeding specialisation of their creators, with deposit feeders producing more sediment disturbance than filter feeders. The mud shrimps *Callianassa subterranea* and *Jaxea nocturna* are both deposit feeders creating dynamic burrow systems which redistribute sediments both onto the mud surface and within the burrows themselves as backfill. The filter feeder *Upogebia stellata*, also a mud shrimp, simply creates a retreat in the mud, while the echiuran worm *Maxmuelleria lankesteri* feeds by reaching out of its burrow to skim off the sediment surface with its proboscis (Nickell *et al.* 1997). *J. nocturna* feeds at a slower rate than *C. subterranea* but displaces sediment from greater than 90 cm depth compared with only 25–30 cm for the latter. The filter feeding *U. stellata* excavates to 30 cm but only to establish a burrow in the first place, while the worm *M. lankesteri* takes the surface layer of sediment and deposits it up to 80 cm below.

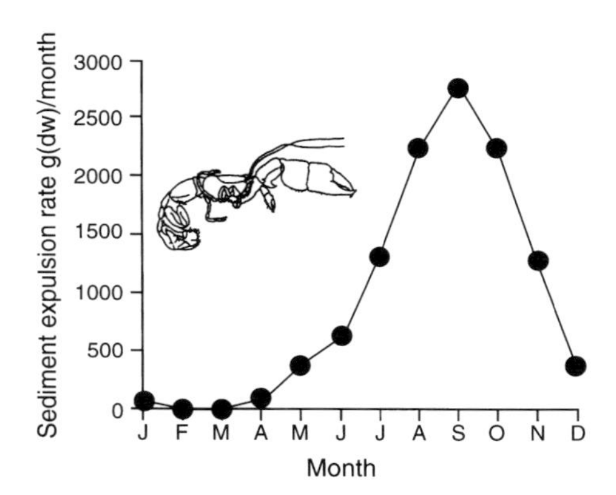

Fig. 7.3. An estimated annual pattern of sediment turnover by the mud shrimp *C. subterranea*. (Adapted from Rowden and Jones 1997.)

Where burrowers make spoil heaps on the mud surface, estimates have been made based on observed burrow densities of the amount of sediment that is being disturbed over time. Rowden and Jones (1997) calculated that the mud shrimp *C. subterranea* alone was displacing 11 kg (dry weight) of sediment per square metre, per year, mainly during the second half of the year (Fig. 7.3). A laboratory study of the same species calculated the displacement for a site in the North Sea at 15.5 kg (dry weight) m^{-2} $year^{-1}$ (Stamhuis *et al.* 1997). In addition to the movement of sediment, the creation of burrows enhances the surface area of the interface between the sediment and the water. It was estimated that in a coastal site in the south of England that, with a density of up to 5,000 m^{-2}, the surface area of burrows of the polychaete *Nereis (Hediste) diversicolor* could increase the total sediment surface area by as much as 300% (Davey 1994, Davey and Watson 1995).

The consequent effect of the burrows on nutrient recycling can be marked. Davey and Watson (1995) found that soluble ammonium and silicate fluxes were up to 100 times higher in locations of high *Nereis* (*Hediste*) densities compared with sites with few or no worms. The effects of the burrowing activities of the bivalve mollusc *Yolida limatula* and the polychaete *N. (H.) diversicolor*, show that they accelerate carbonate dissolution, contributing towards the productivity of the shallow water deposits in which they live.

The influence of marine sediment burrowers on nitrification potentials (i.e. the oxidation of NH_4+ to NO_3) was measured by Mayer *et al.* (1995). Nitrification potential (NP) is a measure of the ability of a sediment to oxidise NH_4+, when NH_4+ and O_2 are not limiting, and is an index of the abundance and activity of nitrifying bacteria. These authors found that the burrow walls of the bivalve *Macoma baltica*, the amphipod *Leptocheirus plumulosus*, and polychaetes *Macroclymene zonalis, Pectinaria gouldii, Loximia medusae* and *Diopatra cuprea* had NPs 2–20 times greater than that of adjacent sediment of the same depth. The influence of burrow ventilation on this effect was indicated by the

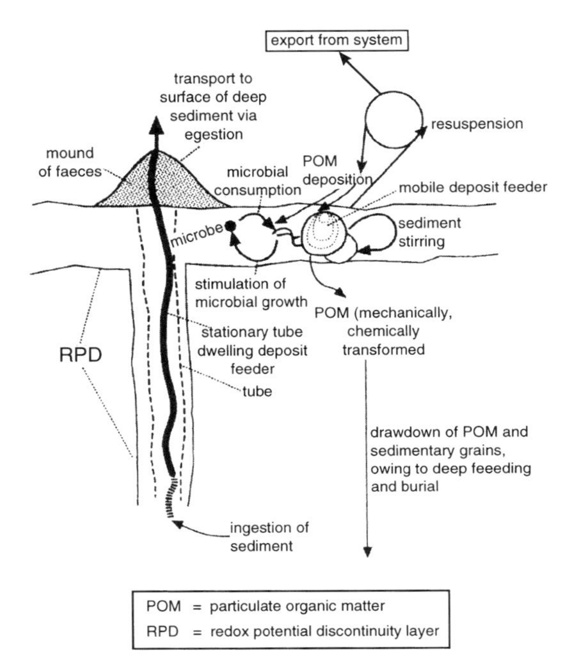

Fig. 7.4. Generalised cross-section of sediment showing the effect of stationary and mobile marine deposit feeding invertebrates on the position of the RPD boundary on the dominant oxidative processes above it, the reducing processes below it, and on movement of POM. (Adapted from Levington 1995.)

correlation between NP enhancement and the duration of ventilation by the occupants.

This nitrification is largely carried out by a specific group of chemautotrophic bacteria that require O_2 as well as the NH_4. In undisturbed sediments the process is generally restricted to the upper 1–6 mm due to the anoxic conditions in the mud. The presence of burrowing organisms in marine sediments, particularly ones that ventilate their burrows, introduces oxygen to much deeper levels of the sediment with major effects on the ecosystem. This is ultimately because oxygen has a high affinity for electrons, generating a high redox potential. This is illustrated by release at of least 40% of the energy in glucose through aerobic respiration compared with only 7% through anaerobic fermentation (Scott Turner 2000). In undisturbed sediments this high energy gain is limited to the upper few millimetres, leaving a huge reserve of potential energy below a sharply defined *redox potential discontinuity* layer (RPD) (Fig. 7.4). What burrowers do through the ventilation of their burrows is to draw down electron acceptors, especially oxygen, deep below the RPD layer, so short circuiting the slow relay of electron donors extending up through the sediment. So, without the burrowers, the RPD is a horizontal boundary only 4–6 mm below the sediment surface. Burrowing invertebrates extend this much deeper into the sediment (Levington 1995).

But, questions Scott Turner (2000), how can this account for the very high densities of burrowers such as lugworms (*Arenicola marina*) in estuarine mud,

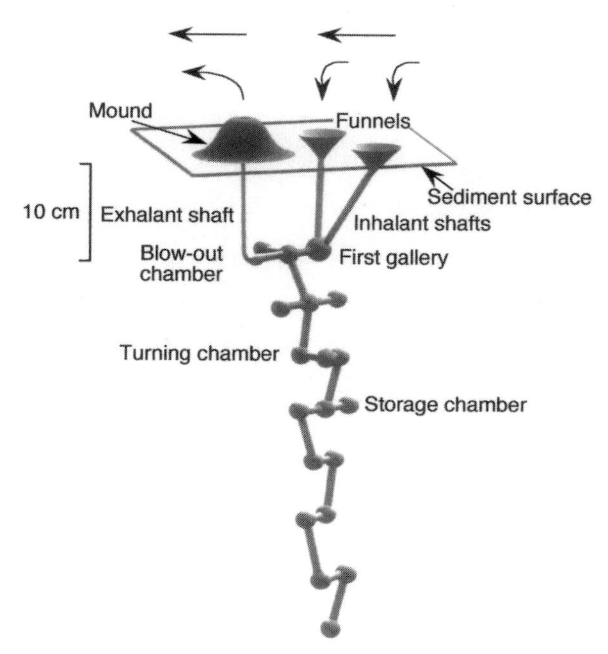

Fig. 7.5. Diagrammatic representation of the burrow system of *C. truncata* in a sandy seabed. The effect of the water current passing over the mound is to induce flow through the burrow system as shown. (Adapted from Zeibis *et al.* 1996a.)

since these chemical reactions are occurring outside the bodies of the organisms that induced them, that is, an expression of their external physiology. The answer, he explains, is that the energy is unlocked by the aerobic bacteria, and it is their greatly enhanced productivity that drives an enriched food chain of diatoms and nematodes, which also boosts the lugworms' nutrition.

Observations on the mud shrimp *Callianassa truncata* show that the extent to which burrowers alter the penetration of oxygen into marine sediments. Although only 2 cm long, the organism creates a complex burrow system that penetrates down about 50 cm. The burrow enhances oxygen penetration in two ways, one through ventilation induced by the architecture of the burrow openings, the other through the ventilation behaviour of the mud shrimp itself (Fig. 7.5) (Zeibis *et al.* 1996*a*).

One or more of the *C. truncata* burrow openings are in the form of cone-shaped funnels, while another is through the top of a 4 cm mound of excavated sediment. Water currents passing over these generate a pressure difference, drawing water in through the cone openings and out through the top of the mound. This increases the penetration of oxygen into the sediment from about 4 mm to about 4 cm. The oxygen penetration into the sediment surrounding the burrow is also assisted by the increase in pore volume of the

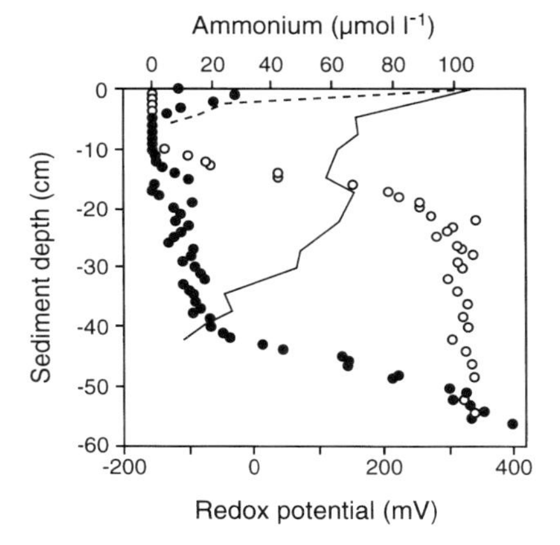

Fig. 7.6. Redox and ammonium profiles in sediments inhabited by *C. truncata* (respectively, continuous line, • circles), and not inhabited (respectively, dashed line, o circles) in relation to sediment depth, showing an effect of occupation nearly to 50 cm (the bottom of the burrow). (Adapted from Zeibis *et al.* 1996a.)

sediment, which results from the general substrate disturbance caused by the burrowing. This has the effect of increasing the oxic sediment volume by a factor of 4.8 compared with that below a smooth sediment surface (Levington 1995; Zeibis *et al.* 1996*b*). The additional ventilation behaviour of the shrimp pushes the measurable increase in oxygen availability (redox potential boundary discontinuity layer) to a depth of greater than 400 mm, that is, essentially to the bottom of the burrow (Fig. 7.6) (Zeibis *et al.* 1996*a*). With a density of 120 mud shrimp burrows per square metre, it can be seen what a substantial difference these burrowers can make to the habitat.

Stabilising aquatic sediments is another significant effect of animal constructions (Dudgeon 1994). Tube-building polychaetes, *Owenia fusiformis* (Fager 1964) and 1 mm long nematodes occurring at densities of over 300 to 10 cm^{-2}, in mucus tubes penetrating 1cm into the marine sediment (Nehring *et al.* 1990), have both been found to have this effect. Similar effects appear to occur in freshwater habitats. For example, Gardarsson and Snorrason (1993) found that high densities of the chironomid *Tanytarsus* in an Icelandic lake, created a covering of the lake floor of occupied and abandoned tubes, and of faecal pellets that resulted in reduced re-suspension of sediment particles. Statzner *et al.* (1999) showed that silk spun by net spinning caddis larvae *Hydropsyche siltalai* could, in the typical range of abundance shown by natural populations, increase the critical shear stress of gravel to water flow by a factor of 2.

7.4.2 Terrestrial soil habitats

The role of terrestrial bioturbators echos that of their marine counterparts, with soil disturbance, soil ventilation and diet all contributing to nutrient recycling. In addition, soil burrowers may contribute to soil drainage. Lavelle (1997) considers the factors controlling the rate of release of nutrients from soil organic matter through microbial activity and identifies three guilds of soil invertebrates each operating at an increasing scale of time and space. The first are termed the *micro-foodweb* organisms, predators on the bacteria and fungi. The second are the *litter transformers*, arthropods normally ingesting purely organic matter. The third level organisms are identified as ecosystem engineers which, in this context are defined as macrofauna that physically change the structure of the soil by their activities; in practice this refers chiefly to earthworms and termites.

Lavelle (1997) stresses the role of the ecosystem engineers not simply in a direct digestive role but also because, through bioturbation and gallery formation, they alter soil porosity for water penetration and gas movement. Therefore they may stimulate soil productivity in a similar way to that described by Scott Turner (2000) for marine sediments.

The effects of bioturbators on soil chemistry and structure in terrestrial substrates are recorded in studies on earthworms, termites, and burrowing vertebrates. The combined effect of bioturbation and digestion on carbon mineralisation was demonstrated for the earthworm *Aporrectodea caliginosa* by comparing the labelled CO_2 emission from soil with added ^{14}C-labelled beach leaf litter with worms present and absent (Wolters and Schaefer 1993). Carbon mineralisation was enhanced both by the experimental mixing of the litter with the soil and by the addition of $(NH_4)_2\ SO_4$ when compared with untreated controls. However, the ^{14}C mineralisation by earthworms exceeded that of both of these treatments (Fig. 7.7).

In more tropical environments, the role of earthworms as decomposers is frequently supplemented by that of termites (Lavelle 1997). Their burrowing activity can also result in the creation of extensive subterranean galleries and numerous openings to them through the soil surface, so that the termites

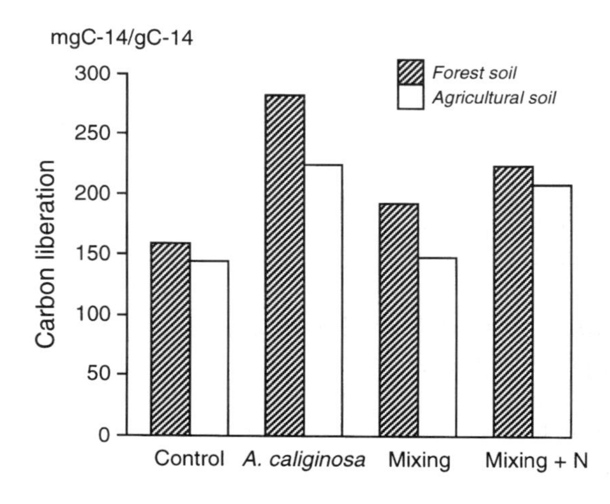

Fig. 7.7. The combined effect of bioturbation and digestion by the earthworm *A. caliginosa* on $^{14}CO_2$–C liberation from forest and agricultural soil compared with, (left) undisturbed control soil and (right) soils after mixing alone, or mixing with addition of $(NH_4)_2\ SO_4$. (Adapted from Wolters and Schaefer 1993.)

can emerge close to foraging sites (Wood and Sands 1978; Darlington 1985). Burrow aperture densities of *Syntermes molestus* and *Syntermes spinosus* in a Central American rain forest site were found to be a minimum of 35 m^{-2} $year^{-1}$, which is enough to have a significant effect on soil aeration and drainage (Martius and Weller 1998). Similar burrows and galleries created by the termite *Gnathitermes tubiformans* were shown in a North American desert ecosystem to enhance drainage and reduce soil erosion. The effect of the presence of the termites was most noticeable in the partially vegetated areas which comprise 80% of this habitat. Elkins *et al.* (1986) conclude that, through these effects on water movement, the termites help to maintain vegetation cover.

Termites by their foraging and soil excavation can change the nutrient and mineral properties of the soil. Wood (1988), reviewing the effects of activities such as *Macrotermes* and *Odontotermes* species on soils, documents various effects in addition to the movement of soil particles through deep excavation and the enhancement of soil porosity. These include the redistribution of organic matter through fungus cultivation and the release of cations such as Ca^+, K^+, and P^+ from faecal-derived organic matter. Similar effects result from the activities of ants. Petal (1978) documents numerous examples of the effects of ant colonies on soil minerals and pH. Soils taken from the mounds of *Lasius* and *Formica* species at a site in Germany were found to be richer in certain minerals including oxides of Na, Mg, and K than soils in the spaces between the mounds (Dean *et al.* 1997).

7.5 Patterning of landscape and patch formation

Bioturbators provide evidence from aquatic and terrestrial habitats of environmental modification, but whether and how burrowers or nest builders generate habitat complexity needs further examination. The study of the patterned habitat referred to as 'mima prairie' provides a case study of this.

7.5.1 Mima prairie

Burrows are of fixed location at least over the short term, and possibly over long periods of time. Consequently the influences of these structures, whether created by an individual or a colony, is localised, which may result in a patch of disturbed or altered environment. This can be seen, for example, from the research on pocket gophers, reviewed by Reichman and Seabloom (2002). The question is whether that localisation and endurance has a significant effect upon the overall habitat of the rodents.

Pocket gophers *Thomomys* (Geomyidae) are subterranean herbivores, which can occur at high densities, making extensive burrow systems while foraging for roots. The excavated soil is either back-filled within the burrow system or brought to the surface, creating a mound. The amounts of soil moved can be very large, ranging from 3.4 up to 57.4 m^{-3} ha^{-1} $year^{-1}$. Burrows can underlie 7.5% of the surface and mounds cover 5–8% of the area.

These rodents are territorial, maintaining spacing by vibration signals through the soil. The interaction between the foraging and territorial behaviour leads to complex spatiotemporal patterns of disturbance, which are aggregated at a scale smaller than that of a single territory and over short time periods. However, at a larger scale the mounds are regular in their distribution as a result of the buffers between territories. The result on the surface is a mosaic of patches creating spatial heterogeneity at a density of hundreds per hectare. With the shifting pattern over time, this can generate further diversity through repeated plant colonisation and succession (Reichman and Seabloom 2002).

A habitat of grassland plains typified by rather evenly spaced mounds, generally several metres wide and 1 m or more high is found in various parts of the world and referred to as 'mima prairie'. Examples are found in North and South America (Cox 1984; Cox and Roig 1986), and Africa (Lovegrove and Siegfried 1989). At least five hypotheses have been proposed to explain this landscape including one of biogenic activity: (1) Particle sorting by frost action, (2) water erosion, (3) wind deposition, (4) seismic activity, and (5) biogenic by fossorial rodent or social insect. Cox and Gakahu (1986) tested predictions of these five hypotheses in relation to the ratio of gravel and pebbles on mound tops and edges compared with inter-mound areas in a mima landscapes in four North American locations. Seven predictions for rodent translocation were upheld whereas other hypotheses were rejected or at best weakly upheld.

The mounds of mima grassland in California can reach 15 m in diameter and nearly 0.8 m in height. They have a radial symmetry that argues against erosion or wind deposition and are generally occupied by pocket gophers *Thomomys* sp. Cox (1984), experimentally tested whether it was the gophers that were responsible for the mounds by placing metal nuts as markers in gopher tunnels at the edges of mounds. Recovery of nuts using a metal detector showed that movement was clearly onto rather than away from the mound, evidence of their movement by the gophers.

Similar landscape effects to those produced by burrowing rodents have been shown to result from the activities of termites, alone or in combination with small mammals. Mounds with a mean diameter of 28 m in Cape Province, South Africa, referred to as *heuweltjies* have been shown to have a distribution that deviates significantly from random towards uniformity. These shallow mounds are occupied both by the termite *Hodotermes viator* and by the common molerat (*Cryptomys hottentotus*). Lovegrove and Siegfried (1989) conclude that intraspecific competition within these two species is sufficient to account for the uniform distribution of the mounds. Another South African study, on mounds with a mean diameter of 17 m, showed current evidence of occupation by *C. hottentotus* but only in the upper portions, and with no trace fossil evidence of their activity. C^{14} dating of the mounds gives dates of 4,000 and 5,500 years ago for their formation, at two different sites by what could have been *H. viator* termites (Moore and Picker 1991). No evidence in support of the physical origin of the mounds was found in either of these studies.

It seems quite possible that in some habitats there remain large or at least significant landscape features, the biogenic origin of which remains unrecognised. Two possible examples have indeed been reported from Australia (Noble 1993). These are 10 m diameter circular features, 10–20 cm high with a central depression, and shallow mounds of 30 m diameter. Neither type was occupied, but evidence supported the view that the smaller was the creation of the now locally extinct malleefowl *Leipoa ocellata*. The larger one was possibly the work of a locally extinct marsupial, the burrowing bettong *Bettongia lesueur* (Potoroidae). Evidence did not support hypotheses for the physical origins of these features (Noble 1993). Sattaur (1991) reasons that *Odontotermes* colonies, through their activities over long periods of time, are responsible for some very large landscape features in southern Africa. These are bands of ridges and gullies which, although only about 2 m high, are up to 1 km long. The positive relationship shown between the extent of landscape modifications and their persistence would suggest that structures of this scale are at least several hundred years old (Whitford and Kay 1999).

7.6 Habitat modification and the promotion of species diversity

Mima prairie shows that habitat modification by colonies, or individual organisms can generate patchiness. This is evident to us because it is at a scale that we can immediately appreciate. However, there seems no reason in principle why comparable effects should not be produced at a much smaller scale with equally important effects. This section assesses whether patchiness is a frequent if not general consequence of the actions of animal builders and whether it promotes species diversity.

7.6.1 The balance of effects favouring increased diversity

A mature colony of *Macrotermes michaelseni* consists of over a million individuals living in a mound 4 m high and foraging on the surrounding dead grasses. In doing this, termites redirect nutrient flow, and alter the landscape.

Dangerfield *et al.* (1998) draw attention to the possibility that such species, as allogenic ecosystem engineers, may have important ecological effects. However, much less obvious architecture may also deserve empirical study. High densities of web-spinning spiders using 'ballooning' dispersal are amongst the first colonisers of nutrient-poor glacial moraines in the Arctic. Hodkinson *et al.* (2001) consider that these spiders may perform an important role in the subsequent development of plant and animal communities because their webs stabilise dusty substrata, attract moisture, and in particular, bring in nutrients through trapping large numbers of chironomid midges. In looking for ecological influences of animal builders, we should include examination of microhabitats created by the actions of small organisms.

The ability of freshwater tube builders to pioneer a process of species diversification was shown experimentally by the enrichment of a whole stream with phosphorous and/or nitrogen (Hershey *et al.* 1988). This resulted in the tube-building larva of the chironomid *Orthocladius rivulorum* developing faster and growing larger compared with unfertilised control areas. Examination of the surface of their tubes showed a rapid growth of diatoms, which gut examination revealed to be the source of food for the chironomids. The diatoms proved to be a monoculture of *Hannaea arcus* forming a distinct and spatially discrete community compared with those found on stones and other surfaces in the area. The tubes therefore represent a distinct type of microhabitat patch.

Petaloproctus socialis is a highly gregarious inshore polychaete, which forms aggregations of up to 100 worms cementing the sediment between them to produce a solid mass of vertical tubes. The worms live head downward in these tubes and are direct deposit feeders of their inshore habitat. Wilson (1979) found that these concretions supported a dense and varied infauna of over 160 other species, predominantly polychaetes. Comparison between this and neighbouring sites without *P. socialis* tube aggregations showed the former was much more rich in macrofauna (Shannen-Wiener diversity of 4.08 for the *P. socialis* habitat; one of the highest recorded for a temperate benthic community, compared with 2.87 where it was absent). Wilson (1979) concludes that in areas where physical disturbance is frequent, the tube aggregations provide a refuge and therefore encourage species diversity.

The effect of the fixed larval cases of the net-spinning caddis larva *Cheumatopsyche* on macro invertebrate colonisation was demonstrated experimentally by comparing colonisation of stones that were either completely denuded by scrubbing or already colonised by caddis larvae (Diamond 1986). Denuded stones were found to have been colonised by fewer other species of various taxa in the 7 day study period than those with caddis cases attached, regardless of whether the cases were occupied or not. How these caddis cases bring about additional habitat complexity is not clear, but Diamond (1986) speculates that local colonisation by caddis larvae leading to macrofaunal diversification could contribute to habitat patchiness.

A similar effect on immigration is created by the presence of mucous tubes of amphipod and copepod Crustacea, and of polychaete worms colonising seagrass *Thalassia testudinum* on the Florida coast (Peachey and Bell 1997). Seagrass meiofauna diversity was greater with a greater percentage cover of mucous tubes. In a field experiment using a conditioned glass substrate, greater meiofaunal numbers immigrated onto substrata bearing mucous tubes of the harpacticoid copepod *Dactylopodia tisboides* than onto substrata with no tubes.

Small-scale patchiness does therefore seem to be a general consequence of the activities of burrowers and of builders, and these patches are colonised by a variety of other species. Animals creating much larger patches appear to produce similar effects. The sand tilefish (*Malacanthus plumieri*), for example, a highly territorial species of the Caribbean shores of Colombia, digs a burrow in the sandy substratum just beyond the coral zone of the fore-reef. By itself the burrow would not greatly alter the character of the landscape, however, each fish builds over its burrow a mound of coral rubble, stones and shells with mean dimensions of $1.33 \times 1.02 \times 0.22\ m^3$. These amount to artificial patch reefs spaced at intervals over an otherwise monotonous sandy environment. They enhance local biodiversity by attracting a wide variety of invertebrate species including crustaceans and echinoderms, and a total of 32 fish species some only as juveniles, of which 66% were carnivorous, the rest herbivores and detritivores (Fig. 7.8) (Büttner 1996).

Evidence of the effects of beavers on local biodiversity indicates that the consequences may be mixed. Several studies report a diversification of species resulting from their activities, although, the populations of some species can be adversely affected. For example, beaver foraging is reported to diminish the diversity of tree species by increasing the importance of conifers (Donkor and Fryxell 1999). The impoundment of a running stream has, on the one hand, the effect of increasing the diversity and biomass of lentic species, chironomidae in particular, while causing a decline in lotic species. A general increase of invertebrate numbers leads to an increase in fish populations such as trout (*Salmo* spp.) and char (*Salvelinus* spp.). The felling of trees and the flooding of their roots caused by the dam, both creates dead timber which provides habitats for certain insects. In addition, meadows created by the tree-felling promote insect populations, which in turn benefit nesting birds and create browsing areas for deer and moose. However, the damming of the river may result in the destruction of some spawning areas and the blocking of fish migration paths (Rosell and Parker 1996).

The occupation of beaver lodges, and dams may only last 4–5 years (Rosell and Parker 1996), leading to the occurrence of a patchy and transitory habitat

Fig. 7.8. Schematic illustration of a nest of the sand tilefish (*M. plumieri*), showing some of the characteristic fauna associated with it: (1) *Stegastes partitus*, (2) *Centropyge argi*, (3) *Holacanthus tricolor* juv., (4) *Pseudogramma gregoryi*, (5) *Chromis insolata* juv., (6) *Apogon* sp., (7) *Serranus baldwini*, (8) *Gnatholepis thompsoni*, (9) *Eucidaris tribuloides*, (10) Amphiuridae, (11) *Cantharus karinae*, (12) Xanthidae, (13) Majidae, (14) *Munida angulata* c.f., (15) *Odontodactylus brevirostris*, (16) Sabellidae. (Adapted from Büttner, H. (1996). *Bulletin of Marine Science*, 58, 248–260.)

in which ecological changes are complex. Dam construction by beavers in a habitat in Finland, characterised by already stable lake and pond habitats, changed the dominant duck species inside 2 years from mallard (*Anas platyrhynchos*) to teal (*Anas crecca*). This occurred solely through an increase in the latter species apparently due to the teal's ability to feed on planktonic cladocerans (Nummi and Pöysä 1997).

The mucus capture nets of pteropod molluscs and houses of Appendicularia illustrate how animal built structures can both increase biodiversity and alter nutrient fluxes even in the relatively uniform environment of marine mesopelagic zone. Houses of the appendicularian *Oikopleura* are no more than about 100 mm in length but, abandoned and occupied houses may together reach densities of 1000 m^{-3}, sufficient to alter the spatial heterogeneity of their mid-water habitat (Alldredge 1976*a*). The abandoned, giant houses of the genus *Bathochordaeus*, which reach 350 mm across, occur at much lower densities but attract a variety of organisms that contribute to their decomposition including bacteria, protists, and copepods. Up to an order of magnitude more metazoa are found in these discarded houses than in the surrounding water. Over 90% of the house-colonising metozoa are copepods, a number exhibiting morphology typical of benthic rather than open ocean feeders (Steinberg *et al.* 1994). This community of organisms, through their breakdown of the house remains, contribute to the recycling of carbon (Steinberg *et al.* 1997), however, this breakdown is not fast enough to prevent the sedimentation of *Bathychordaeus* house debris and mucus aggregates from the webs of pteropods such as *Limacaria retroversa* having an important influence upon the removal of captured fine particles such as pikoplankton from the mid waters and their deposition as ocean floor sediment (respectively, Noji *et al.* 1997; Silver *et al.* 1998).

7.6.2 Patch dynamics and species diversity

Patchiness in ecosystems is regarded as an indicator of heterogeneity, and is evidence of some agent of disturbance maintaining the dynamic mosaic (Denslow 1985; White and Pickett 1985). Agents such as fire or storms may create patches on a large scale, a falling tree in a tropical forest a much smaller patch, the burrow of a mud shrimp, a smaller patch still (Loucks *et al.* 1985). A feature of smaller patches is the relatively large perimeter in relation to patch area. This allows them to be rapidly exploited or colonised by lateral migrations from the surroundings, or to be colonised by edge species that are adapted to microhabitats created by both the disturbed and undisturbed communities (Sousa 1985). A habitat like mima prairie, which has a relatively small-scale patchiness, therefore offers a lot of edge habitat to support diversity.

Reichman *et al.* (1993) demonstrate that the disturbance to the surface around the burrow openings of the pocket gopher *Geomys bursarius*, through the deposition of burrow spoil, creates an edge effect wave in the tall grass prairie vegetation. The wave is revealed in the change in the plant biomass at different distances from the burrow entrance. The disturbance of the soil inhibits plant growth but, just beyond this point, the vegetation flourishes in conditions where light has full access on one side and where soil minerals or organic nutrients from animal excreta are washed out of the recently excavated soil. This zone of vigour then has an inhibitory role on the vegetation just beyond it, while beyond that the vegetation recovers. The disturbed patch therefore has more than a simple effect upon habitat heterogeneity.

Burrowers offer a constant source of disturbance, with new rodent burrows appearing while others become inactive (Reichman and Seabloom 2002). This further supports diversity by allowing fugitive species a constant source of new opportunities (Sousa 1985) and perpetuating the presence of patches in all states of succession. Klaas *et al.* (2000) looked at the spatiotemporal pattern of surface disturbance induced by spoil from pocket gopher mounds. They found that, at a scale of <20 m there was a clustered pattern of mound production due to digging activity within a single burrow system. But, at a scale of >20 and <30 m the pattern of mound production changed little over time. These authors suggest that the high variability in rates of mound production over short distances may act to enhance the establishment of a more diverse ecological community. Once established, colonising species have a high probability of colonising a fresh mound nearby. Species requiring an undisturbed habitat can still persist.

Disturbed soil patches may also enhance within-species variation in plants. Different plant species are characteristic of the highly disturbed soil areas created by the mounds of the banner-tail kangaroo rat (*Dipodomys spectabilis*), compared with the undisturbed areas in between. In addition, because many of the annual plant species in the habitat favour, but do not exclusively occupy disturbed soil patches, they locally show great within-species life-history variation (Guo 1996). Finally, small-scale patches are also known to promote diversity by facilitating the coexistence of species with similar habitat requirements. This is because first arrivals at a patch have an advantage over those arriving later so one or a few species are unable to colonise all patches (Tokeshi and Townsend 1987).

In some habitats more than one species of ecosystem engineer may be sympatric. This may well be a widespread phenomenon, but is very little studied. That the consequences of this may be important is indicated by the investigation of Cardinale *et al.* (2002) into the more general question of whether organisms in species-diverse communities do better together than on their own. Their study is rare in being an empirical test of this problem, but is useful in the context of this book in choosing as a model, the feeding rates of three species of net spinning caddis. In testing the species both alone and with all three together, they found that in the mixed assemblage, the total consumption of suspended particles was a remarkable 66% greater, with all three species benefiting significantly. The factor identified as bringing about this mutually beneficial effect was topographical complexity created by the feeding nets. With all three species together, stream beds had greater complexity, increasing turbulent mixing of water between the stream bed and the overlying water column resulting in a greater overall capture efficiency by the nets.

7.6.3 Direct effects of burrowers and builders on local resources

The effects of animal builders on community diversity can be exerted by two different routes. One is the direct effect of the architecture or building activity, which this chapter has already shown can be profound. The other is the effect on the control of local resources exercised by a resident population of builders themselves. The definition 'physical ecosystem engineers' (Lawton and Jones 1995; Jones *et al.* 1997), seeks to emphasise the importance of physical alteration of the environment by these organisms on other species by excluding from it the direct effects that one species may have on others through, for example, direct competition for food. This is a helpful distinction, however, Hansell (1993) pointed out that ecosystem engineers, by modifying habitats to their own advantage, may come to enjoy a security of tenure in a habitat which helps them to exert greater direct control over resources such as food. Distinguishing these two influences, the physical ecosystem engineering effects and the direct organism-to-organism effects, and understanding the interaction between them is important and currently insufficiently studied. A good case study of the direct control exercised over resources by a local population of builders is that of the competition for the seed crop by desert dwelling rodents and ants.

Seed storage is widespread in small burrowing rodents (Vander Wall 1990), but it is in Cricetidae (mice, hamsters, and gerbils) and Heteromyidae (kangaroo rats *Dipodomys*, kangaroo mice *Microdipodops*, and pocket mice *Perognathus*) where specialist seed storers are found (Brown *et al.* 1979*a*). The most highly developed seed storage is found in the specialised, arid habitat dwellers, the kangaroo rats (*Dipodomys*), where seed storage is primarily an adaptation to overcome the sporadic nature of seed production. Shaw (1934) describes the burrow system of *Dipodomys ingens* containing 35 L of seeds for an animal of about 150 g. Storage on this scale potentially brings them into competition with other rodents, and with seed storing ants which in some habitats significantly reduce seed populations (Westoby *et al.* 1982) and influence the abundance and distribution of some plants (Risch and Carroll 1986). Kangaroo rat species may also influence seed availability by limiting seed production. Kerley *et al.* (1997) established through the study of captive *Dipodomys ordii* that, as the grasses are growing, these kangaroo rats cut and consume a large proportion of the tillers (seed head stems).

In some North American desert habitats up to six species of seed-eating rodents may be found coexisting. However, a strong association is shown between the size of rodent and the size of seed collected, which probably reduces interspecific competition (Fig. 7.9) (Brown and Lieberman 1973). This is further supported by evidence that overlap of foraging preferences between species is greater in productive habitats, while in less productive habitats, ecologically similar species are excluded, creating less diversity. Brown (1973) also found that in dune habitats, where the number of desert rodent species could vary between one and five, there was always a regular body size separation between species, with the difference in size inversely related to the number of species present. This implies that where resources limit the number of species, body size differences are greatest. Desert ant species may also be competitors of each other. Species with similar sized workers and foraging characters tend not to coexist (Brown *et al.* 1979*a*).

Exclusion plots from which both seed collecting ants and rodents have been removed generate higher seed biomass in soil samples than when either or both are present (Brown *et al.* 1979*b*). Rodents increase 20% in numbers of individuals and 29% in biomass when ants are excluded, and ant colonies

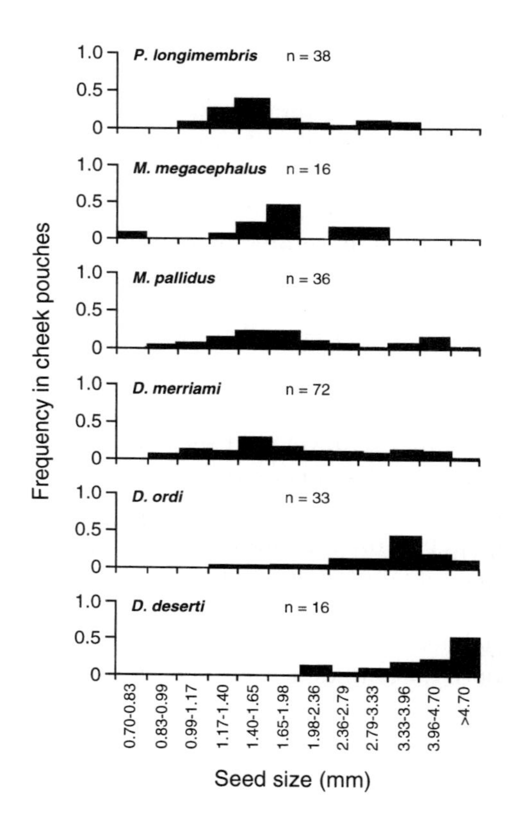

Fig. 7.9. Frequency distribution of sizes of natural seeds in the cheek pouches of six species of sympatric heteromyid rodents. (Adapted from Brown and Lieberman 1973.)

increase by 71% after rodent exclusion when compared with controls. All this provides evidence of strong competition between the two taxa. However the relationship is more complex. The long-term effects are different for rodents and ants. In the absence of ants, the rodents continue to show benefit but, without rodents, the ant populations gradually decline. This is apparently because with only ants present, the more vigorous, larger seeded plants favoured by the rodents, begin to displace the less vigorous, smaller seeded species harvested by the ants (Davidson *et al.* 1984).

7.7 Mutualisms and associations

Table 1.1 demonstrates that the most prevalent function of animal-built structures is the control of conditions within a home, providing a buffer against the hazards of the physical environment or predation pressure outside. This protected environment attracts other species, again contributing to local species diversity.

Where associating organisms simply take advantage of the living space provided by builders, we should anticipate facultative exploitation by species that are also found in other equivalent, protected spaces. The polychaete *Phyllochaetopterus socialis* aggregates to form dense tangled masses of tubes, an architecturally complex environment. Nalesso *et al.* (1995) examining these for any associated organisms of greater than 1 mm length, produced a list of 68 species of mostly Crustacea and Mollusca. The great majority of these were either found or known to occur in other substrata, so the observed associations appeared to be largely facultative.

This may of course be a stimulus to speciation by bringing together organisms that otherwise rarely meet. In particular, it could stimulate the evolution of a more intimate relationship between a builder and other species. This section examines the role of builders in stimulating mutualistic associations. In particular it tests the prediction that architecture, as distinct from the builder, will in general foster facultative associations. Obligate associations on the other hand should evolve where the presence of the squatter impacts directly upon the fitness of the builder; this could be in ways quite unrelated to the architecture itself, for example, predation upon or parasitism of the population of builders, or competition with them for resources. Here we would predict instances where evolutionary arms races have given rise to highly evolved relationships including obligate dependencies upon the builder species.

7.7.1 Facultative squatters in social insect nests

The complex nest architecture of social insects provides a particularly powerful test of whether architecture alone can stimulate the evolution of mutualistic relationships with animal architects and whether those dependencies are predominantly facultative or obligate.

Ants and termites have been estimated to represent about one-third of the total animal biomass of Amazonian rainforest (Fitkau and Klinge 1973). Many of these species make large structures that either endure as burrow systems in the ground or as above-ground mounds or arboreal nests. These structures remain intact while the colony remains active and may be sufficiently durable to persist after the colony has died out, providing potentially important refuges for other species. In steppe areas of Wyoming, subterranean termites *Reticulitermes tibialis* will, in the autumn when ants become less active, intrude into the nest space excavated by the harvester ant *Pogonomyrmex occidentalis* (Crist and Friese, 1994).

On examination of an abandoned arboreal nest of *Nasutitermes tatarendae* the day after the base of the tree in which it was located was inundated by seasonal Amazonian flood, Martius *et al.* (1994) found it to contain several thousand individuals belonging to 77 arthropod taxa, many apparently terrestrial refugees. Martius (1994) found that over 60% of arboreal nests of

Nasutitermes corniger in Amazonian forest, whether occupied or abandoned, contained colonies of the ant *Dolichoderus bispinosus*. Nests of *Anoplotermes* species, normally located at the base of tree trunks, extended up the trunk during the seasonal flood to create bulbous epigeic nests. Colonies of terrestrial termites, *Coptotermes* and *Rhinotermes* as well as other terrestrial arthropods make use of these nests during the floods.

In the Cameroon rainforest of West Africa, abandoned or occupied mushroom-shaped mounds of *Cubitermes* occur at a density of about 125 ha^{-1} (Dejean and Durand 1996). Mounds of *Cubitermes fungifaber* and *Cubitermes banksi* apparently provide important habitats for the establishment of new colonies of other termites, mostly *Cubitermes*. Incipient colonies of squatter *Cubitermes* occurred in 72% of examined *Cubitermes* mounds, along with at least some individuals of more than 25 other termite species.

The nests of these two *Cubitermes* species are also apparently of importance as refuges for colonies of ants. In a sample of 725 termitaries, Dejean and Durand (1996) found 799 ant colonies of 37 genera which covered 151 species, 11 of them undescribed. This, claim the authors, is among the highest diversity of ant species described for any habitat and illustrates how a rather invariant nest architecture can provide a breeding habitat for a diverse range of ants. Of course, not just animals benefit from such refuges. Nests of the paper wasp *Polistes hebraeus* on La Reunion Island were found to harbour 52 species of fungi of 31 genera (Fouillaud and Morel 1995).

7.7.2 Leaf rolls as generalised shelters

The experimental creation of simple refuges shows that they are quickly exploited by opportunistic facultative species. Larvae of the birch tube-maker (*Acrobasis betulella*) (Lepidoptera, Pyralidae) roll up young birch leaves to provide shelters for themselves when not feeding on neighbouring leaves. Cappuccino (1993) found that individual leaf rolls could be occupied by larvae of more than one instar or even by larvae of a guild of leaf 'sandwich-makers' belonging to other lepidopteran families. Creating artificial *A. betulella* shelters demonstrated that these attracted retreat-dwelling species more than control leaves. Artificial leaf sandwiches created by joining two leaves together, were colonised by larvae of several sandwich-maker species as well as *A. betulella*, and also attracted other arthropods such as spiders. Kudo (1994) found that leaves rolled by larvae of a tropical moth *Olethrentes mori* provided a preferred moulting refuge for aggregations of the homopteran bug *Elasmucha putoni*.

Martinsen *et al.* (2000) measured the impact of the presence of rolled cottonwood (*Populus* sp.) leaves made by the lepidopteran larvae *Anacampsis niveopulvella* on both the diversity and abundance of other arthropod species.

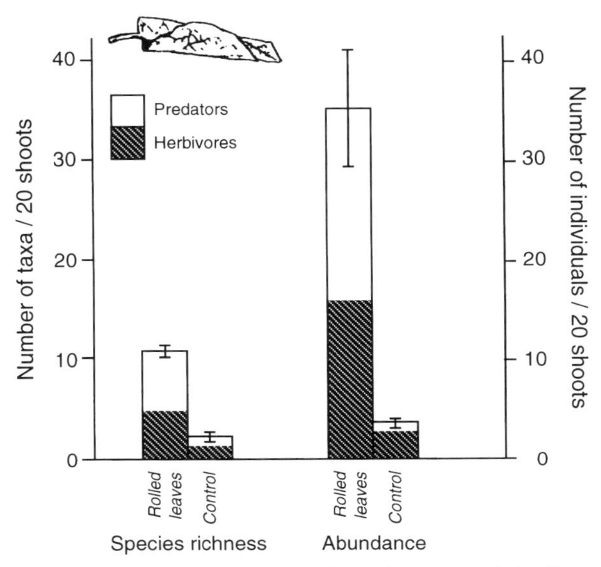

Fig. 7.10. The presence of leaf rolls on cottonwood trees (*Populus* sp.) significantly increased both species richness and abundance compared with adjacent shoots without a leaf roll (control). Vertical lines represent ± 1 SE. Overall means are broken down into herbivores and predators. (Adapted from Martinsen *et al.* 2000.)

By experimentally manipulating the availability of rolled leaves, they observed four times greater species richness (seven orders of insects as well as spiders and mites) (Fig. 7.10), and seven times greater abundance on cottonwood shoots with rolled leaves than without. Many of the arthropod colonists were found to be predators, and the provision of artificial rolls rather than leaf rolls showed that in the main the rolls were providing shelter not food (Martinsen *et al.* 2000).

7.7.3 Specialised associations with animal builders

There are a number of burrowing animals that are commonly found to have another organism sharing the burrow with them. The exact nature of these relationships is often not understood, but they show a situation where obligate relationships might evolve because the burrower and the squatter live in close association and have the opportunity to interact either competitively or mutualistically.

The grapsid crab *Chasmagnathus granulata* is an abundant inhabitant of estuarine environments in the south-western Atlantic where it excavates semi-permanent burrows in the soft intertidal sediments. These burrows seem to act as nurseries for recently settled megalope and juvenile *C. granulata*.

However, the juveniles also occupy the inhalent siphon holes of the stout razor clam (*Tagelus plebius*), which constructs deep permanent burrows (Gutiérrez and Iribarne 1998). The presence of these siphon holes may allow crab larvae to colonise poorly structured habitats where adult crab burrows are currently rare or absent. So, in this case there may to be a degree of dependency of the crab upon the razor shell, but it is apparently opportunistic, and the mollusc obtains no benefit or known penalty.

In the steppe habitat of China and Tibet, two species of snowfinch (*Montifringilla*) are dependent upon the burrows of pikas (*Ochotona*) for nest sites (de Schauensee 1984). The burrows of prairie dogs (*Cynomys*) seem to be critical for the survival of burrowing owls. (*Athene cunicularia hypugaea*). Butts and Lewis (1982) found that in Oklahoma, 66% of adult owls' nests in 'towns' of the black tailed prairie dog (*Cynomys ludovicianus*), that the owls made little modification to the dog burrows in which they build their nests, and their numbers declined rapidly in abandoned burrow systems. Burrowing owls nesting in non-prairie dog areas are apparently occupying marginal habitats possibly through competitive exclusion from the limited optimal areas. In these cases, it appears that there may be nothing special about the mammal burrows other than the absence of other holes or cavities, and again the mammals themselves are not known to be affected by the intrusion.

Many examples can be found where the builders are exploited, without benefit to themselves. The most common victims are social arthropods, where the builders themselves or the resources they gather become a target for specialist predators, parasites, or scroungers. An example is webs of the social spider *A. eximius*. This is an Amazonian forest species, the webs of which are two or more metres across, composed of multiple sheet webs with knock down threads. These accumulate large amounts of dry leaves in their interior, used by the spiders as refuges and nurseries (Brach 1975). *Crematogaster* c.f. *victima* ants have been observed to forage in these webs, grabbing prey knocked down by the silk threads and driving off the spiders with the secretion of their venom gland (Fowler and Venticinque 1996). Spiders of the genus *Argyrodes* specialise to varying degrees in stealing prey caught in the webs of other spiders. Some *Argyrodes* species remove prey that is of little interest to the host spider (Cangialosi 1990), but the stealing of prey by *Argyrodes antipodanus* from the webs of *Nephila plumipes* has been shown experimentally to reduce the growth rate of the host (Grostal and Walter 1997).

Argyrodes species vary in their specialisation on particular spider hosts. *Argyrodes globosus* is recorded by Henaut (2000) on the webs of five species belonging to three different genera, although showing a preference for webs of a certain size and of a particular species. *A. antipodanus* is not host specific, but prefers orb webs of *Nephila* and *Cyrtophora*, which are surrounded by scaffold threads (Grostal and Walter 1999). *Argyrodes ululans* does however see to specialise on the large web complexes of the social spider *Anelosimus eximius* (Cangialosi 1990).

The expectation that opportunities for the evolution of obligate associations between architects and other species are potentially great where the architects live socially, in large long-lived structures is also supported by the wide range of arthropods found in symbiotic relationships with ants alone. This includes numerous species from at least 10 orders of insects. Among the Coleoptera alone there are 35 families containing species dependent on ant colonies in some way; some, such as detritivores, may be beneficial to nest hygiene, while others may harm the colony by soliciting food from ant workers or eating ant larvae (Hölldobler and Wilson 1990).

A number of species of social bees or wasps have become parasitic in the nests of near relatives. *Vespula austriaca* parasitizes the nests of *Vespula rufa*, exploiting its force of workers to raise its brood composed exclusively of reproductives (Wilson 1971), a relationship known as *inquilinism*. Similar parasitic relationships are found in bumble bees where, for example, several *Psithyrus* species occur as inquilines in the nests of *Bombus* species, and in *Polistes* wasps where *Polistes sulcifer* is an inquiline in the nests of *Polistes gallicus*. In some of these examples, the inquiline is known to be a sibling species of the host. This phenomenon is referred to as *Emery's rule* and seems to indicate that the pathway to exploitation was facilitated by the parasitising species so closely resembling the host that nestmate recognition defences were not fully effective.

Social insects of differing kinds may exploit the defences of other colonies by nesting in association with them. These aggregations may be mutually beneficial. Jeanne (1978) describes aggregations of up to 23 nests of the wasp *Polybia rejecta*, where each colony apparently benefits by the collective defence that they can mount against nest predators. Sakagami *et al.* (1989) found 67 nests of the stingless bee *Trigona moorei* all located within an arboreal *Crematogaster* ant nest.

The collective stinging power of wasp or ant colonies is also exploited by a number of bird species that place their nests in close proximity to them. Hansell (2000) reviews 28 species of birds from nine families reported as nesting in association with the nests of social wasps. For a number of these bird species the association may be facultative but for the red-cheeked cordon bleu *Uraeginthus bengalus* the association seems to be essential, although the wasp species chosen may vary and apparently gains no reciprocal benefit.

A few examples do show mutually beneficial specialist associations between builders and cohabiting species. Burrows dug by shrimps of the genus *Alpheus* are attractive refuges for gobiid fish. Karplus (1987) reviews evidence of 29 gobiid species associated with 13 different *Alpheus* species. Although this indicates that these relationships may not be obligate at the species level, they are mutualistic. Burrow digging is virtually the exclusive role of the shrimp, which may also groom the fish. In return the shrimp may obtain food items as a result of grooming and, through contact communication and sometimes even visual signals from the fish, anticipate the approach of danger.

The above examples support the view that specialised symbiotic relationships between builder and another organism do generally involve direct interactions between the two parties which are beyond the role of the architecture itself. None of these exemplifies a specialised mutualistic association between the two species, but this is the nature of the association between ants and termites that cultivate fungus (Section 1.3.3), and arguably between ants and some epiphytic plants in the creation of 'ant gardens'.

The ponerine ant species *Pachycondyla goeldii* builds an arboreal carton nest in which it plants seeds of epiphytic species such as bromeliads. The roots of the seedlings reinforce the ant nest wall (Corbara 1996). Mutualistic relationships such as these to create so-called 'ant gardens' may be facilitated by the plants providing special nutritive bodies attached to hard-coated seeds, which encourage ants to bring them back to the nest and then abandon them (Hölldobler and Wilson 1990). Kaufmann *et al.* (2001) observed that all 16 species of plant epiphytes on a giant bamboo (*Gigantochloa*) in a Malaysian forest site were growing in ant carton nests. This, they reasoned, was because the hard smooth surfaces and relatively short life (15–20 years) of the bamboo stems renders them difficult for plants to colonise without the aid of the ant nests. One of the ant species (*Crematogaster* sp.), was only found to build nests on bamboo and, when offered a choice of seeds, only brought those of epiphitic species back to the nest. None of these plants appeared to provide food for the ants, so their benefit to the ants seems to be that of stabilising the nest with their roots.

7.8 Conclusions

7.8.1 Is the keystone concept supported?

It is now possible to assess whether habitat 'modifiers' (Mills *et al.* 1993), in the form of animal architects, give weight to the keystone concept and, if so whether they form a particularly important category of them.

The marine bivalve *Limaria hians* deserves the status of a keystone species according to Hall-Spencer and Moore (2000). *Limaria* species are unusual among bivalves in enveloping themselves in a nest of byssus threads. When occurring at high densities, *L. hians* can create massive solid reefs, distinct from the surrounding seabed. Hall-Spencer and Moore (2000) recorded 284 species of marine organisms associated with these reefs including Protozooa, Macroalgae, and a wide range of invertebrates. Although the authors do not compare this with the diversity in the immediate surrounding areas, the argument of keystone status is that species diversity on the byssus reef is very much greater.

Burrowers too have been invoked as keystone species. The plateau pika (*Ochotona curzioniae*) is, for example, regarded as a keystone species by Smith and Foggin (1999). Through soil disturbance it brings about an increase in plant diversity, while the burrows themselves provide nesting sites for several species of birds, notably snowfinches (*Montifringilla* species) (Schauensee 1984). The pikas themselves are important prey of mustelids, in particular the stepps polecat (*Mustela eversmanni*), whose dependence upon the pikas is similar to that of the black-footed polecat (*Mustela nigriceps*) on the prairie-dog (*Cynomys*) (Schaller 1998).

Gopher tortoises (*Gopherus polyphemus*) that inhabit the longleaf pine (*Pinus palustris*) habitats of south-western United States are known to have 360 'commensal' species associated with their burrows (Jackson and Milstrey 1989), ranging from vulnerable vertebrate species such as the eastern indigo snake (*Drymarchon coaris couperi*), to numerous species of arthropods.

The size of the tortoises in an undisturbed site was found to be greater than that in a disturbed site, reflecting the greater life expectancy of the tortoises (Guyer and Hermann 1997). Larger tortoise burrows endure longer than smaller ones, and have a half life of 12–24 years. Given the slow growth rate and population replacement rate in this species, it could be claimed that its importance to the conservation of diversity of this ecosystem justifies the label keystone species.

Multiple effects of habitat modification brought about by the red-naped sapsucker (*Saphyrapicus nuchalis*) doubly qualify it as a keystone species according to Daily *et al.* (1993). It is a primary excavator of nest cavities in aspens (*Populus tremuloides*) infected with the heartwood fungus *Fomes igniarius.* Abandoned nest cavities provide essential nest sites for two obligate secondary cavity nesters, the tree and the violet-green swallow (*Tachycineta bicolor* and *Tachycineta thalassina*). In addition, the sapsuckers obtain the food that gives them their name by drilling sap wells in spruce *Picea*, aspen *Populus,* and willow *Salix,* which then provide feeding sites used by over 40 other species including hummingbirds, chipmunks, and butterflies.

In the view of Brown and Heske (1990) the guild of kangaroo rats (*Dipodomys* species) should be regarded as keystone species. This is supported by studies on *Dipodomys* species in the United States using exclusion pens over a 20 year period. Exclusion of the three *Dipodomys* species from Chihuahuan Desert shrub habitat transformed it over a period of 12 years, through loss of selective predation of seeds and reduction of soil disturbance, into a habitat of tall grasses colonised by eight new grassland rodent species.

But how much of this status is due to soil disturbance by the kangaroo rats and how much to seed predation? After *Dipodomys* exclusion, only 20–30% of the energy of seed production was taken over by the community of small granivorous rodents, creating a substantial redirection of energy in the ecosystem. In the years since 1996 this has been transformed again by the arrival of the pocket mouse (*Chaetodipus baileyi*), which is similar in size to *Dipodomys* species and consumes as much energy as the long absent kangaroo rat species (Ernest and Brown 2001). Whether the soil disturbance of this species is equivalent to that of *Dipodomys* has not been established but seed predation is clearly a major factor in control by *Dipodomys* species over this ecosystem.

The keystone concept initially received much attention particularly in the field of environmental conservation, where the promise to identify some small component of a complex community which, if protected, would support the rest of the biological diversity, held out hope of simpler, more effective management decisions. The survey of Hall-Spencer and Moore (2000) on the keystone status of the byssus reefs formed by the bivalve *L. hians*, is essentially an argument for greater protection for this habitat. However critics, while generally acknowledging that some species may exert powerful influences on ecosystems, have expressed dissatisfaction with the concept as originally proposed by Paine (1969).

This disagreement comes not from an assessment of whether the removal of a species has a substantial effect on the ecosystem, so much as on the definition of the term keystone. Mills *et al.* (1993) and Bond (2001) consider that there is lack of clarity in what measure should be used as an indicator of keystone status and of data to make such an evaluation. Other authors have put forward refinements to the original definition of Paine (1969) to allow the term to be strengthened. Power *et al.* (1996) added to the criterion of large effects resulting from the removal of keystone species, that these effects should be disproportionately large relative to their abundance. Jordan *et al.* (1999), examining the role of keystone species in foodweb dynamics, conclude that keystones are characterised more by making network flows less reliable than by the likelihood of species loss. However, their emphasis on trophic interactions seems unsuited to assessing the full impact of habitat modifiers.

Ceballos *et al.* (1999) compared sites where prairie dogs were active with ones from which they had become extinct. They claimed that, through their burrowing, plant grazing activities, and availability as prey, the prairie dogs maintained habitat heterogeneity and supported diversity (Fig. 7.11), so confirming their status as keystone species. Kotliar *et al.* (1999), while acknowledging that prairie dogs do exert some keystone level influences, question their role in supporting diversity. They find that the majority of vertebrate species occurring in association with prairie dogs are in fact facultative and

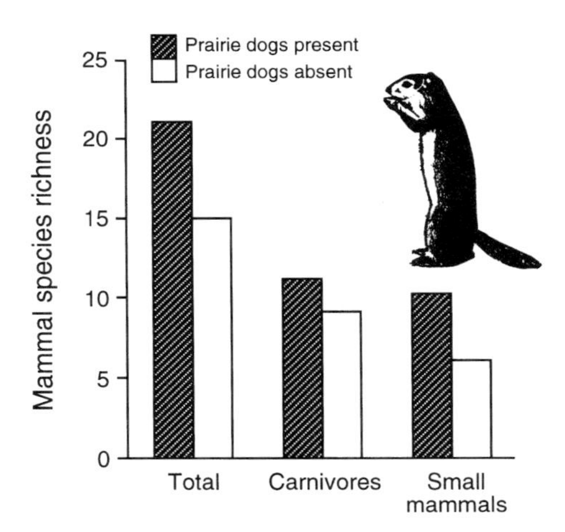

Fig. 7.11. The effect of black-tailed prairie dogs *Cynomys ludovicianus* compared with their absence in North Western Mexico on total species richness of mammals, on carnivores, and on small mammals. (Adapted from Ceballos *et al.* 1999.)

opportunistic, while only a few species like the black-footed ferret are exceptional in this respect. As this chapter has shown, this is indeed the expected outcome of physical habitat alteration, so a criterion that requires builders and burrowers to support a variety of obligately dependent species would deny most of them the status of keystone species.

Kotliar (2000) uses the prairie dog *Cynomys* example to argue that embracing the qualification of proportional effect of dominant species and greater than proportional effect for keystone species (Power *et al.* 1996) is inadequate. This is because there may well not be a simple linear relationship between abundance and overall community importance (Fig. 7.12). She advocates a third criterion be added to the definition: keystone species should perform roles not shown by other species or processes. She urges the abandonment of any conceptual dualism between keystone and non-keystone species, and a more detailed examination of the circumstances such as abundance, spatial and temporal distribution under which a species can exert keystone-level influence. This echoes the conclusion of Bond (2001), who looks to more empirical data to revive the value of the keystone concept. This is an approach that would encourage the importance of builders in ecosystems to be understood more fully.

7.8.2 Burrowers and builders; burrows and buildings

This chapter has provided evidence of the influence of builders and burrowers on ecosystems which, starting from the changing of the physical and chemical properties of the substratum, feed through into the relative abundance of microorganisms, plants, and other animals. Examples of large effects are apparent in both terrestrial and aquatic ecosystems. However, the extent

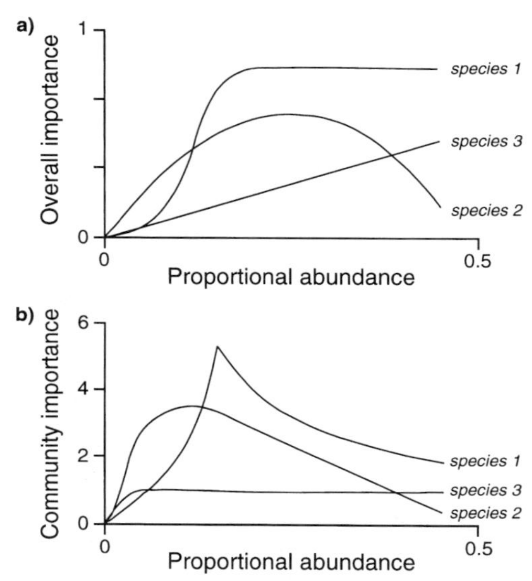

Fig. 7.12. Theoretical curves illustrating possible relationships between abundance and importance. (Adapted from Kotliar 2000.) (a) Power *et al.* (1996) differentiate between dominant and keystone species on the basis that whereas a dominant species increases in importance in proportion to its abundance (species 3), the importance of keystone species will increase at a greater rate (species 1 and 2). (b) Kotliar (2000) argues that relationships between abundance and community importance may be more complex (e.g. curves for species 1, 2, and 3). Species 1, which might represent prairie dog colonies, supports few species when at low population level. As its population increases, threshold values for the support of several new species are reached, but further increase in colony size dilutes community importance.

of the influence of these ecosystem engineers is shown repeatedly to depend not simply upon habitat modification itself but on other influences exerted by their activities, most obviously as foragers and as the prey or hosts of other species. Whether, through these combined effects, builders and borrowers should be considered as keystone species seems immaterial. Jones *et al.* (1997) are probably right to limit the definition of physical ecosystem engineering to the effects of physical alteration of the environment, but it is important to recognise that, by virtue of the position the engineers have created for themselves, they may also exert significant direct biological influences on other species.

Hansell (1987*a*, 1993) reasoned that the security offered by the building of nests and burrows gives opportunities for continued development of populations of builders to approach that of the carrying capacity of the environment and associated enhancement of control over local resources including food. This remains unproven, however, this chapter confirms that species which seem to be contributing significantly to the character of a habitat through their building or burrowing, should be regarded as potentially of great importance. Then, as Kotliar (2000) and Reichman and Seabloom (2002) suggest, the nature of the influences should be dissected to look at effects of abundance on degree of control and to separate the influences the concerned species exerts through its building or burrowing from its other effects. The prediction of Hansell (1993) that, through the added protection they afford themselves, builders and burrowers reduce their possibility of extinction, also remains untested, although Brandl and Kaib (1995) do argue that termite species may have been resistant to extinction for reasons largely based on their mound dwelling biology.

Repeated examples in this chapter confirm that increased species diversity is the general outcome of the presence of animal builders, while few examples show the result to have been exclusion. Both these outcomes were proposed by Hansell (1993), and on balance by Jones *et al.* (1997). It is now possible to say in more detail how this might arise. Exclusion is the result of competition for resources, (between ants and rodents for seeds), or habitat destruction (by beaver dams). Diversity comes about first by the creation of protected spaces within nests and burrows. This is well illustrated by examples of arthropods taking over nest space created by *Cornitermes* and *Cubitermes* species and by the creation of patchiness in the environment as shown by bioturbators at all scales. These dependencies appear to be largely facultative, although this needs further study. Nevertheless the creation of these microhabitat niches may be relatively important for the survival of viable populations of a wide variety of opportunist species. Obligate symbiotic relationships seem generally to involve direct interaction between the builders and other species, generally within the confines of the nest or burrow, although the nature of these and their evolution also needs further investigation.

8 Evolution

8.1 Animal architecture as behavioural evidence

The second half of this chapter examines the influence that animal built structures exert on the evolutionary trajectory of the builders, but the chapter begins by assessing what evidence the structure of a building provides on a builder's evolutionary past.

8.1.1 Architectural characters as phylogenetic evidence

The evolution of animal building can be studied by fitting the respective structures onto a phylogeny established on other criteria. For example, Winkler and Sheldon (1993) determined the pattern of evolution of nest design within the swallows (Hirundinidae) by matching the nests of 17 different species to a consensus tree based on DNA–DNA hybridisation. This showed a cavity nesting, then a mud nest building group of species evolving from an ancestral (*plesiomorphic*) soil burrowing state. The initial design of the mud nests was bracket shaped, but from this arose an enclosed nest form, which finally evolved into a retort shape, with the addition of an entrance tube.

So the grouping of nest types (borrowers, cavity nesters, mud builders) in this case mirrors the phylogenetic groupings based on the DNA hybridisation analysis. This also supports what can be understood as a logical progression from the excavation of mud to building with mud, and then increasing complexity in the built structure. So could a reliable phylogeny have been established on the basis of nest characters alone?

There has been controversy over this point, with critics claiming that, as behavioural characters they will have been more labile in evolutionary history than, for example, anatomical characters. If this is the case, then it is predicted that *homoplasy* (mistaken homology) will be more likely to occur where behavioural characters are used in phylogenetic studies than where morphological characters alone are relied upon.

This prediction was tested by De Querioz and Wimberger (1993), drawing upon evidence from 20 studies where both morphological as well as behavioural characters were used, and a further 44 where either behavioural or morphological characters alone were used. They found that morphological characters were no less homoplasious than behavioural ones. Nevertheless, there are some problems with this interpretation, in particular that the behavioural characters chosen in these studies were probably selected because they were judged by researchers to be the most reliable of the behavioural characters available. However, since this will also have been the case for the morphological characters, the authors conclude that both are equally reliable in phylogenetic studies.

Other studies have not been so supportive of the use of characters derived from built structures. Nests of paper wasps of the genus *Mischocyttarus*, for example, have distinctive species-specific nest designs, yet Wenzel (1991) using eight characters of nest architecture and building material, concluded that all had been subject to innovation. The consequence of this was that the nests of some close relatives looked markedly different, while those of species in different lineages were convergently alike.

In a study of the swiftlet genus *Aerodramus* (Apodidae), nest characters also failed to provide reliable information on the phylogeny of a group of 15 species and sub-species (Lee *et al.* 1996). The nest characters failed to map consistently onto a phylogeny derived from mitochondrial DNA. However, the nest characters used were quite simple and only four in number, so their weakness could be ascribed either to the general unreliability of nest-derived characters or to their lack of detail in the data obtained (Hansell 2000). The ovenbirds (Furnariidae) are a family noted for the between-species variation in their generally large and complex nests. Zyskowski and Prum (1999) produced a phylogeny for 50 genera based entirely on 24 nest characters. From this they deduced, for example, that cavity nesting is a plesiomorphic character in the family, but in separate lineages cavity nesting has been lost and replaced with new structures. The authors describe the analysis as 'heuristic', and therefore awaiting support using other characters; nonetheless, they regard the wealth of detail present in the nests of this family as significant evolutionary evidence.

The use of characters provided by nests or other built structures have also been the target of criticism, not because they are behavioural, but specifically because they are not. The argument here is that they are inherently less informative than characters obtained from the building behaviour itself.

Stuart and Hunter (1998) test the relative value of what they term *end product* (built structure) characters against actual behaviour, in a phylogenetic study on blackflies (Simuliidae). In order to resolve the phylogeny of 17 species of *Simulium, Eusimulium,* and some related genera, they used five characters relating to the morphology of the pupal cocoon and a further 14 characters derived from the behaviour exhibited during the spinning of the cocoon.

The five end product characters generated two equally parsimonious trees, with no homoplasy but leaving many unresolved areas. The 14 behavioural characters generated 22 equally parsimonious trees, but not markedly different from one another, and with the tree much more fully resolved (Figs 8.1 and 8.2). Their conclusion is that the cocoon spinning characters are more informative than the end product ones.

The weakness of the end product characters in this study, as the authors point out, can be attributed to their small number relative the behavioural ones. However, a more serious weakness in them is indicated by the relatively poor fit of the end product tree to a third phylogeny, this time derived from well-studied morphological and cytological characters of the organisms themselves. This is, for example, indicated by the supposed monophyletic origin of the boot-shaped cocoon as indicated by the end product tree, while the behavioural

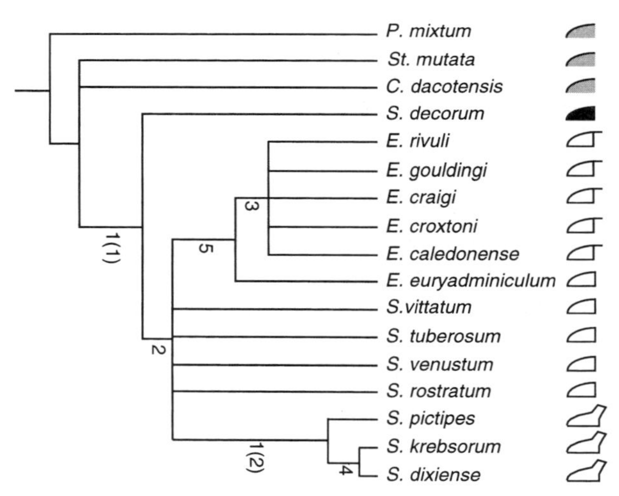

Fig. 8.1. A phylogeny for blackflies based on five end product characters observed in the pupal cocoon are insufficient to resolve substantial areas of a tree incorporating 17 taxa. The tree length is 6 with values of consistency index (CI) = 1.00 and retention index (RI) = 1.00, showing that no character evolved more than once. The cocoon symbols, incorporating the end product characters, represent: hatched = loosely woven, sac-shaped cocoon; open triangle = tightly woven slipper-shaped cocoon with collar; open triangle with extension = as previous design but with anterodorsal projection; black triangle = loosely woven, slipper-shaped cocoon lacking a well-defined collar; open triangle with spout = tightly woven boot-shaped cocoon. (Adapted from Stuart and Hunter 1998.)

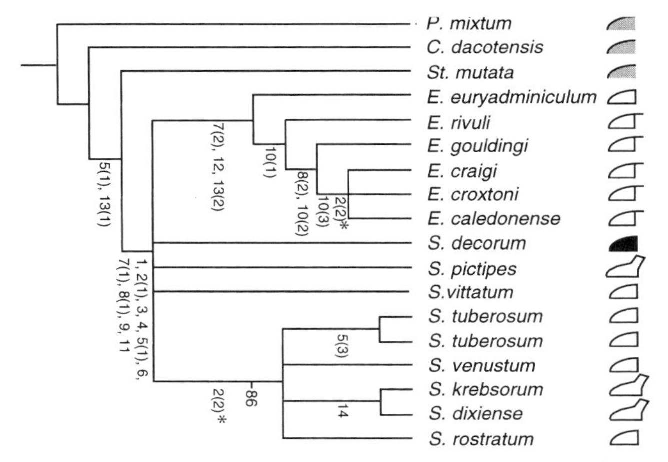

Fig. 8.2. A phylogeny for the same taxa as shown in Fig. 8.1, but using 14 behavioural characters, produces a more highly resolved tree than obtained from end product characters. Asterisks (*) represent a homoplasy in that character. '86' in larger type, is the percentage of trees generated, supporting that node. The tree length is 25, with high vales for CI = 0.92 and RI = 0.96. (Adapted from Stuart and Hunter 1998.)

characters apparently correctly represent it as a holoplasy, covergently arrived at in *Simulium pictipes* and *Simulium dixense*.

Stuart and Currie (2002) further elaborate on the distinction between and relative merits of end product characters and the behaviour that produces them, in a study on the case building of 35 genera of caddis larvae from 10 families. The purpose of the study in this instance is not to test whether end products are good indicators of evolutionary history, but whether end products and the behaviour that generates them give equally good information. In brief, the conclusion of their study is, no. They found several examples across the caddis families where larval cases with the same distinctive design features had been convergently arrived at using different behaviours, while further examples revealed distinctly different designs produced by very similar behaviour.

The conceptual conclusion of Stuart and Currie (2002) is that end product characters cannot be used to assume underlying behaviour, because they are not behaviour. The authors draw an analogy between ontogenetic processes leading to mature morphology of an organism, and behavioural actions leading to end products. Their conclusion is that since development and morphology are different disciplines, end products and building behaviour should similarly be regarded as separate.

It is at one level a truism to say, as Stuart and Currie (2002) do, that an end product is not behaviour, but the product of it. However, end products do contain, amongst other things, behavioural information. As a behavioural record, the end product may indeed be incomplete, with some behavioural actions never represented in it, other behaviours destroyed or distorted, but much may remain, if properly understood. So, for example, a completed structure such as a caddis case expresses no time intervals, nevertheless it may be possible to reconstruct very accurately the *order* in which its components were added, once it has been established directly from the behaviour that materials are always added to the front end.

A current weakness in the use of end product characters in phylogenetic studies is their small number and rather gross level of description. More detailed examination of end products will generally reveal much more behavioural information. Hansell (2000) developed a scheme for the systematic description of bird nests which, although far from detailed, recognised four different nest zones each of which could contain one or more of 33 categories of material. A nest could also be one of eight different shapes, located in one of eight different sites and attached in one of ten different ways.

Examination of partially built structures is a further source of structural characters, and can, for example, provide information on the timing of building events or reveal the presence of features that are lost or obscured at a later stage of construction. Examination of the developmental stages of a structure also provides information on the order in which parts of the structure are built.

Wenzel (1991, 1993), in studies on the evolution of varied nest architecture in vespid wasps, argues the parallel between ontogenetic studies on morphology and the development of a nest through behaviour. In ontogenetic studies it has long been contended that differences expressed early in development generally represent divisions between higher, more anciently derived taxa, while those observed later in development, reveal lower, more recently derived taxa (Fig. 8.3). This provides animal built structures with a powerful tool to distinguish homologous from merely analogous features.

It is important to note in this study Wenzel (1993) describes as 'behavioural', characters that can be determined by examination of the nests alone during stages in their development. This approach is validated with significant success in matching a phylogeny of 28 genera of Polistinae based on these behavioural plus end product criteria with one based solely on the morphology of the wasps themselves. Stuart and Currie (2002) take the view that, since development and morphology are separate disciplines, so by analogy are characters of behavioural development and end products. They cite Wenzel (1993) in support of this view, apparently overlooking the fact that Wenzel's 'behavioural' criteria are derived almost entirely from the morphology of the nests observed during their architectural development.

End products, it seems should be regarded as analogous to the mature organism, while partially completed structures are equivalent to stages of

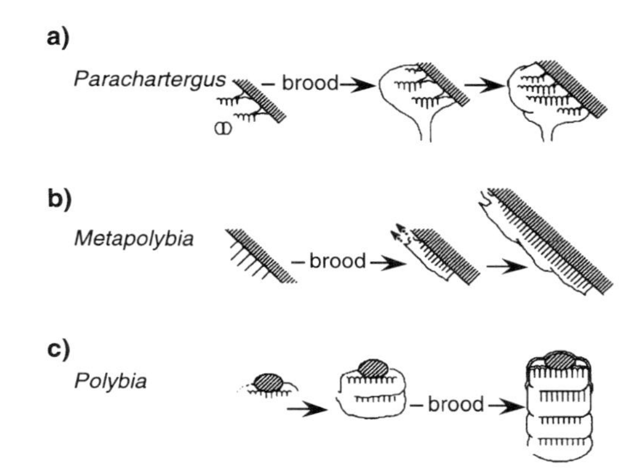

Fig. 8.3. Major designs in eusocial vespid nest architecture, and their pattern of nest development. Each row displays the nest development of a named genus ((a) *Parachartergus*, (b) *Metapolybia*, and (c) *Polybia*) chosen to represent all members of a wider group. The left-most figure in a row represents nest initiation. To its right are shown successive growth stages. (Adapted from Wenzel 1991.)

development. From this I think we should conclude that the process of construction (the behaviour itself) properly equates to the dynamic molecular and cellular phenomena by which the organism becomes manifest. Neither end product nor 'partial product' are the behaviour itself, but both provide evidence of it that can substitute for at least some of the evidence that would be obtained if these ephemeral events could be fully recorded. The fact that they cannot, while end products can be examined at leisure, gives the latter a special advantage.

8.1.2 Ethospecies and ichnogenera

The behavioural record as an end product has occasionally been regarded as sufficiently distinctive that, in the absence of comparable morphological data, species status can nevertheless be invoked; species recognised in this way have been termed *ethospecies*. Emerson (1956), for example, distinguished species of termite of the genus *Apicotermes* on the basis of very distinctive architectural differences in nest design. This concept has not proved popular and, in view of the points already discussed, any claim of specific status based on a few end product characters alone is likely to appear weak. There is however a particular exception to this, *ichnofossils*.

Ichnofossils are fossilised evidence of the nature and actions of their creator, without fossil evidence of the organism itself. Fossil impressions of caddis leaf panel cases from the Oligocene (24–37 m years ago) (Hellmund and Hellmund 2001) record monocot and dicot leaf panels, regularly cut and oriented in a manner closely resembling that of the living genera of Phryganeidae. The Late Cretaceous nest burrow in Fig. 5.4 shows an elegant helical design thought to have been dug by a solitary bee (Genise and Hazeldine 1998). Such records of animal built structures are likely to be very fragmentary but, for some organisms, may be better preserved than the organisms that produced them. Soft-bodied borrowers may leave ichnofossils that cannot be reliably associated with a particular class or even phylum of organism (Hansell 1984). More interestingly, social organisms may leave traces much more substantial than any one of their number, as recorded, for example, in the substantial ground nests of termites.

Genise and Brown (1994) recognise two new *ichnogenera* and five new *ichnospecies* of termites based on trace fossils from the Upper Eocene to the Lower Miocene (that is, from about 40 m to 20 m years ago) in Egypt. This is indicative of the diversity of these insects at that time. More importantly, Hasiotis and Dubiel (1995) described a new termite ichnogenus and ichnospecies from the Late Triassic in Arizona, pushing back the origin of insect social life to around 220 m years ago, before the radiation of angiosperms in the Early Cretaceous, an event which was thought to have been the stimulus for it.

8.2 Building and the key adaptation concept

The concept of a *key adaptation* or *key innovation* identifies a particular evolutionary change as allowing a lineage to utilise resources previously unavailable (Futuyma 1986), which enable it to diversify to exploit the varied aspects of the newly accessible habitat. What is the evidence that innovations in animal building have brought about the extension of habitat range in a particular species, stimulated an episode of rapid speciation, or contributed towards the evolution of greater social complexity? If there have been such innovations, what has characterised them?

I have identified two alterations in nest building behaviour that are associated with the transition from solitary to social behaviour of wasps Hansell (1987a). The first is the transition from an excavated to a constructed nest. It can be regarded as a change in the behaviour directed at the soil, from that of excavation alone, to the use of the material after excavation. This occurred in both the Specidae and Vespidae (Eumeninae).

The second change was in the choice of the building material from mud to paper. This change apparently coincided in the Vespidae with a transition from subsocial to primitively eusocial colony life. This association in time could have been a causal link, either indirectly through the opportunities that light-strong paper offered for attachment of nests in a new range of sites, or directly through its mechanical advantages permitting the construction of much larger structures. I referred to these as respectively innovations of *design* and of *technology* (Hansell 2000).

This is this same dichotomy that Craig (1987*a*,*b*) makes in the context of spider webs, when she asks whether changes in the function of webs has been achieved more easily through changes in web design or the properties of the silk threads. Craig's example, however, shows that a further distinction needs to be recognised in the nature of the technological innovation. When building materials are collected, technological change is the result of *behavioural* innovation alone but, when the materials are self-secreted, any innovation must include ones at the *physiological* level. In this respect I was (Hansell 2000) too simplistic in asking which of the two, design or technological changes, had been the most important in bringing about evolutionary change in birds' nests. Among birds, only the salivary mucus of swifts and swiflets (Apodidae) is a specialised self-secreted building material, nevertheless, I did fail to point out that technological change in bird nest building could have come about either through a change in a collected or secreted material.

8.2.1 Key adaptation, habitat range extension, and speciation

It is difficult to test for an association between architectural innovation and extension of habitat range. Control over the environment through building should assist this, but the effect may be obscured by subsequent speciation over the extended range.

The distribution of ants of the genus *Oecophylla* at least provides a supportive example of this, since the only two species of the genus extend across the entire Old World tropics. *Oecophylla longinoda* occurs across Africa, and *Oecophylla smaragdina* continues through India to Australia. These are weaver ants, capable of using the silk spun by their larvae to create nests by sticking a few leaves together. These ants also have complex social organisation and communication, but Hölldobler and Wilson (1990) regard nest building as a significant contributor to their ecological dominance over such a wide area. Their nest building technique enables each colony to spread itself between dozens of nests and thereby exert territorial control over whole trees or even groups of them.

The evidence that building innovation stimulates speciation has received more attention. Collias (1997), for example, claimed that nest construction in birds had been an important factor contributing to the speciation of passerines, because of the ability to place their relatively small nests in a variety of locations. While not disagreeing with the logic of this argument, this hypothesis remains untested and other aspects of bird biology such as their feeding anatomy or mechanism of song production could equally be regarded as candidates responsible for speciation in passerines (Riakow 1986, Hansell 2000). Social wasps, particularly the Polistinae, provide another example suggestive of an association between nest building diversity and speciation, in this case paper technology providing the opportunity for adopting new nest sites and nest enlargement (Hansell 1996b), but again the hypothesis remains untested.

The problem is to be able to show that species diversification coincided with the innovation, that no other innovation occurred at the same time, and that related species without that innovation failed to show that rate of diversification. Such tests have been applied to some taxa. That by Connor and Taverner (1997) examines the consequences for speciation of the evolution of the leaf mining habit.

A leaf miner creates a burrow inside the leaf in which it spends its larval life. About 10,000 described species show this habit, which has evolved a number of times and is found in four insect orders: Lepidoptera, Diptera, Hymenoptera, and Coleoptera. The argument for leaf mining as a key adaptation is based on the supposition that it provided protection from predators, from the defences of the plant itself, and from physical factors such as desiccation or UV irradiation. To test whether this has proved to be a stimulus for adaptive radiation Connor and Taverner (1997) compared the species diversity in phylogenies of these leaf mining taxa with their non-mining sister groups.

In the Coleoptera, leaf mining is a derived (apomorphic) character relative to external feeding, but its non-mining sister group is equally speciose. In the Diptera, the origins of the mining habit are too complex to make the necessary comparison. In the Hymenoptera, leaf mining seems to have evolved six times, three from external feeding ancestors which can provide comparable sister groups. Data are incomplete but none provides support for the key adaptation hypothesis.

In the Lepidoptera, the leaf mining habit seems to have evolved only once, however it seems to be plesiomorphic to the external leaf feeding habit in the order. In this case the external feeding macrolepidoptera are half as rich again in species as leaf miners, defying the predictions of the key adaptation hypothesis with evidence that in abandoning the leaf mining habit the lineage was in fact liberated from certain constraints, and speciation stimulated.

8.2.2 Building with silk: a key adaptation?

The current scarcity of evidence that construction behaviour has provided a significant stimulus to an expansion of habitat range or speciation may simply reflect a lack of appropriate research or the difficulty in obtaining such data. However, one theme that has constantly recurred in this book has been the varied roles and ecological influences of silk. It also currently provides the strongest case for an innovation in building technology being a key adaptation.

According to Craig (1997) the origin of silk secreting glands in both spiders and insects has taken place through two routes that she labels *systemic* and *surficial*. Craig traces the origins of systemic glands to modifications of crural

	Surficial glands (Epidermal origin)		Systemic glands (Crural origin)	
	Paurometabolous	Holometabolous	Paurometabolous	Holometabolous
Insects	Adults & Larvae **tarsal glands** (Embiidina)	Adults **tarsal glands** (Empiidae: Diptera)	Adults & larvae *salivary glands* (Orthoptera) *colleterial glands* (Mantoidea)	Adults *colleterial glands* (Hymenoptera) Larvae ***salivary glands*** (Lepidoptera) *peritrophic membrane* Adults & Larvae *malphigian tubules* (Coleoptera)
Spiders	Adults & Larvae **piriform, acinous** (Araneomorphae) **cribellar glands** (Deinopoidea)	None	Adults & larvae **tubulliform, ampullate** (Araneomorphae) **pseudoflagellar** (Uloboridae) **flagellar glands** (Araneoidea)	None

Fig. 8.4. Silk producing systems in insects and spiders, appear to have evolved both form glands of surficial (epidermal) origin, and ones of systemic (crural) origin. Glands named in bold are dedicated to the production of silk; glands indicated in italics produce silk as a secondary function. (Adapted from Craig 1997.)

glands present on the ancestor from which the Arthropoda and Onycophora arose. Crural glands still remain in the Onycophora secreting proteinaceous, defensive glues. However, both spiders and insects have evolved new kinds of silk glands from modified epidermal (surficial) structures. The cribellar glands are an example in the spiders, while in insects the tarsal glands of Embiidina provide just one example of glands of surficial origin (Fig. 8.4).

So, silk production was characteristic of the earliest spiders, possibly with the role of creating a shelter or cocoon to protect the eggs or to provide a shelter for the silk producer. Later the threads leading to the shelter probably became used to communicate the approach of potential prey by a sit and wait predator, as can now be seen in the array of threads that radiate from the burrows of a spider such as *Segestria*. With the development of silk capture devices came the evolution of a complex collection of glands each producing a specialised secretion (Craig 1997).

Craig (1997) considers the secretion of fibrous proteins in insects to have been a primitive feature of the Hexapoda, initially exhibited by adults in a reproductive context, as seen, for example, in the Thysanura. Secretions that are recognisably silks have evolved multiple times through modification of a variety of different structures in both adult and larval forms (Figs 8.5 and 8.6). In the hemimetabolous Embiidina, for example, silk glands in the tarsi have

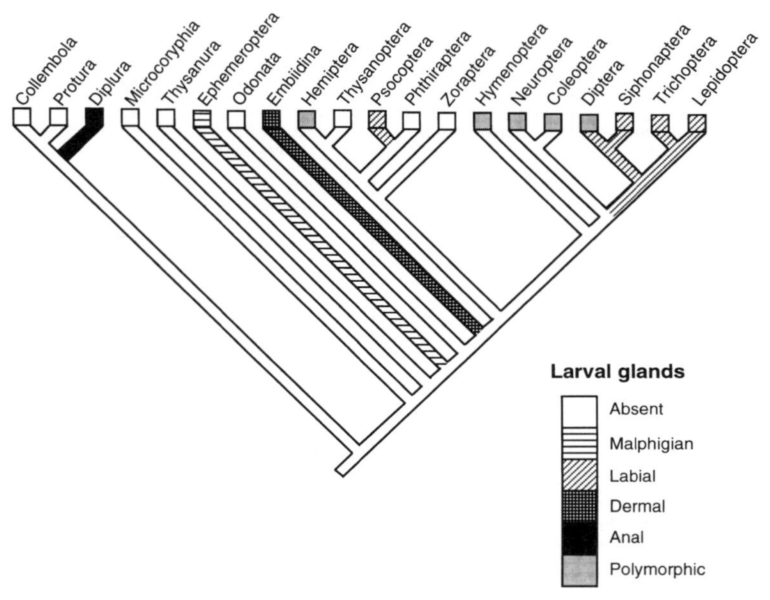

Fig. 8.5. The phylogeny of silks produced by hexapod larvae. Silk production is common among hexapod larvae, but originates from a number of different glands. (Adapted from Craig 1997.)

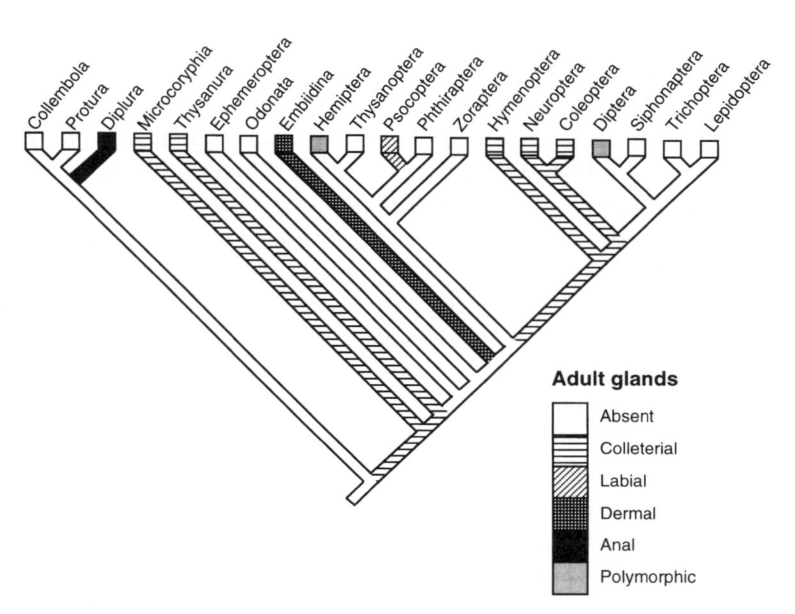

Fig. 8.6. The phylogeny of silk produced by hexapod adults. Silk production among adult hexapods is less common than among larvae. (Adapted from Craig 1997.)

evolved to create protective galleries. In the holometabolous orders various glands evolved in larvae chiefly to create pupal cocoons, although in bees and wasps they may secondarily strengthen the whole nest, while in the ant genera *Oecophlylla* and *Dendromyrmex* the larval silk is used directly in nest construction. In Siphonaptera, Trichoptera, and Lepidoptera it is paired labial glands that produce silk, generally for pupal cocoons but also for other structures, such as portable protective cases, prey capture devices, or communal silk tents.

The pattern of evolution of silk has evidently been complex. It is a dominant feature of the spiders and is a major aspect of the biology of certain insect orders, yet its production has evidently been repeatedly lost during evolution (Craig 1997), possibly due to the costs of silk production. However, one practical consequence of the repeated evolution of the innovation of silk in different biological context is the opportunity to look for evidence of it as a key adaptation in any of its incarnations. One kind of evidence is that species with webs have shown a stronger tendency to speciate than near relatives without webs. The spider genus *Tetragnatha* provides such evidence.

Tetragnatha is an interesting genus in the Araneoidea; while some of its species construct a web of orb design with sticky droplet capture principle, others (characterised by long leg spines) build no webs at all. Both types of species occur on Hawaii. Gillespie and Croom (1995) compared the pattern of specialisation in Hawaii between the spiny-legged and web spinning clades, using morphological and mitochondrial DNA characters to reconstruct the phylogenies. Their prediction was that habitat specialists should be able to diverge within an island through the action of disruptive selection, while for generalists, isolation on separate islands would be necessary. They found that in the spiny-leg clade, regarded as a habitat generalist, the most closely related species was never found on the same island and was likely to be found on a widely separated island in the group, strongly suggesting allopatric speciation occurring between islands only.

Species of the web spinning clade have not been formally described, so species were provisionally assigned to populations that showed distinct sets of unique characters. The differentiation of these species could be recognised between volcanoes on the same island, and possibly even on a single volcano. Gillespie and Croom (1995) attribute this ability to speciate within a much shorter geographical separation than the spiny-leg clade, at least in part, to habitat specialisation associated with web use.

Additional evidence of the influence of web building on spider evolution comes from an examination of the consequences of innovations in the type of silk produced. The obvious target for study is the innovation of the adhesive capture thread in spiders, which was the replacement for the cribellate capture thread.

In this context, the key adaptation hypothesis predicts that adaptive radiation should follow shortly upon the emergence of the adhesive capture thread. Bond and Opell (1998) statistically demonstrate that the sticky thread producing clade Araneoidea is significantly more species diverse than the cribellar silk producing Deinopoidea, and that this diversification process arose at the point of origin of the araneoid group. It is therefore reasonable to conclude in this case that the new capture principle did act as a key innovation. However, the way in which it acted is less clear since, the change in the nature of the adhesive principle was also associated at the same time with a lowering of the UV reflectance of the thread and a change in web orientation from horizontal to vertical. Either way, it involves a physiological change in technology.

Bond and Opell (1998) believe that the association of these events with the onset of species radiation indicates a rapid evolutionary response to newly available resources and consequent selection pressures. But can such a complex of complimentary adaptations be properly called a 'key' innovation? The authors believe that they can, arguing that significant new capabilities that lead to speciation are important in evolution even though their causes may be the outcome of the coincidence of a complex of character changes.

Changes in web design may also have contributed to habitat diversification and speciation in spiders; the two have certainly been associated with one another. The orb web design is a feature of spiders that vary considerably in size, but across that size range they differ in their design and mode of operation. Large orb web builders build webs with threads under high tension in order to catch large, fast moving prey. On the other hand the small-bodied species build orbs with threads under low tension designed to catch slow flying or light prey (Craig 1987*a*) (Section 5.4.4). There has been an evolutionary trend in the Araneoidea towards smaller size, and it is the smaller sizes that have shown a departure from the orb design to produce a variety of reduced or modified designs for use in a variety of habitats. In the view of Craig (1987*a*,*b*), whereas an orb web design is necessary for the structural efficiency of the high tension webs, it is not for the low tension webs. This has provided the opportunity for design innovations to operate in new habitats, without the necessity of corresponding innovations in the web material. An example is the webs of *Wendilgarda* species, which take the form of species-typical arrays of vertical threads hanging down into the water which bear sticky threads at their lower end just above the water surface. These are webs specialised in the capture of water striders (Gerridae) or similar prey (Eberhard 2000). In this instance therefore it appears to be design innovation that has been the principal route of evolutionary change.

This raises the question of whether the spider web building result can be generalised to all other kinds of builders, or only more narrowly to builders of prey capture devices. It could be argued that the architecture designed to capture prey is likely to lead more directly to intraspecific competition, specialisation, and then speciation than architecture designed for personal protection (i.e. that home building species will tend to speciate less rapidly than trap builders). Home building, by reducing extremes of risk, makes a wider range of habitats tolerable but leads only indirectly to intraspecific competition.

A model group to test whether silk has been a stimulus to speciation in trap builders and/or home builders is the Trichoptera. If architecture for prey capture is a stronger speciating force than that for protection, we would predict that net spinning caddis larvae have shown a greater tendency to speciate, or a greater ability to speciate over shorter distances, than have the case builders. Unfortunately the picture is still incomplete, although net spinning appears to have contributed to speciation in that group.

Plague (1999) consider the process that led to speciation in net spinning caddis flies. These nets depend for their effectiveness upon the filtration of moving water, but are found at one extreme in the fast flowing water of small streams, at the other in broader rivers where the current is very slow. Net spinners are claimed to have arisen initially in more rapid flowing water later occupying slower rivers, evolving finer nets and speciating in the process. Plague (1999) offer three hypotheses to explain the dispersal and speciation of net spinning caddis into slower water, greater food concentration, avoidance of predators, or competitive displacement. Of the three, he favour the third, enhanced intraspecific competition.

He suggests that this could have led to speciation through a process of temporal reproductive isolation. This would occur where less competitive individuals within the population were forced into the slower flowing water, which is either less nutrient rich or for which their mesh size is too large to be efficient. Either way, they would experience a slower developmental rate compared with individuals in the faster flowing water, leading to separation of emergence times between the two, hence reduced gene flow, possibly leading to speciation of a population of increasingly specialised slow water net spinners.

Consideration has not been given to the possible role of larval cases in the speciation of case building caddis. Predation pressure seems to have been influential on case design (Section 6.2.2), but predation threats in any one habitat are likely to be diverse, quite apart from other possible roles for the cases. The prediction of a lower tendency to speciate in case building than net spinning caddis does seem worth testing.

The possible role of silk as a key innovation extends beyond its effect on silk producers to secondary silk users, of which a particularly important group is the birds. Here is another possible source of evidence that silk has contributed towards speciation among home builders. Collias (1997) takes the view that nest building in birds, together with small body size and flight could have made an important contribution to the radiation seen particularly in the passerines. An especially important material in the construction of nests by small birds is arthropod silk (Hansell 2000).

The importance of silk in bird nest construction has been emphasised by various authors (Collias and Collias 1984; Hansell 1984). Hansell (2000) located arthropod silk as a structural material in species belonging to 56% of passerine bird families, and in the hummingbirds (Trochilidae), although its usefulness is confined to small birds with a threshold-like effect which apparently excludes species with nests weighing over about 30 g. The innovation of a Velcro fabric of silk in combination with plant materials allows easy nest construction (Section 3.3.1), however it is the role of silk in the attachment of the nest that seems to offer the greatest potential as a key innovation by extending the range of possible nest sites and therefore nesting habitats. Nests of the goldcrest *Regulus regulus* are hung in a framework of silk strands lashed between fine twigs, those of the long-tailed hermit hummingbird *Phaethornis superciliosus* are bound onto the tip of a leaf with silk, and that of the origma *Origma solitaria* hung by silk strands from nodular mineral encrustations on the roof of caves (Hansell 2000). It would be worth while examining more systematically the possible role of silk use in the nests of small birds as a key adaptation.

8.3 Does building facilitate social evolution?

I proposed (Hansell 1987*a*) that nest building could act to facilitate the evolution of eusociality in the hymenoptera, because nest building produces a more regulated and therefore more K-selecting environment. I further predicted (Hansell 1993, 1996*b*) that innovations in building behaviour, by bringing about a significant advance in the ability to control the environment, could lead to an extension of habitat range for that taxon. This prediction was not specifically linked to one of consequent speciation, however, an innovation giving rise to the invasion of new habitats might subsequently facilitate adaptive radiation.

The ability of builders and burrowers to alter habitats is extensively examined in Chapter 7, but this has important evolutionary consequences too. An organism may alter its habitat in a manner that endures beyond its own lifespan; it may also alter conditions for a wide variety of organisms. However, more specifically it may alter the local environment for later generations of conspecifics, even for its own descendants. In 1993, I predicted that a species in which a builder exerts substantial control over an environment that persists across generations, might be less affected by environmental fluctuations, and therefore less vulnerable to extinction than equivalent non-building species. More recently, other authors (e.g. Odling-Smee *et al.* 1996) have explored the evolutionary consequences of this form of material inheritance. They make some predictions that are not intuitively obvious, and which are examined in more detail at the end of this chapter.

8.3.1 Nests and eusociality in insects and other arthropods

The concept of *inclusive fitness* (Hamilton 1964) provided the first theoretical explanation of how intraspecific cooperation could evolve through an individual sacrificing personal fitness in order to enhance the fitness of a relative. Hamilton's (1964) papers also raised the possibility that the repeated independent evolution of eusociality in the Hymenoptera had been facilitated by their haplodiploid system of inheritance.

The emergence of social groups through evolution involves the balance between the decision of individuals to stay together or to leave the group. The decision of an individual to remain on the natal nest and help rear kin is the so-called *subsocial* route to eusociality in the Hymenoptera and some other insects. However, an alternative evolutionary path to social living, is the so-called *parasocial* one, in which individuals enhance personal fitness through mutual cooperation, subsequently evolving eusociality through the development of a caste system (Wilson 1971).

Evidence is available to confirm that in the Hymenoptera, both routes have been taken. Mutualistic associations can be sustained even when individual fitnesses in the group are very unequal, provided that individuals are better off in the group than out. Such associations are now well known in, for example, social wasps such as *Polistes* (Ito 1993). However coefficients of relatedness in social insect colonies have now been found to be often very much lower than anticipated by Hamilton (1964).

Strassmann *et al*. (1994) found coefficients of relatedness of no more than 0.1, compared with the maximum of 0.75 for full sisters, in colonies of *Liostenogaster flavolineata* (Stenogastrinae), even though the colonies are in many respects eusocial. Arathi and Gadagkar (1998) found that in *Ropalidia marginata* (Polistinae), in which colonies may be founded by associations of relatives or non-relatives, cooperation was as good and brood care as effective in nests founded by two non-nest-mate females, as two nest-mate females. This, they conclude illustrates that kinship alone is insufficient to explain the evolution of hymenopteran eusociality, and consequently ecological factors must play a significant role. The purpose of this section is to determine whether the nest (or other built structure) constitutes one of these ecological factors and, if so, how it might operate to stimulate the evolution of eusociality.

Hansell (1987*a*) and Myles (1988) independently considered the possible effects of nest building insects on their subsequent evolution of social organisation. They pointed out the value of the nest itself as a reason for an individual to stay in the home colony rather than leave to found a new one, the former author largely in the context of the social hymenoptera, the latter for the evolution of social complexity and caste formation in termites.

The nest-related model of Hansell (1987*a*) for the evolution of eusociality supposes that, by reducing environmental variation for nest occupants, the population increases such that it is regularly near the carrying capacity of the environment. This leads to increased difficulties for nest leavers in founding new colonies, and so an increased tendency to stay in the natal colony. An alternative nest-related hypothesis has been advanced for the evolution of eusociality. Here the reason for an individual remaining in or joining a group is seen as the ability to share the benefits of, or possibly inherit, the existing nest compared with paying the cost of creating a new one (Myles 1988).

These two hypotheses, (*habitat saturation / lack of opportunity*, and *cost of nest/ foundation*) are in principle distinguishable experimentally, although in practice there may be some uncertainty in the definition or in the demonstration. Availability of nest sites is an issue of habitat saturation that can be tested by observing the tendency of individuals to leave the home colony when additional nest sites are made available. However, the tendency of individuals of a burrowing species to found new colonies in the wet season when burrowing is easier, might be a problem of seasonal habitat saturation or costs of nest foundation. Food availability might also be a factor that influences independent nest foundation, and could be tested by food provisioning; however, it is more difficult to be sure that the food shortage is a consequence of high intraspecific competition resulting from a high steady population sustained by the protection provided by their nests, or to some other cause.

In practice these fundamental distinctions are often not made; instead various ecological factors that relate to the nest, are proposed as possible influences on the benefit to an individual of staying or leaving. An articulation of this can be abbreviated as the *expansible nest site* hypothesis (Alexander *et al.* 1991), although the hypothesis is in fact a composite, which predicts that the evolution of eusociality will be facilitated in circumstances where: development to adulthood is slow, where the nest site is defensible, long lasting, expansible, and nest location food rich.

Each of these variables is worthy of independent investigation, but an interesting nest-related one is the scope for expanding the nest. Although in Hansell (1987*a*) I predicted that nest building would tend to facilitate the evolution of social living, I also argued that building a nest could constrain social evolution in a situation where inadequate building technology prevented enlargement of the nest. This explanation was invoked specifically to account for the colony sizes of 10 or less, found in the primitively eusocial vespid wasps, the Stenogastrinae, compared with the colonies of thousands of individuals found among the Polistinae and Vespinae.

In some stenogastrine species, the cells of the brood comb are made predominantly of mineral soil particles (e.g. *Stenogaster*: Hansell and Turillazzi 1991) or varying proportions of mineral and organic particles, depending upon species (e.g. *Liostenogaster*: Turillazzi 1999). In some other stenogastrine species, the cell walls are composed of a kind of paper, consisting largely of plant fragments, much of it rotted woody material. These nests are fragile and crumbly as well as being thicker and heavier than the much stronger paper found in the walls of brood cells of comparable sized polistine species. This had earlier led Hansell (1987*a*) to hypothesise that poor quality nest paper, unsuited to building large nests, might have acted as a constraint on the evolution of large colonies in the Stenogastrinae, while no such constraint had acted upon the Polistinae.

This nest material constraint hypothesis to explain the limited colony sizes in the Stenogastrinae has since been challenged. Turillazzi (1989) explains that the lack of large colonies in Stenogastrinae in terms of a complex of factors, including the absence of adequate social regulatory mechanisms or nest defence behaviour. Wenzel (1991) also points out that predation pressure could act to limit colony size, while the fragile nest material in the subfamily appears to be an apomorphic rather than a plesiomorphic condition.

Severe predator pressure, physical conditions, and inter or intraspecific competition have been widespread influences on the evolution of societies through extended parental care (Clutton-Brock 1991). This, through help provided by offspring to assist the parents in further reproductive efforts, typifies the subsocial route, which in some insects has led to eusociality. Building behaviour can be one aspect of parental care and is in some taxa, such as the birds, for example, almost the sole reason for building.

In the amphipod crustacean *Leptocheirus pinguis*, larvae which initially hatch in the maternal burrow, may remain and extend the system each with its own side branch. This tendency to stay may be due to the risks of predation or possibily the hazards of substrate disturbance in shallow water (Thiel *et al.* 1997). Similar behaviour by the young of terrestrial burrowing organisms have, however, been attributed to the costs of burrowing. In the scorpion *Heterometrus fulvipes*, subsocial communities develop during the dry season as offspring remain in, and extend the deep burrow of their mother. Such burrow systems may last at least 2 years, by which time there are multiple entrances and food sharing among the group. Shivashankar (1994) attributes the tendency of young to stay in particular to the cost of digging in hard soil in the dry season. Building costs are invoked by Cohen and Cohen (1976) to explain the cooperative building of grass-lined burrows by groups of 100 or more desert cockroach *Arvenivaga apacha* within the mounds of kangaroo rats, although their function is attributed to protection against cold and water penetration.

In the case of these cockroaches, it is not known whether the groups form through the subsocial or parasocial route. However, multifemale nest foundation in the wet tropics by paper wasp species such as *Ropalidia fasciata*, *Polistes versicolor*, and *Mischocyttarus angulatus* are seen by Ito (1993) as generally mutualistic, brought about by the need for frequent nest rebuilding after predation or tropical storms, a cost that can be shared if females live in groups.

Habitat saturation has been invoked as a stimulus to eusociality where there is evidence of a shortage of specialised nest sites. Duffy (1996), for example, has attributed the occurrence of a fully eusocial shrimp *Synalpheus gregalis* in colonies of over 300, with a single reproductive female and degrees of relatedness equivalent to full sibs, to compete for the scarce sponges in which they live. This, it is argued, has led juvenile shrimps to stay and defend rather than leave and found. This is supported by experimental evidence of their ability to discriminate between nest mates and intruders, and respond aggressively to the latter (Duffy *et al.* 2002).

This sponge-dwelling shrimp does, however, exhibit all those characteristics predicted by the expansible nest site hypothesis (Alexander *et al.* 1991) to foster eusociality.

The habitat saturation hypothesis has been empirically tested by the supply of additional nest sites. The allodapine bee *Exoneura bicolour*, nests in the pithy interiors of dead plant stems. Females are quite long-lived and all are capable of egg-laying; they are however, only facultatively social and so may nest alone. Bull and Schwarz (1996), in an experiment where nest site availability was manipulated, failed to show that, where opportunities for independent foundation occurred, cooperative nesting was reduced. He concluded that some other mutual advantage was derived from cooperation, possibly protection of the brood from predators. The provision of vacant nests to the facultatively eusocial stenogastrine wasp *L. flavolineata* was, found to provoke only a small proportion of helper individuals on nearby nests to become foundresses (Field *et al.* 1998).

Similar experiments on ants have produced inconsistent results. Herbers and Banschbach (1999) found that the addition of hollow twigs to the leaf litter, to provide nest sites similar to those used by the ant *Myrmica punctiventvis* had no effect on the subdivision of existing colonies, although such an effect was reported by Herbers (1986) for the ant *Leptothorax longispinosus*.

The nest is a marked and even spectacular feature of many eusocial arthropods. There are reasons for believing that the eusociality in these organisms may be promoted by the nest, but the strength of this influence is still unclear. Furthermore, evidence to test how such an influence might be exerted is fragmentary and inconclusive.

8.3.2 Rodent burrows and the evolution of eusociality

Parental care and social living are features of a wide range of rodents, however, eusociality is only shown by two genera of fossorial rodents, both of the family Bathyergidae: *Heterocephalus* and *Cryptomys*. In *Heterocephalus glaber* and *Cryptomys damaensis*, the social group contains only one reproductive female and one reproductively effective male. The remainder of individuals are classified as frequent or occasional workers. These species occur in habitats of low, unpredictable rainfall, feeding on patchily distributed subterranean plant tubers. Jarvis *et al.* (1994) explained the cooperation in these species as the outcome of a need to dig through hard soil to locate the widely and patchily distributed food plants. This promotes staying within the group which in turn, enhances relatedness, in turn supporting cooperation. This has been termed the *aridity food-distribution hypothesis*.

Two other *Cryptomys* species are found in areas of higher rainfall and evenly dispersed food, and are not eusocial, so conforming to the predictions of this hypothesis. However, another species of this genus, *Cryptomys mechowi* lives in a seasonally wet habitat, yet exhibits full eusociality. Wallace and Bennett (1998) claim that this contradicts the aridity of food-distribution hypothesis, and instead propose *dry season soil hardness* as the mechanism that discourages leaving in the dry season and favours extensive cooperative burrowing in the wet season to locate the dispersed food plants.

Both of these hypotheses implicate burrowing activity as a key selection pressure in the evolution of eusociality. Burda *et al.* (2000) however challenge the view that social living in the Bathyergidae is an adaptation to arid habitats at all. They point out that the other three genera of the family are also found in arid areas yet are all solitary. This, in their view, indicates that social life in the Bathyergidae is plesiomorphic and that the fossorial way of life could in fact select for a solitary existence. This means that the explanation for eusociality in *Heterocephalus* and some *Cryptomys* is, under this hypothesis, one of phylogenetic constraint; a failure for whatever reason to respond to selection pressures for a more solitary existence.

8.3.3 Spider silk spinning and the evolution of social living

Out of more than 30,000 species of spiders only a few dozen of them are known to live a predominantly social existence. The extent of their cooperation extends across prey capture, feeding, brood care, and building. In spite of their relative scarcity, social spiders occur in at least 7 different families, with indications that the habit has arisen 12 or more times independently (Avilés 1995).

The nature of the social organisation varies somewhat between species, but none has evolved full eusociality. The reasons for this are unclear, but the speculative and ingenious explanation of Craig (2003) is one of physiological constraint: spiders lack the flexible developmental pathways necessary for eusociality to evolve. This results from the hormonal regulation of silk production, which needs to continue uninterrupted during the growth phase of the life cycle, limiting evolutionary innovation in development generally.

Archaearanea wau (Theridiidae) exemplifies those spider species that exhibit a permanent cooperative social existence. It occurs in colonies of several hundred that build and maintain massive webs. Leaves caught in the tangle of threads can extend a few metres above the sheet web, and act as retreats for the spiders. Lubin (1995) found that although individuals take on more tasks as they grow, they never show job switching as typified in the age-related job specialisation of eusocial insects, and all adult females can reproduce. Another feature that distinguishes these social spiders from social insects is a strong tendency for inbreeding, leading to high degrees of within-colony relatedness and female-biased sex ratios.

A variety of ecological explanations have been offered for the evolution of social living in spiders. These include advantages not strictly related to building behaviour, such as collective brood care. The question for this chapter is whether built structures in particular have made their contribution. In this context it is important to note that the majority of cooperative spiders build not only prey capture devices but also refuges or homes, giving two different consequences of building either of which might favour evolution towards greater social cooperation.

Various advantages of group hunting, have been invoked for the evolution of sociality in spiders, in particular, the ability to capture large prey, individual economy in web construction, and more effective prey attraction. Observations of cooperative capture and handling of large prey by social spiders have invited comparisons with social vertebrate predators that benefit from hunting in this way. This hypothesis does not in itself require that the evolution of spider cooperative hunting has anything to do with web construction, although some studies do make this connection.

Anelosimus eximius (Theridiidae) is a species which lives in large groups of thousands of individuals of varied ages. The web consists of a sheet of silk with a network of overhead knock-down threads. Cooperation is shown in both prey capture and parental care. Attacks on prey are carried out simultaneously by several individuals, enabling insect prey much larger than an individual spider to be caught, but adult females also tend each other's egg sacs and occasionally regurgitate food to offspring that are not their own (Christenson 1984).

Cooperative handling of prey may have been a selection pressure for the evolution of social living in *Aebutina binotata* (Dictynidae). It lives in colonies with a median number of about 40, but with a capture web of limited size. The

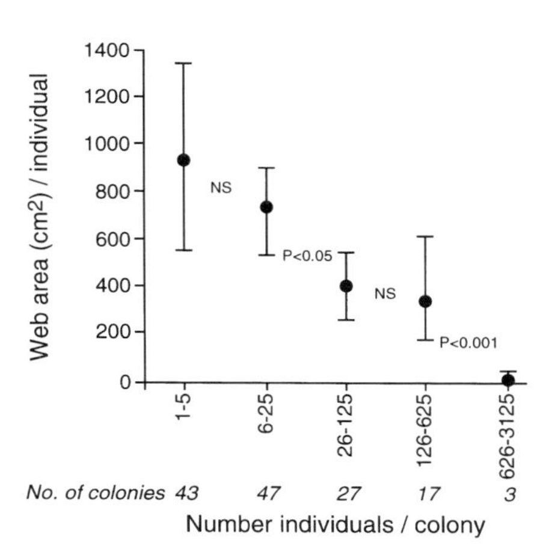

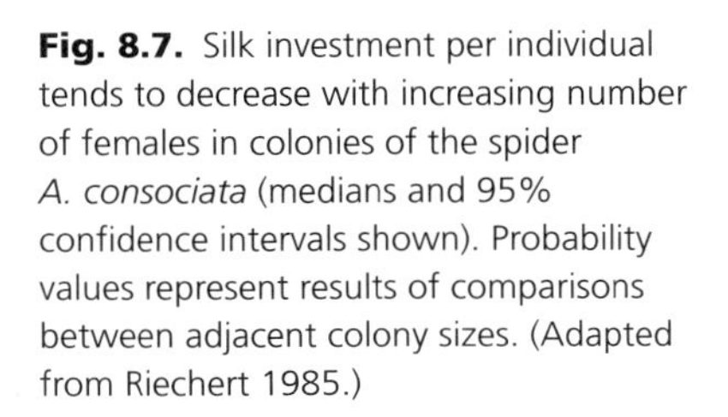
Fig. 8.7. Silk investment per individual tends to decrease with increasing number of females in colonies of the spider *A. consociata* (medians and 95% confidence intervals shown). Probability values represent results of comparisons between adjacent colony sizes. (Adapted from Riechert 1985.)

web is made of cribellate silk and does no more than cover the upper surface and screen the lower surface of the leaf under which the spiders all shelter. Cooperation occurs in web construction, although it is largely undertaken by adult females, and there is also cooperation in brood care and prey capture. Avilés (1993) however, asserts that two particular collaborative activities are of importance in the evolution of their social system, one of which concerns prey capture. The first is the difficulty in moving large prey captured on the upper leaf surface into the sheltered space below it; the second is a nomadic lifestyle that necessitates repeated reconstruction of the web, a cost of reconstruction explanation.

The testing between the two hypotheses of cooperative prey capture and cost of web construction, Riechert (1985) in a study of *Agelana consociata*, dismisses the cooperative prey capture argument on the grounds that the captured prey in this species are all small. In addition she shows that as group size goes up, so individual prey capture rates and female egg production go down, confirming that frequency of prey capture is in this instance not the selective advantage. On the other hand, as group size goes up, so web area per individual (i.e. individual costs of web production) go down (Fig. 8.7). This, she reasons, amounts to an important saving in an environment where heavy rains during 6 months of the year result in web rebuilding 2 out of every 5 days.

The contention of the prey attraction hypothesis is that, where the web is used to attract prey, a collective effort will more than compensate for the shared prey. Reflection of UV wavelengths from the web is stronger in families of cribellate spiders than in ecribellae species. The UV reflectance of these webs is apparently able to attract insect prey (Section 1.4). If web aggregation attracted additional prey, it could have led to communal living. In support of

Fig. 8.8. Communal feeding in the social spider *Tapinillus* sp. (Photo Leticia Avilés 1994.)

this hypothesis, the majority of social spiders are found among the cribellate families, although *Anelosimus* (Theridiidae), and *Agelana* (Agelanidae) are exceptions to this.

These examples are supportive of an association between social living spider species and web building, but insufficient to confirm a causal link between the latter and the former. However a particularly suggestive example of some link between the two comes from a species of *Tapinillus*. It is apparently unique among the family of lynx spiders (Oxypodidae) in two respects: first it builds a web and second it lives in colonies of several dozen individuals with cooperative prey capture (Fig. 8.8) (Avilés 1994). There is, however, a challenge to this web-related hypothesis in the form of two genera where species do not spin capture webs yet live socially (Avilés 1995); they are *Diaena* (Thomisidae) and *Delena* (Sparassidae).

Diaena species cooperate in silking leaves together to make an elaborate communal nest, on the surface of which they forage individually (Avilés 1995), while *Delena cancerides*, lives in groups under the bark of dead trees. Prey capture in the field has not been observed in this species but, in a laboratory situation, prey capture is cooperative but does not involve the use of a web (Rowell and Avilés 1995). These examples are indeed evidence counter to that of web involvement in the evolution of spider sociality, but what is being examined here is not whether web construction in particular had been a predisposing factor in social evolution, but whether *any* building behaviour has had that influence. This, in the case of all social spiders, includes not just webs but also homes. Home-related hypotheses for the evolution of eusociality in spiders include, as with web building, the cost of building, but also the greater effectiveness of a collective over an individual shelter.

Whitehouse and Jackson (1998) identify both shortage of suitable websites and collective extended maternal care as possible advantages of social life in *Argyrodes flavipes*. It is an unusual member of its genus, for not being klepto-parasitic and being permanently social, living in small colonies of around four adults in which all females may breed. However, no experiment appears to have been conducted on this or other social spider species to determine if the provision of web or nest sites promotes emigration from the home colony.

Protection from adverse weather conditions is one of a number of possible advantages cited by Downes (1994) for cooperation in colonies of *Phryganoporus candidus* (Desidae). These are initially founded by a single female, but grow with the addition of young and possibly by merging of colonies to reach as much as 200 or more individuals. They cooperatively construct a chambered silk tunnel covered by cribellate capture web. Avilés and Gelsey (1998) specul-ate that the transition from periodic to perpetual colony life in cooperative spiders could have been facilitated by the continued occupation of established or proven nest sites. In support of this nest-security hypothesis, they cite evid-ence from *A. eximius* that, although the probability of female reproduction decreases with colony size, offspring survival increases.

The causes for the evolution of social living spiders may not be simple or the same across all species. However, silk structures, either in the form of webs or nests, have the potential to shift the balance of individual decision-making towards staying in rather than leaving the colony in a variety of ways, and seem certain to have exercised that influence in at least some of the spider lineages achieving permanent social life.

These studies, not only on social spiders, but also on eusocial insects and fossorial rodents, emphasise the need for more experimental evidence on the role of nest construction and other building behaviour on the evolution of social living. Nonetheless they also encourage the belief that, since the ecological consequences of building are varied, there may be no simple predominant way in which building can facilitate the evolution of greater sociality.

8.4 Niche construction, ecological and cultural inheritance

8.4.1 Ecological inheritance

A point that has been repeatedly raised in this book is that by altering its external environment, an organism may change a variety of selection pressures acting upon it. Caddis larvae with longer cases may not simply be better protected from a fish predator but will have altered their mobility and respiratory environment. Selection for an improved extended phenotype (Dawkins 1982) is therefore likely, as with any other adaptation, to alter selection pressures acting at a variety of gene loci. This reasoning also needs to be extended beyond the builder itself, since the structure might outlive the builder and influence the environment experienced by the next or subsequent generations of the builders' descendants. This is particularly obvious for prominent ecosystem engineers like burrowers (see Sections 7.4 and 7.5). Their offspring may not simply inherit genes for building behaviour but may inherit some remnants or effects of the extended phenotypes of their ancestors. This additional pathway of inheritance has been termed *ecological inheritance* (Laland *et al.* 2000).

The orb web of a spider is an ephemeral structure, and has a life expectancy that may be only of a few days or even hours. Nevertheless, although this structure is not passed to the builder's descendants, any young spider is likely to find itself foraging in a habitat already populated by other spiders and their webs, creating circumstances that could exert selection pressures upon it. Burrows or constructed houses are generally more durable, and therefore provide abundant opportunities for ecological inheritance. This was the basis for the hypotheses examined in Section 8.3 of the influence of building in the evolution of social behaviour. These hypotheses compare the fitness consequences of inheriting the home nest with those of leaving. An additional assumption of these is that resource availability beyond the home nest may itself have been modified by the presence of other conspecific nest builders.

Laland *et al.* (1996) and Odling-Smee *et al.* (2003) used a deliberately simplified population genetic model to assess the possible evolutionary consequences of an environmental inheritance pathway brought about by niche construction. This can be summarised as occurring when an organism modifies its relationship with the environment either by perturbing it at its current location, or by relocating to expose itself to different factors (Odling-Smee *et al.* 2003). This covers a very broad range of ecological influences, but within its boundaries lie those exerted by the builders and burrowers.

This model incorporates two loci and imagines that the capacity of a population for niche construction is influenced by the frequency of alleles at a locus **E** (influences the organisms effect in changing the *environment*). The model defines the amount of a *resource* **R** in the environment to be dependent on niche construction by past and present generations of the organism. **R** is a function of the frequency of alleles at the **E** locus over a number of recent generations. It could represent one of a variety of effects, for example, the availability of food or presence of a predator. The amount of this resource in the environment in turn determines the contribution made to the organism's

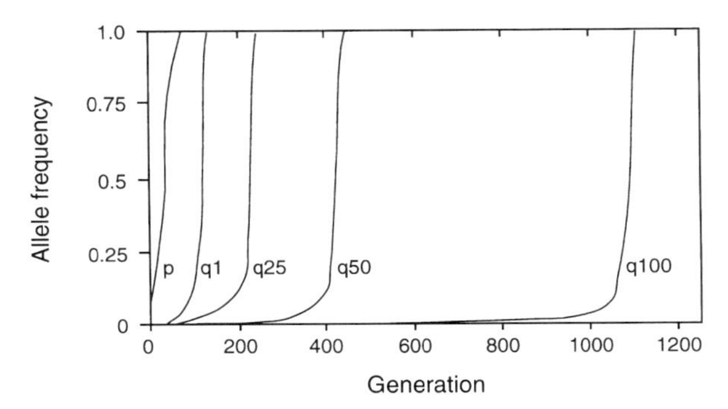

Fig. 8.9. The consequence of the spread of allele *E*, on selection favouring allele *A*, depends upon the number of generations during which *E* needs to act in order to alter the environment (**R**) sufficiently to change selection pressure acting at the **A** locus. The figure plots the frequencies of *E*(p) and *A*(q) against time for 1, 25, 50, and 100 generations. (Adapted from Laland *et al.* 1996.)

fitness by a second locus, **A**. Consequently the influence on the habitat exerted by **E** alleles, creates feedback onto locus **A**. The population is assumed to be isolated, randomly mating and diploid, with individuals defined by two diallelic loci. These are at **E** (alleles *E* and *e*) and at **A** (alleles *A* and *a*).

This model makes some significant predictions - one of particular interest being the time lag with which changes in the frequency of *E* would impact upon *A*. In the case of the spider's web, where there is no environmental inheritance, the effect is essentially immediate and does not extend beyond it. However, the greater the number of previous generations that are necessary to alter **R** through environmental inheritance, the greater the time lag in the effect on *A* (Fig. 8.9), creating *evolutionary inertia*.

This evolutionary inertia is due to two factors. The first is that, when selection favouring allele *E* begins, it takes several generations for the resource to accumulate (or indeed deteriorate) sufficiently to exert selective influence upon locus **A**. The second is that allele *A*, although it will eventually be favoured when the altered level of resource begins to have an effect, it will initially continue to decline. As a result, its frequency will take some generations under the new selection regime simply to return to its original frequency, indeed it could be lost from the population altogether before selection favouring it takes effect.

Less obviously, the feedback time lag also generates evolutionary momentum. This is seen when selection at the **E** locus stops or reverses. This does not prevent the resource from continuing to accumulate for several generations, with the result that evolutionary change at the **A** locus will continue in the original direction.

Variants of this model make other important predictions (Odling-Smee *et al.* 2003), for example, that, depending upon the interaction between the effects of niche construction and sources of external selection, niche construction will be powerful enough to create new evolutionary outcomes generating and sustaining new polymorphisms or eliminating ones that are otherwise stable. The model can also make predictions about co-evolutionary relationships, since selection at the **A** locus of one organism may be subject to influences originating from the **E** locus of another (Laland *et al.* 1999; Odling-Smee *et al.* 2003).

8.4.2 Cultural inheritance

The lack of developmental studies on animal building has already been remarked upon in this book. This leaves us with little information on the role of novel learned building behaviour let alone its social transmission. Laland *et al.* (1996, 1999) point out that this too can be a route for changes in selection pressure on locus **A**, through an influence on changes of the distribution of resource **R**; the opening of foil lids of milk bottles by British tits (Paridae) could, for example, have brought to bear new selection pressures on their digestive system (Odling-Smee *et al.* 1996).

Learning may come about through individual trial and error or particularly in higher vertebrates through social learning. In non-humans transmission in this case is generally horizontal (Laland *et al.* 2000) (i.e. within the same generation) however, it is in some cases vertical (i.e. across generations), as in the case of chimpanzees learning to crack nuts with a hammer and anvil (Matsuzawa 1994). This gives three routes by which construction behaviour of one generation may influence the next: *genetic*, social learning (*cultural transmission*), and ecological inheritance (*niche construction*) (Laland *et al.* 2000) (Figs 8.10 and 8.11).

Human niche construction is now the most pervasive and influential on the planet (Vitousek *et al.* 1997). It is also overwhelmingly cultural in its transmission, as is evidenced by the elaborate social and institutional mechanisms for the transmission of this capability. Does this allow us to make any particular predictions about how and at what pace we have experienced, and will exhibit, evolutionary change?

Laland *et al.* (2000) predict that the effects of niche construction generally have been to mitigate selection pressures, and thus allow populations to maintain greater levels of genetic variation at those loci that effectively became shielded from selection. It should therefore be a feature of all niche constructors, but of humans in particular, that they are very resistant to genetic evolution in response to changing environments, but retain the capacity for dramatic evolutionary change. This should for example be observed in

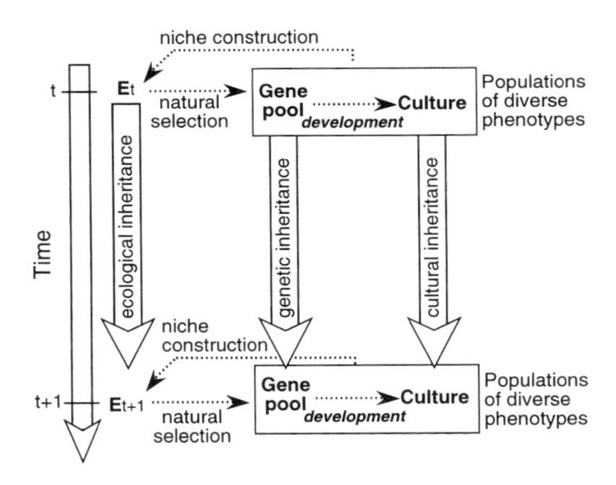

Fig. 8.10. This representation of the causal relationship between biological evolution and cultural change illustrates three possible routes through which one generation can inherit influences from the previous one: ecological, genetic, and cultural. Through ecological inheritance, niche construction behaviour of any generation is able to interact with a habitat already modified by the equivalent activities of its ancestors. (Adapted from Laland *et al.* 2000.)

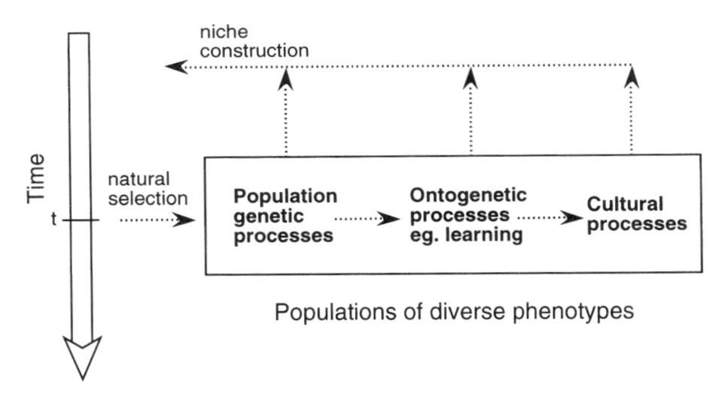

Fig. 8.11. A more detailed representation of the box *'populations of diverse phenotypes'* in Fig. 8.10. It shows how adaptation under natural selection in populations of complex organisms, such as humans, acts upon genetic, ontogenetic, and cultural processes, all of which may bring about a shift in niche construction. (Adapted from Laland *et al.* 2000.)

the extreme case that selection pressures led to the abandonment of niche construction behaviour. If this occurred to a niche constructor of 'keystone' significance, it could result in, as Laland *et al.* (2000) put it 'a cascade of evolutionary events that realign ecosystems'.

8.4.3 Human evolution and niche construction

The heavy reliance in human development of social learning across generations is in marked contrast with that of even our nearest living relatives, raising the question of the nature of the selection pressures enhancing this channel of inheritance. Laland *et al.* (2000) cite current theory as being in agreement that genetic transmission is favoured when environmental change is slow, giving genetic innovation time to respond to it. By contrast individual learning is favoured in circumstances of rapid environmental change because it can constantly update behavioural responses. Transmission of learned behaviour through the vertical path (across generations) is however, not favoured during time of rapid environmental change because what applied to the previous generation may no longer be applicable, but is best suited to circumstances of intermediate rates of change. This poses a problem in explaining why, in the evolution of the hominids, there appears to have been a trend towards increasing dependence on vertical transmission while initially rates of environmental change were unaltered.

Niche construction could provide at least a partial explanation for this, since our ancestors by their increasing impact upon the world could have made it less unpredictable. Vertical transmission of learned information then closes a positive feedback loop in which transmitted information assists in the creation of a more predictable world (not just by environmental control exerted through niche construction, but also of acquired knowledge) (Laland *et al.* 2000). This is a cycle that, through cultural tradition, rewards anticipation and advanced planning, but if not balanced by a degree of innovation and its cultural transmission, could lead to increased cultural conservatism.

Cultural traits have been found to exist not only in chimpanzees but also in other great apes and some monkeys. This is using a fairly strict definition of cultural transmission to include the elements of: innovation, transmission from the innovator to another, standardisation of the pattern, and its persistence in the absence of the innovator. This makes it likely that primate culture preceded the separation of the hominids from their common ancestry with the great apes (McGrew 1998). The concept of innovation as an individual act accords with our own notions of 'authorship' and 'intellectual property rights'. Reynolds (1993) wishes to dispel this; he stresses the social context of most human craft-based work, and the almost certain social context of related niche construction and tool making behaviours in our hominid ancestors.

Fruth and Hohmann (1996) consider the obverse of this: How does building behaviour create a social environment that stimulates other cultural innovations? Their example is the possible consequence of nest building by chimpanzees. All the four great ape species also build nests. In chimpanzees, these are constructed either as day nests or night nests. They hypothesise that nests became advantageous to these species when their body sizes became too large for them to rest comfortably in the trees. Initially the function of the nest could have been to protect valuable feeding sites, but it evolved to provide protection from predators at night and the practical but beneficial experience of a good night's sleep. More important from the point of view of human evolution however, is their hypothesis for the role of nests in the evolution of hominid intelligence. This extrapolation derives from the social night nesting exhibited by chimpanzees. The social environment that this creates not only encourages social construction behaviour but the possibility for the exchange of other kinds of information—a possible context for the development of language (Fruth and Hohmann 1996).

It remains to assess what role inherited niche construction played in human evolution. The speculation of Laland *et al.* (2000), that vertical cultural transmission may have increased in its effectiveness as our ancestors made the environment a more predictable place, only concerns itself with ecological inheritance of a very general kind, habitat modification by fire, for example. But what role if any, was played by the passing of individual possessions across generations?

Before the abandonment of a nomadic way of life within the last few tens of thousands of years, the scope for the inheritance of possessions must have been limited. A possible class of inherited objects would be tools, but the life expectancy of even the stone tools of our recent human ancestors is unlikely to have given them sufficient utilitarian value to merit transfer across generations, but as objects of sexual attraction the situation might have been different. Of course it is possible to substitute in Figs 8.10 and 8.11 the words sexual selection for natural selection. Bowerbird displays represent transmission of sexually selected traits through both genetic and cultural inheritance; the question here is what role has the inheritance of sexually selected objects themselves had on human evolution?

It has been noted that some early human stone implements show evidence of skill in manufacture but not of effectiveness as tools. One hypothesis that could account for this is that they are display objects that may assist in attracting mates or enhancing social position (Kuhn and Stiner 1998). If this was the case, could they have significantly influenced the attractiveness of someone inheriting them or gaining possession of them? The answer may be very little; on the other hand, as I have indicated in Section 5.5, our enthusiasm for the arts does raise legitimate biological questions. A major component of artistic endeavour is the fabrication of *objects* of desire. As 'heirlooms' these have more than economic significance. How ancient is that significance and was sexual selection the motor?

Bibliography

Acuña, J.L. (2001). Pelagic tunicates: why gelatinous? *American Naturalist*, **158**, 100–107.

Aichorn, A. (1989). Nestbautechnik des Schneefinken (*Montifringilla n. nivalis* L.). *Egretta*, **32**, 49–63.

Aitken, K.E.H., Wiebe, K.L., and Martin, K. (2002). Nest site reuse patterns for a cavity-nesting bird community in interior British Columbia. *Auk*, **119**, 391–402.

Alcock, J. (1972). The evolution and use of tools by feeding animals. *Evolution*, **26**, 464–473.

Alexander, R.D., Noonan, K.M., and Crespi, B.J. (1991). The evolution of eusociality. In *The biology of the naked mole-rat* (eds. P.W. Sherman, J.U.M. Jarvis, and R.D. Alexander), pp. 3–44. Princeton University Press, Princeton, NJ.

Alldredge, A. (1976*a*). Discarded appendicularian houses as sources of food, surface habitats, and particulate organic matter in planktonic environments *Limnology and Oceanography*, **21**, 14–23.

Alldredge, A. (1976*b*). Appendicularians. *Scientific American*, **235** (1), 94–102.

Alstad, D.N. (1987*a*). Particle size, resource concentration, and the distribution of net spinning caddisflies. *Oecologia*, **71**, 525–531.

Alstad, D.N. (1987*b*). A capture-rate model of net spinning caddisfly communities. *Ecologia*, **71**, 532–536.

Anderson, C. and Franks, N.R. (2001). Teams in animal societies. *Behavioural Ecology*, **12**, 534–540.

Andrade, D.V. and Abe, A.S. (1997). Foam nest production in the armoured catfish. *Journal of Fish Biology*, **50**, 665–667.

Arathi, H.S. and Gadagkar, R. (1998). Cooperative nest building and brood care by nestmates and non-nestmates in *Ropalidia marginata*: implications for the evolution of eusociality. *Oecologia*, **117**, 295–299.

Atkinson, R.J.A. (1986). Mud-burrowing megafauna of the Clyde sea area. *Proceedings of the Royal Society of Edinburgh,* **90B**, 351–361.

Atkinson, R.J.A. and Taylor, A.C.T. (1991). Burrows and burrowing behaviour of fish. In *The environmental impact of burrowing animals and animal burrows* (eds. P.S. Meadows and A. Meadows) *Symposium of the Zoological Society of London,* **63**, 133–155. Clarendon Press, Oxford.

Atkinson, R.J.A., Froglia, C., Arneri, E., and Antolini, B. (1998). Observations on the burrows and burrowing behaviour of *Brachynotus gemmellari* and on the burrows of several other species occurring on *Squilla* grounds off Ancona, Central Adriatic. *Scientia Marina,* **62**, 91–100.

Avilés, L. (1993). Newly-discovered sociality in the neotropical spider *Aebutina binotata* (Dictynidae?). *Journal of Arachnology,* **21**, 184–193.

Avilés, L. (1994). Social behaviour in a web-building lynx spider, *Tapinillus sp.* (Araneae: Oxyopidae). *Biological Journal of the Linnean Society,* **51**, 163–176.

Avilés, L. (1995). Causes and consequences of co-operation and permanent-sociality in spiders. In *The Evolution of social behaviour in insects and arachnids* (eds. J.C. Choe and B.J. Crespi), pp. 476–498. Cambridge University Press, Cambridge.

Avilés, L. and Gelsey, G. (1998). Natal dispersal and demography of a subsocial *Anelosimus* species and its implications for the evolution of eusociality in spiders. *Canadian Journal of Zoology,* **76**, 2137–2147.

Balasingh, J., Koilraj, J., and Kunz, T.H. (1995). Tent construction by the short-nosed fruit bat *Cynopterus sphinx* (Chiroptera: Pteropodidae) in Southern India. *Ethology,* **100**, 210–229.

Bailey, W.J. and Roberts, J.D. (1981). The bioacoustics of the burrowing frog *Heleioporus* (Leptodactylidae). *Journal of Natural History,* **15**, 693–702.

Bankover, V., Christov, R., Kujumgiev, A., Marcucci, M.C., and Popov, S. (1995). Chemical composition and antibacterial activity of Brazilian propolis. *Zietschrift fur Naturforschung, C,* **50**, 167–172.

Barber, I., Nairn, D., and Huntingford, F.A. (2001). Nests as ornaments: revealing construction by male sticklebacks. *Behavioural Ecology,* **12**, 390–396.

Bass, M. and Cherrett, J.M. (1994). The role of leaf-cutting ant workers (Hymenoptera: Formicidae) in fungus garden maintenance. *Ecological Entomology,* **19**, 215–220.

Beaver, R.A. (1989). Insect-fungus relationships in the bark and ambrosia beetles. In *Insect-fungus interactions* (eds. N. Wilding, M. Collins, P.M. Hammond, and J.F. Webber), pp. 121–143. Academic Press, London.

Beck, B.B. (1980). *Animal tool behaviour*. Garland STPM, New York.

Belić, M.R., Skarka, V., Deneubourg, J.L., and Lax, M. (1986). Mathematical model of honeycomb construction. *Journal of Mathematical Biology,* **24**, 437–449.

Bierans de Haan, J.A. (1922). Uber den Bauinstinkt Einer Kocherlarve (*Limnophilus marmoratus* Curtis). *Bijdragen tot de dierkunde,* **22**, 321–327.

Biere, J.M. and Uetz, W. (1981). Web orientation in the spider *Micrathena gracicis* (Araneae: Araneidae). *Ecology,* **62**, 336–344.

Biermann, C.J. (1993). *Essentials of pulping and papermaking.* Academic Press, London.

Bink, F.A. (1985). Bouw van het hibernaculum door de Grote ijsvogelvlinder (*Limenitis populi* (Linnaeus)) (Lepidoptera: Nymphalidae). *Entomologische Berichten,* **45**, 115–117.

Bird, F.L. and Poore, G.C.B. (1999). Functional burrow morphology of *Biffarius arenosus* (Decapoda; Callianassidae) from southern Australia. *Marine Biology,* **134**, 77–87.

Bissett, J.L. and Moran, V.C. (1967). Life history and cocoon spinning behaviour of the South African Mantispid (Neuroptera, Mantispidae). *Journal of the Entomological Society of South Africa,* **30**, 82–95.

Blackledge, T.A. and Wenzel, J.W. (2000). The evolution of cryptic spider silk: a behavioural test. *Behavioural Ecology,* **11**, 142–145.

Boesch, C. (1991). Teaching among wild chimpanzees. *Animal Behaviour,* **41**, 530–532.

Boesch, C. (1993). Aspects of transmission of tool-use in wild chimpanzees. In *Tools, language and cognition in human evolution* (eds. K.R. Gibson and T. Ingold), pp. 171–183. Cambridge University Press, Cambridge.

Boggs, D.F., Kilgore, D.L., and Birchard, G.F. (1984). Respiratory physiology of burrowing mammals and birds. *Comparative Biochemistry and Physiology,* **77**, 1–7.

Bonabeau, E., Theraulaz, G., Deneubourg, J.-L., Aron, S., and Camazine, S. (1997). Self-organisation in social insects. *Trends in Ecology and Evolution,* **12**, 188–193.

Bonabeau, E., Dorigo, M., and Theraulaz, G. (1999). *Swarm intelligence.* Oxford University Press, Oxford.

Bond, J.E. and Coyle, F.A. (1995). Observations on the natural history of an *Ummidia* trapdoor spider from Costa Rica (Araneae, Ctenizidae). *Journal of Arachnology,* **23**, 157–164.

Bond, J.E. and Opell, B.D. (1997). The functional significance of medially divided cribella in the spider genus *Mallos* (Araneae, Dictynidae). *Bulletin of the British Arachnological Society,* **10**, 239–241.

Bond, J.E. and Opell, B.D. (1998). Testing adaptive radiation and key innovation hypotheses in spiders. *Evolution,* **52**, 403–414.

Bond, W. (2001). Keystone species—Hunting the snark? *Science,* **292**, 63–64.

Borgia, G. (1985*a*). Bower quality, number of decorations and mating success of male satin bowerbirds (*Ptilonorhynchus violaceus*): an experimental analysis. *Animal Behaviour,* **33**, 266–271.

Borgia, G. (1985*b*). Bower destruction and sexual competition in the satin bowerbird (*Ptilonorhynchus violaceus*). *Behavioural Ecology and Sociobiology,* **18**, 91–100.

Borgia, G. (1993). The cost of display in the nonresource based mating system of the satin bowerbird. *American Naturalist*, **44**, 734–743.

Borgia, G. (1995*a*). Complex male display and female choice in the spotted bowerbird: specialised functions for different bower decorations. *Animal Behaviour,* **49**, 1291–1301

Borgia, G. (1995*b*). Threat reduction as a cause of differences in bower architecture, bower decoration and male display in two closely related bowerbirds, *Chlamydera nuchalis* and *C. maculata*. *Emu*, **95**, 1–12.

Borgia, G. (1997). Comparative behavioural and biochemical studies of bowerbirds and the evolution of bower-building. In *Understanding and protecting our biological resources* (eds. M.L. Reaka-Kudla, D.E. Wilson, and E.O. Wilson), pp. 263–276. Joseph Henry Press, Washington, DC.

Borgia, G. and Gore, M.A. (1986). Feather stealing in the satin bowerbird (*Ptilonorhynchus violaceus*): male competition and the quality of the display. *Animal Behaviour,* **34**, 727–738.

Borgia, G. and Mueller, U. (1992). Bower destruction, decoration stealing and female choice in the spotted bowerbird *Chlamydera maculata*, *Emu*, **92**, 11–18.

Borgia, G. and Presgraves, D.C. (1998). Coevolution of elaborated male display traits in the spotted bowerbird: an experimental test of the threat reduction hypothesis. *Animal Behaviour,* **56**, 1121–1128.

Borgia, G., Pruett-Jones, S., and Pruett-Jones, M. (1985). The evolution of bower-building and the assessment of male quality. *Zeitschrift für Tierpsychologie,* **67**, 225–236.

Borgia, G., Kaatz, I.M., and Condit, R. (1987). Flower choice and bower decoration in the satin bowerbird *Ptilonorhynchus violaceus*: a test of hypotheses for the evolution of male display.*Animal Behaviour,* **35**, 1129–1139.

Brach, V. (1975). The biology of the social spider *Anelosimus eximius*. *Bulletin of the Southern California Academy of Science*, **74**, 37–41.

Brandl, R. and Kaib, M. (1995). Diversity of African termites: an evolutionary perspective. *Mitteilungen der Deutsche Gesellschaft fur Allggemeine und Angewandte Entomologie*, **10**, 653–660.

Brandt, M. and Mahsberg, D. (2002). Bugs with a backpack: the function of nymphal camouflage in the West African assassin bugs *Paredocla* and *Acanthaspis* spp. *Animal Behaviour*, **63**, 277–284.

Broadley, A. and Stringer, A.N. (2001). Prey attraction by larvae of the New Zealand glowworm, *Arachnocampa luminosa* (Diptera: Mycetophilidae). *Invertebrate Biology*, **120**, 170–177.

Brouwer, L. and Komdeur, J. (2004). Green nesting material has a function in mate attraction in the European starling. *Animal Behaviour*, **67**, 539–548.

Brown, C.R. and Brown, M.B. (1986). Ectoparasitism as a cost of coloniality in cliff swallows (*Hirundo pyrrhonota*). *Ecology*, **67**, 1206–1218.
Brown, C.R. and Brown M.B. (1991). Selection of high quality host nests by parasitic cliff swallows. *Animal Behaviour*, **41**, 457–465.
Brown, J.H. (1973). Species diversity of seed-eating desert rodents in sand dune habitats. *Ecology*, **54**, 775–787.
Brown, J.H. and Heske, E.J. (1990). Control of a desert-grassland transition by a keystone rodent guild. *Science*, **250**, 1705–1707.
Brown, J.H. and Lieberman, G.A. (1973). Resource utilisation and coexistence of seed-eating desert rodents in sand dune habitats. *Ecology*, **54**, 788–797.
Brown, J.H., Reichman, O.J., and Davidson, D.W. (1979*a*). Granivory in desert ecosystems. *Annual Review of Ecologyand and Systematics*, **10**, 201–227.
Brown, J.H., Davidson, D.W., and Reichman, O.J. (1979*b*). An experimental study of competition between seed-eating desert rodents and ants. *American Zoologist*, **19**, 1129–1143.
Brown, S.A. (2004). The influence of benthic hydrodynamics on the ecology of trichoptera larvae. Unpublished Ph.D. thesis. University of Glasgow.
Bruce, M.J., Herberstein, M.E., and Elgar, M.A. (2001). Signalling conflict between prey and predator attraction. *Journal of Evolutionary Biology*, **14**, 786–794.
Bull, N.J. and Schwarz, M.P. (1996). The habitat saturation hypothesis and sociality in an allodapine bee: cooperative nesting is not 'making the best of a bad situation'. *Behavioural Ecology and Sociobiology*, **39**, 267–274.
Burda, H., Honeycutt, R.L., Begall, S., Locker-Grütjen, O., and Scharff, A. (2000). Are naked and common mole-rats eusocial and if so, why? *Behavioural Ecology and Sociobiology*, **47**, 293–303.
Butler, J.M. and Roper, T.J. (1996). Ectoparasites and sett use in European badgers. *Animal Behaviour*, **52**, 621–629.
Büttner, H. (1996). Rubble mounds of sand tilefish *Malacanthus plumieri* (Bloch, 1787) and associated fishes in Colombia. *Bulletin of Marine Science*, **58**, 248–260.
Butts, K.O. and Lewis, J.C. (1982). The importance of prairie dog towns to burrowing owls in Oklohoma. *Proceedings of the Oklahoma Academy of Sciences*, **62**, 46–52.
Byrne, R.W., Corp, N. and Byrne, J.M.E. (2001). Estimating the complexity of animal behaviour: how mountain gorillas eat thistles. *Behaviour*, **138**, 525–557.
Camargo, J.M.F. de (1970). Nikhos e biologia de algumas espécies de Meliponideos da regiâo de Pôrto Velho, Território de Rondônia, Brasil. *Revista Biologia Tropical* (San José, Costa Rica), **16**, 207–239.
Camazine, S., Deneubourg, J.-L., Franks, N.R., Sneyd, J., Theraulaz, G., and Bonabeau, E. (2001). *Self-organisation in biological systems*. Princeton University Press, Princeton.

Campbell, M.K. (1999). *Biochemistry, 3rd Edition*. Harcourt Brace, Orlando., FL.

Cangialosi, K.R. (1990). Life cycle and behaviour of the kleptoparasitic spider, *Argyrodes ululans* (Araneae, Theridiidae). *Journal of Arachnology*, **18**, 347–358.

Cappuccino, N. (1993). Mutual use of leaf shelters by lepidopteran larvae on paper birch. *Ecological Entomology*, **18**, 287–292.

Caraco, T. and Gillespie, R.G. (1986). Risk-sensitivity: foraging mode in an ambush predator. *Ecology*, **67**, 1180–185.

Cardinale, B.J., Palmer, M.A., and Collins, S.L. (2002). Species diversity enhances ecosystem functioning through interspecific facilitation. *Nature*, **415**, 426–429.

Carpenter, J.M. (1991). Phylogenetic relationships and the origin of social behaviour in the Vespidae. In *The social biology of wasps* (eds. K.G. Ross and R.W. Matthews), pp. 7–32. Cornell University Press, Ithaca, NY.

Ceballos, G., Pachero, J., and List, R. (1999). Influence of prairie dogs (*Cynomys ludovicianus*) on habitat heterogeneity and mammalian diversity in Mexico. *Journal of Arid environments*, **41**, 161–172.

Chapin, J.P. (1954). The birds of the Belgian Congo, Part 4. *Bulletin of the American Museum of Natural History*, **75B**, 1–846.

Chappell, J. and Kacelnik, A. (2002). Tool selectivity in a non-primate, the New Caledonian crow (*Corvus moneduloides*). *Animal Cognition*, **5**, 71–78.

Chauvin, G., Vannier, G., and Gueguen, A. (1979). Larval case and water balance in *Tinea pellionella*. *Journal of Insect Physiology*, **25**, 615–619.

Chilton, J., Choo, B.S., and Popov, O. (1994). Morphology of some three-dimensional beam grillage structures in architecture and nature. Natuerliche Konstructionen 9, Evolution of Natural Structures, *Third International Symposium of Sonderforchungsbereich* **230** (Stuttart), 19–24.

Chmiel, K., Herberstein, M.E., and Elgar, M. (2000). Web damage and feeding experience influence web site tenacity in the orb-web spider *Argiope keyserlingi* Karsch. *Animal Behaviour*, **60**, 821–826.

Choe, J.C. (1994). Ingenious design of tent roosts by Peter's tent-making bat, *Uroderma bilobatum* (Chiroptera: Phyllostomidae). *Journal of Natural History*, **28**, 731–737.

Choe, J.C. (1997). A new tent roost of Thomas' fruit-eating bat, *Artibeus watsoni* (Chiroptera: Phyllostomidae), in Panama. *Korean Journal of Biological Science*, **1**, 313–316.

Christenson, T.E. (1984). Behaviour of the colonial and solitary spiders of the theridiid species *Anelosimus eximius*. *Animal Behaviour*, **32**, 725–734.

Christy J.H., Backwell, P.R.Y., Goshima, S., and Kreuter, T. (2002). Sexual selection for structure building by courting fiddler crabs: an experimental study of behavioural mechanisms. *Behavioural Ecology*, **13**, 366–374.

Christy, J.H., Backwell, P.R.Y., and Schober, U. (2003*a*). Interspecific attractiveness of structures built by courting male fiddler crabs: experimental evidence of a sensory trap. *Behavioural Ecology and Sociobiology*, **53**, 84–91.

Christy, J.H., Baum, J.K., and Backwell, P.R.Y. (2003*b*). Attractiveness of sand hoods built by courting male fiddler crabs, *Uca musica*: test of a sensory trap hypothesis. *Animal Behaviour*, **66**, 89–94.

Clark, L. (1990). Starlings as herbalists: countering parasites and pathogens. *Parasitology Today*, **6**, 358–360.

Clark, L. and Mason, J.R. (1985). Use of nest material as insecticidal and anti-pathogenic agents by the European starling. *Oecologia*, **67**, 169–176.

Clark, L. and Mason, J.R. (1988). Effect of biologically active plants used as nest material and the derived effect to starling nestlings. *Oecologia*, **77**, 174–180.

Clutton-Brock, T.H. (1991). *The evolution of parental care*. Princeton University Press, Princeton, NJ.

Cohen, A.C. and Cohen, J.L. (1976). Nest structure and micro-climate of the desert cockroach, *Arenivaga apacha* (Polyphagidae, Dictyoptera). *Bulletin of the Southern California Academy of Sciences*, **75**, 273–277.

Cole, M.R. (1998). *Regulation of nest construction behaviour and nest development in vespine wasp, with special reference to Dolichovespula norwegica* and *D. sylvestris*. Unpublished Ph.D. thesis, University of Glasgow.

Cole, M.R., Hansell, M.H., and Seath, C.J. (2001). A quantitative study of the physical properties of nest paper in three species of Vespine wasps (Hymenoptera, Vespidae). *Insectes Sociaux*, **48**, 33–39.

Coleman, S.W., Patricelli, G.L., and Borgia, G. (2004). Variable female preferences drive complex male displays. *Nature*, **248**, 742–745.

Colgin, M.A. and Lewis, R.V. (1998). The minor ampullate silk proteins contain new repetetive sequences and highly conserved non-silk-like "spacer" regions. *Protein Science*, **7**, 667–672.

Colin, P.L. (1973). Burrowing behaviour of the yellowhead jawfish *Opisthognathus aurifrons*. *Copeia*, **1973**, 84–90.

Collias, E.C. and Collias, N. E. (1973). Further studies on development of nest building behaviour in a weaverbird. *Animal Behaviour*, **21**, 371–382.

Collias, N. (1986). Engineering aspects of nest building by birds. *Endeavour*, **10**, 9–17.

Collias, N.E. (1997). On the origin and evolution of nest building by passerine birds. *The Condor*, **99**, 253–269.

Collias, N.E. and Collias, E.C. (1962). An experimental study of the mechanism of nest building in a weaverbird. *Auk*, **79**, 568–595.

Collias, N.E. and Collias, E.C. (1964*a*). The development of nest building behaviour in a weaverbird. *Auk*, **81**, 42–52.

Collias, N.E. and Collias, E.C. (1964*b*). The evolution of nest building in weaverbirds (Ploceidae). *University of California Publications in Zoology*, **73**, 1–162.

Collias, N.E. and Collias, E.C. (1984). *Nest building and bird behaviour*. Princeton University Press, Princeton, NJ.

Connor, E.F. and Taverner, M.P. (1997). The evolution and adaptive significance of the leaf-mining habit. *Oikos*, **79**, 6–25.

Conrad, K.F. and Robertson, R.J. (1993). Clutch size in eastern phoebes (*Sayornis phoebe*). I. The cost of nest building. *Canadian Journal of Zoology*, **71**, 1003–1007.

Copeland, M. and Crowell, S. (1937). Observations and experiments on the case building instinct of two species of Trichoptera. *Psyche, Cambridge*, **44**, 125–131.

Corbara, B. (1996). Arboreal nest building and ant-garden initiation by a ponerine ant. *Naturwissenschaften*, **83**, 227–230.

Cott, H.B. (1940). *Adaptive colouration in animals*. Methuen, London.

Cox, G.W. (1984). The distribution and origin of mima mound grasslands in San Diego County, California. *Ecology*, **65**, 1397–1405.

Cox, G.W. and Gakahu, G. (1986). A latitudinal test of the fossorial rodent hypothesis of Mima mound origin. *Zeitschrift für Geomorphologie*, **30**, 485–501.

Cox, G.W. and Roig, V.G. (1986). Argentina mima mounds occupied by ctenomid rodents. *Journal of Mammalogy* **67**, 428–432.

Coyle, F.A. (1986). The role of silk in prey capture by non-araneomorph spiders. In *Spiders: webs behaviour and evolution* (ed. W.A. Shear), pp. 269–305. Stanford University Press, Stanford.

Craig, C.L. (1986). Orb-web visibility: the influence of insect flight behaviour and visual physiology on the evolution of web designs within Araneoidea. *Animal Behaviour* **34**, 54–68.

Craig, C.L. (1987*a*). The ecological and evolutionary interdependence between web architecture and the silk spun by orb weaving spiders. *Biological Bulletin of the Linnean Society*, **30**, 135–162.

Craig, C.L. (1987*b*). The significance of spider size to the diversification of spider-web architectures and reproductive modes. *American Naturalist*, **129**, 47–68.

Craig, C.L. (1989). Alternative foraging modes of orb web weaving spiders. *Biotropica*, **21**, 257–264.

Craig, C.L. (1990). Effects of background pattern on insect perception of webs spun by orb-weaving spiders. *Animal Behaviour*, **39**, 135–144.

Craig, C.L. (1991). Physical constraints on group foraging and social evolution: observations on web spinning spiders. *Functional Ecology*, **5**, 649–654.

Craig, C.L. (1997). Evolution of arthropod silks. *Annual Review of Entomology*, **42**, 231–267.

Craig, C.L. (2003). *Spiderwebs and silk*. Oxford University Press, Oxford.

Craig, C.L. and Bernard, G.D. (1990). Insect attraction to ultraviolet-reflecting spider webs and web decorations. *Ecology*, **7**, 616–623.

Craig, C.L., Okubo, A., and Andreasen, V. (1985). Effect of spider orb-web and insect oscillations on prey interception. *Journal of Theoretical Biology*, **115**, 201–211.

Craig, C.L., Weber, R.S., and Bernard, G.D. (1996). Evolution of predator-prey systems: spider foraging plasticity in response to the visual ecology of prey. *American Naturalist*, **147**, 205–229.

Craig, C.L., Hsu, M., Kaplan, D,. and Pierce, N.E. (1999). A comparison of the composition of silk proteins produced by spiders and insects. *International Journal of Biological Macromolecules*, **24**, 109–118.

Craig, C.L., Riekel, L., Herberstein, M.E., Weber, R.S., Kaplan, D,. and Pierce, N.E. (2000). Evidence of diet effects on the composition of silk proteins produced by spiders. *Molecular Biology and Evolution*, **17**, 1904–1913.

Crist, T.O. and Friese, C.F. (1994). The use of ant nests by subterranean termites in two semi-arid ecosystems. *American Midland Naturalist*, **131**, 370–373.

Currey, J.D. (1970). *Animal skeletons. Studies in biology no. 22*. Edward Arnold, London.

Daanje, A. (1957). Die Blattrolltechnik von *Apoderus coryli* L. und *Attelabus nitens* Scop. (Coleoptera, Attelabinae). *Behaviour*, **11** , 85–155.

Daanje, A. (1964). Ueber die Ethologie und Blattrolltechnik von *Deporaus betulae* L. und ein Vergleich mit den anderen blattrollenden Rhynchitinen und Attelabinen. *VerhandlungenKoninklijke Nederlandse Akademie Wetenschappen*, **61**, 3–215.

Daily, G.C., Ehrlich, R,. and Haddad, N.M. (1993). Double keystone bird in a keystone species complex. *Proceedings of the National Academy of Science USA*, **90**, 592–594.

Dangerfield, J.M., McCarthy, T.S., and Ellery, W.N. (1998). The mound-building termite *Macrotermes michaelseni* as an ecosystem engineer. *Journal of Tropical Ecology*, **14**, 507–520.

Darchen, R. (1962). Observation direct du dévelopment d'un rayon de cire. Le role des chaines d'abeilles. *Insectes Sociaux*, **9**, 103–120.

Darden, T.R. (1972). Respiratory adaptations of a fossorial mammal, the pocket gopher (*Thomomys bottae). Journal of Comparative Physiology*, **78**, 121–137.

Darlington, J.P.E.C. (1984). Two types of mound built by a termite *Macrotermes subhyalinus* in Kenya. *Insect Science and its Application*, **5**, 481–492.

Darlington, J.P.E.C. (1985). The structure of mature mounds of the termite *Macrotermes michaelseni* in Kenya. *Insect Science and its Applications*, **6**, 149–156.

Darlington, J.P.E.C., Zimmerman, P.R., Greenberg, J., Westberg, C., and Bakwin, P. (1997). Production of metabolic gasses by nests of the termite *Macrotermes jeanneli* in Kenya. *Journal of Tropical Ecology*, **13**, 491–510.

Darwin, C. (1871). *The descent of man* (2nd Edition 1922). John Murray, London.

Darwin, C. (1881). *The formation of vegetable mould through the action of worms with observations on their habits.* John Murray, London.

Davey, J.T. (1994). The architecture of the burrow of *Nereis diversicolor* and its quantification in relation to sediment water exchange. *Journal of Experimental Marine Biology and Ecology*, **179**, 115–129.

Davey, J.T. and Watson, P.G. (1995). The activity of *Nereis diversicolor* (Polychaeta) and its impact on nutrient fluxes in estuarine waters. *Ophelia*, **41**, 57–70.

Davidson, D.W., Inouye, R.S., and Brown, J.H. (1984). Granivory in a desert ecosystem: Experimental evidence for indirect facilitation for ants and rodents. *Ecology*, **65**, 1780–1786.

Dawkins, R. (1982). *The extended phenotype*. Freeman, Oxford.

Daws, A.G., Bennet-Clark, H.C., and Fletcher, N.H. (1996). The mechanism of tuning of the mole cricket singing burrow. *Bioacoustics*, **7**, 81–117.

Dawson, W.D., Lake, C.E. and Schumpert, S.S. (1988). Inheritance of burrow building in *Peromyscus*. *Behaviour Genetics*, **18**, 371–382.

Dean, W.R.J., Milton, S.J., and Klotz, S. (1997). The role of ant nest-mounds in maintaining small scale patchiness in dry grasslands in Central Germany. *Biodiversity and Conservation*, **6**, 1293–1307.

Dejean, A. and Durand, J.L. (1996). Ants inhabiting *Cubitermes* termitaries in African rain forests. *Biotropica*, **28**, 701–713.

Deneubourg, J.L. (1977). Application de l'ordre par fluctuations a la description de certains étapes de la construction du nid chez les termites. *Insectes Sociaux*, **24**, 117–130.

De Neve, L. and Soler, J.J. (2002). Nest-building activity and laying date influence female reproductive investment in magpies: an experimental study. *Animal Behaviour*, **63**, 975–980.

Denny, M. (1976). The physical properties of spider's silk and their role in the design of orb webs. *Journal of Experimental Biology*, **65**, 483–506.

Denny, M.W. (1993). *Air and water. The biology and physics of life's media.* Princeton University Press, Princeton, NJ.

Denslow, J.S. (1985). Disturbance-mediated coexistence of species. In *The ecology of natural disturbance and patch dynamics* (eds. S.T.A. Pickett and P.S. White), pp. 307–323. Academic Press, Orlando, FL.

De Querioz, A. and Wimberger, P.H. (1993). The usefulness of behaviour for phylogeny estimation: levels of homoplasy in behavioural and morphological characters. *Evolution*, **47**, 46–60.

Diamond, J.M. (1982). Evolution of bowerbirds' bowers: animal origins of the aesthetic sense. *Nature*, **297**, 99–102.

Diamond, J. (1986). Animal art: variation in bower decorating style among male bowerbirds *Amblyornis inornatus*. *Proceedings of the National Academy of Sciences, USA*, **83**, 3042–3046.

Diamond, J. (1987). Bower building and decoration by the bowerbird *Amblyornis inornatus*. *Ethology*, **74**, 117–204.

Diamond, J. (1988). Experimental study of bower decoration by the bowerbird *Amblyornis inornatus*, using coloured poker chips. *American Naturalist*, **131**, 631–653.

Diamond, J.M. (1986). Effects of larval retreats of the caddisfly *Cheumatopsyche* on macroinvertebrate colonisation in piedmont, USA streams. *Oikos*, **47**, 13–18.

Donkor, N.T. and Fryxell, J.M. (1999). Impact of beaver foraging on structure of lowland boreal forests of Algonquin Provincial Park, Ontario. *Forest Ecology and Management*, **118**, 83–92.

Downes, M.F. (1994). The nest of the social spider *Phryganoporus candidus* (Araneae: Desidae): structure, annual growth cycle and host plant relationships. *Australian Journal of Zoology*, **42**, 237–259.

Downing, H.A. (1994). Information analysis by the paper wasp, *Polistes fuscatus*, during nest construction (Hymenoptera, Vespidae). *Insectes Sociaux*, **41**, 361–377.

Downing, H.A. and Jeanne, R.L. (1986). Intra-and interspecific variation in nest architecture in the paper wasp *Polistes* (Hymenoptera, Vespidae). *Insectes Sociaux*, **33**, 422–443.

Downing, H.A. and Jeanne, R.L. (1988). Nest construction by the paper wasp, *Polistes*: a test of stigmergy theory. *Animal Behaviour*, **36**, 1729–1739.

Downing, H.A. and Jeanne, R.L. (1990). The regulation of complex building behaviour in the paper wasp, *Polistes fuscatus* (Insecta, Hymenoptera, Vespidae). *Animal Behaviour*, **39**, 105–124.

Drury, W.H. (1962). Breeding activities, especially nest building of the yellowtail (*Ostinops decumanus*) in Trinidad, West Indies. *Zoologica*, **47**, 39–58.

Dubost, G. (1968). Les mammifères souterrains. *Revue Écologie et de Biologie du Sol*, **1**, 99–133.

Dudgeon, D. (1987). A laboratory study of optimal behaviour and the costs of net construction by *Polycentropus flavomaculatus* (Insecta: Trichoptera: Polycentropidae). *Journal of Zoology, London*, **211**, 121–141.

Dudgeon, D. (1994). The functional significance of selection of particles by aquatic animals during building behaviour. In *The biology of particles in aquatic systems 2nd Edition* (ed. R.G. Wotton), pp. 289–312. Lewis Publishers, Boca Raton, FL.

Duffy, J.E. (1996). Eusociality in a coral reef shrimp. *Nature*, **381**, 512–514.

Duffy, J. E., Morrison, C.L., and Macdonald, K.S. (2002). Colony defence and behavioural differentiation in the eusocial shrimp *Synalpheus regalis*. *Behavioural Ecology and Sociobiology*, **51**, 488–495.

Dworschak, P.C. (1998). The role of tegumental glands in burrow construction by two Mediterranean Callianassid shrimp. *Sneckenbergiana Maritima*, **28**, 143–149.

Dworschak, P.C. and Rodrigues, S. de A. (1997). A modern analogue for the trace fossil *Gyrolithes*: burrows of the thalassinidean shrimp *Axianassa australis*. *Lethaia*, **30**, 41–52.

Eberhard, W. (1969). Computer simulation of orb-web construction. *American Zoologist*, **9**, 229–238.

Eberhard, W.G. (1974). The natural history and behaviour of the wasp *Trigonopsis cameronii* Kohl (Sphecidae). *Transactions of the Royal Entomological Society of London*, **125**, 295–328.

Eberhard, W.G. (1975). The 'inverted ladder' orb web of *Scoloderus* sp. and the intermediate orb of *Eustala* (?) sp. Araneae: Araneidae. *Journal of Natural History*, **9**, 93–106.

Eberhard, W.G. (1981*a*). Construction behaviour and the distribution of tensions in orb webs. *Bulletin of the British Arachnological Society*, **5**, 189–204.

Eberhard, W.G. (1981*b*). The single line web of *Phoroncidia studo* Levi (Araneae: Theridiidae): A prey attractant? *Journal of Arachnology*, **9**, 229–232.

Eberhard, W.G. (1992). Web construction by *Modisimus* sp. (Araneae, Pholcidae). *Journal of Arachnology*, **20**, 25–34.

Eberhard, W.G. (2000). Breaking the mould: behavioural variation and evolutionary innovation in *Wendilgarda* spiders (Araneae Theridiosomatidae). *Ethology, Ecology and Evolution* **12**, 223–235.

Edmonds, D.T. and Vollrath, F. (1992). The contribution of atmospheric water vapour to the formation and efficiency of a spider's capture web. *Proceedings of the Royal Society of London*, **248**, 145–148.

Edwards, R. (1980). *Social wasps - their biology and control.* Rentokil Limited, W. Sussex.

Eisner, T. and Eisner, M. (2000). Defensive use of a faecal thatch by a beetle larva (*Hemisphaerota cyanea*). *Proceedings of the National Academy of Science*, **97**, 2632–2636.

Eisner, T, and Nowicki, S. (1983). Spider web protection through visual advertisement: role of the stabilimentum. *Science*, **219**, 185–187.

Elgar, M.A., Allan, R.A., and Evans, T.A. (1996). Foraging strategies in orb-spinning spiders: Ambient light and silk decorations in *Argiope aetherea* Walckenaer (Araneae: Araneoidea). *Australian Journal of Ecology*, **21**, 464–467.

Elkins, N.Z., Shabol, G.V., Ward, T.J., and Whitford, W.G. (1986). The influence of subterranean termites on the hydrological characteristics of a Chihuahuan desert ecosystem. *Oecologia*, **68**, 521–528.

Emerson, A.E. (1938). Termite nests—A study of the phylogeny of behaviour. *Ecological Monographs*, **8**, 248–284.

Emerson, A.E. (1956). Ethospecies, ethotypes, taxonomy, and the evolution of *Apicotermes* and *Allognathotermes* (Isoptera, Termitidae). *American Museum Novitates*, **1771**, 1–31.

Emlen, J.T. (1954). Territory, nest building, and pair formation in the cliff swallow. *Auk*, **71**, 16–35.

Endler, J.A. (1981). An overview of the relationship between mimicry and crypsis. *Biological Journal of the Linnaean* Society, **16**, 25–31.

Epstein, M.C. (1995). False-parasitised cocoons and the biology of Aididae (Lepidoptera: Zygaenoidea). *Proceedings of the Entomological Society of Washington*, **97**, 750–756.

Ernest, S.K.M. and Brown, J.H. (2001). Delayed compensation for missing keystone species by colonisation. *Science*, **292**, 101–104.

Ernst, W.H.O. and Sekhwela, M.B.M. (1987). The chemical composition of lerps from the mopane psyllid *Arytaina mopane* (Homoptera, Psyllidae). *Insect Biochemistry*, **17**, 905–909.

Evans, H.E. (1964). Observations on the nesting behaviour of *Moniaecera asperata* (Fox) (Hymenoptera, Sphecidae, Crabroninae), with comments on communal nesting in solitary wasps. *Insectes Sociaux*, **11**, 71–78.

Evans, H.E. (1966). *The comparative ethology and evolution of the sand wasps.* Harvard University Press, Cambridge, MA.

Evans, H.E. and West Eberhard, M.J. (1970). *The wasps.* The University of Michigan Press, Ann Arbor, MI.

Evans, M.R. and Thomas, A.L.R. (1997). Testing the functional significance of tail streamers. *Proceedings of the Royal Society of London B*, **264**, 211–217.

Fager, E.W. (1964). Marine sediments: effects of a tube-building polychaete. *Science*, **143**, 356–359.

Farrow, G.E. (1971). Back-reef and lagoonal environments of Aldabra atoll distinguished by their crustacean burrows. *Symposium of the Zoological Society of London*, **28**, 455–500.

Fenaux, R. (1986). The house of *Oikopleura dioica* (Tunicata, Appendicularia): structure and functions. *Zoomorphology*, **106**, 224–231.

Field, J.P., Foster, W., Shreeves, G., and Sumner, S. (1998). Ecological constraints on independent nesting in facultatively eusocial hover wasps. *Proceedings of the Royal Society of London B*, **265**, 973–977.

Fischer, R. and Meyer, W. (1985). Observations on rock boring by *Alpheus saxidomus* (Crustacea: Alpheidae). *Marine Biology*, **89**, 213–219.

Fisher, R.A. (1930). *The genetical theory of natural selection*. Clarendon Press, Oxford.

Fitkau, E.J. and Klinge, H. (1973). On biomass and trophic structure of the central Amazonian Rain Forest Ecosystem. *Biotropica*, **5**, 2–14.

Fitzgerald, T.D. and Clark, K.L. (1994). Analysis of leaf-rolling behaviour of *Caloptilia serotinella* (Lepidoptera: Gracillariidae). *Journal of Insect Behaviour*, **7**, 859–872.

Fitzgerald, T.D. and Willer, D.E. (1983). Tent-building behaviour of the Eastern tent caterpillar *Malacosoma americanum* (Lepidoptera: Lasiocampidae). *Journal of the Kansas Entomological Society*, **56**, 20–31.

Fitzgerald, T.D., Clark, K.L., Vanderpool, R., and Phillips, C. (1991). Leaf shelter-building caterpillars harness forces generated by axial retraction of stretched and wetted silk. *Journal of Insect Behaviour*, **4**, 21–32.

Flood, P.R. (1991). Architecture of, and water circulation and flow rate in, the house of the planktonic tunicate *Oikopleura labradoriensis*. *Marine Biology*, **111**, 95–111.

Flood, P.R. and Deibel, D. (1998). The appendicularian house. In *The biology of pelagic tunicates* (ed. Q. Bone), pp. 105–124. Oxford University Press, Oxford.

Flood, P.R. and Fiala-Médioni, A. (1982). Structure of the mucous feeding filter of *Chaetopterus variopedatus* (Polychaeta). *Marine Biology*, **72**, 27–33.

Fouillaud, M. and Morel, G. (1995). Fungi associated with the nests of the paper wasp *Polistes hebraeus* (Hymenoptea: Vespidae) on La Reunion Island. *Environmental Entomology*, **24**, 298–305.

Fowler, H.G. and Venticinque, E.M. (1996). Interference competition and scavenging by *Crematogaster* ants (Hymenoptera: Formicidae) associated with the webs of the social spider *Anelosimus eximius* (Araneae: Theridiidae) in the Central Amazon. *Journal of the Kansas Entomological Society*, **69**, 267–269.

Frank, C.L. (1988*a*). Influence of moisture content on seed selection by kangaroo rats. *Journal of Mammalogy*, **69**, 353–357.

Frank, C.L. (1988*b*). The effects of mouldiness level on seed selection by *Dipodomys spectabilis*. *Journal of Mammalogy*, **69**, 358–362.

Franks, N.R. and Deneubourg, J.-L. (1997). Self-organising nest construction in ants: individual worker behaviour and the nest's dynamics. *Animal Behaviour*, **54**, 779–796.

Franks, N.R., Wilby, A., Silverman, B. and Tofts, C. (1992). Self-organising nest construction in ants: sophisticated building by blind bulldozing. *Animal Behaviour*, **44**, 357–375.

Frith, C.B., Borgia, G. and Frith, D.W. (1996). Courts and courtship behaviour of Archbold's bowerbird *Archboldia papuensis*. *Ibis*, **138**, 204–211.

Frith, C.B. and Frith, D.W. (1994). Courts and seasonal activities at them by male tooth-billed bowerbirds, *Scenopoeetes dentirostris* (Ptilonorynchidae). *Memoirs of the Queensland Museum*, **37**, 121–145.

Frith, C.B. and Frith, D.W. (2000*a*). Attendance levels and behaviour at bowers by male golden bowerbirds, *Prionodura newtoniana* (Ptilonorhynchidae). *Memoirs of the Queensland Museum*, **45**, 317–341.

Frith, C.B. and Frith, D.W. (2000*b*). Home range and associated sociobiology and ecology of male golden bowerbirds, *Prionodura newtoniana* (Ptilonorhynchidae). *Memoirs of the Queensland Museum*, **45**, 343–357.

Frohlich, D.R. and Parker, F.D. (1985). Observations on the nest-building and reproductive behaviour of a resin-gathering bee: *Dianthidium ulkei* (Hymenoptera: Megachilidae). *Annals of the Entomological Society of America*, **78**, 804–810.

Fruth, B. and Hohmann, G. (1996). Nest building behaviour in the great apes: the great leap forward? In *Great ape societies* (eds. W.C. McGrew, L.F. Marchant, and T. Nishida), pp. 225–240. Cambridge University Press, Cambridge, MA.

Fukuyama, K. (1991). Spawning behaviour and male mating tactics of a foam-nesting treefrog *Rhacophora schlegelii*. *Animal Behaviour*, **42**, 193–199.

Funmilayo, O. (1979). Food consumption, food preferences and storage in the mole. *Acta Therologica*, **24**, 379–389.

Futuyma, D.J. (1986). *Evolutionary biology, 2nd Edition*. Sinauer, Sunderland, MA.

Gadagkar, R. (1991). *Belonogaster, Mischocyttarus, Parapolybia*, and the independent-founding *Ropalidia*. In *The social biology of wasps* (eds. K.G. Ross and R.W. Matthews), pp. 149–190. Cornell University Press, Ithaca, NY.

Galloway, J. (1988). Nature's second-favourite structure. *New Scientist*, 31 March, 36–39.

Gardarsson, A. and Snorrason, S.S. (1993). Sediment characteristics and density of benthos in Lake Myvatn, Iceland. *Verhandlungen der Internationalen Vereinigung fur Theoretische und Angewandte Limnologie*, **25**, 452–457.

Garnier-Sillam, E. and Harry, M. (1995). Distribution of humic compounds in mounds of some soil-feeding termite species of tropical rainforests: its influence on soil structure stability. *Insectes Sociaux*, **42**, 167–185.

Genise, J.F. and Brown, T.M. (1994). New trace fossils of termites (Insecta: Isoptera) from the late Eocene-early Miocene of Egypt, and the reconstruction of ancient isopteran social behaviour. *Ichnos*, **3**, 155–183.

Genise, J.F. and Hazeldine, P.L. (1998). 3-D reconstruction of insect trace fossils: *Ellipsoideichnus meyeri* Roselli. *Ichnos*, **5**, 167–175.

Gerber, C., Badertscher, S., and Leuthold, R.H. (1988). Polyethism in *Macrotermes bellicosus* (Isoptera). *Insectes Sociaux*, **35**, 226–240.

Gilby, A.R., McKellar, J.W., and Beaton, C.D. (1976). The structure of lerps: carbohydrate, lipid, and protein components. *Journal of Insect Physiology*, **22**, 689–696.

Gillespie, R.G. and Croom, H.B. (1995). Comparison of speciation mechanisms in web-building and non-web-building groups within a lineage of spiders. In *Hawaiian biogeography: Evolution on a hot spot archipelago* (eds. W.L. Wagner and V.A. Funk), pp. 121–146. Smithsonian Institution, Washington.

Gilmer, R.W. (1972). Free floating mucus webs: a novel feeding adaptation for open ocean. *Science, N.Y.*, **176**, 1239–1240.

Gilmer, R.W. and Harbison, G.R. (1986). Morphology and field behaviour of pteropod molluscs: feeding methods in the families Cavoliniidae, Limacinidae and Peraclididae (Gastropoda: Thecosomata). *Marine Biology*, **91**, 47–57.

Goodwin, D. (1976). *Crows of the world*. British Museum (Natural History), Publication No. 771, London.

Goodwin, D. (1983). *Pigeons and doves of the world*. British Museum (Natural History), Comstock Publishing Associates, Cornell University Press, Ithaca, NY.

Gordon, D.M. (1996). The organisation of work in social insect colonies. *Nature*, **380**, 121–124.

Gordon, J.E. (1976). *The new science of strong materials, 2nd edition*. Penguin, Harmondsworth.

Gordon, J.E. (1978). *Structures, or why things don't fall down*. Penguin, Harmondsworth.

Gorman, M.L. and Stone, R.D. (1990). *The natural history of moles*. Christopher Helm, London.

Gosline, J.M., DeMont, M.E., and Denny, M.W. (1986). The structure and properties of spider silk. *Endeavour, New Series*, **10**, 37–43.

Gosline, J.M., Guerette, P.A., Ortlepp, C.S., and Savage, K.N. (1999). The mechanical design of spider silks: from fibroin sequence to mechanical function. *Journal of Experimental Biology*, **202**, 3295–3303.

Götmark, F. (1993). Conspicuous nests may select for non-cryptic eggs: a comparative study of avian families. *Ornis Fennica*, **70**, 102–105.

Gotts, N.M. and Vollrath, F. (1991). Artificial intelligence modelling a web-building in the garden cross spider. *Journal of Theoretical Biology*, **152**, 485–511.

Grassé, P.-P. (1959). La reconstruction du nid et les coordinations interindividuelles chez *Bellicositermes natalensis* et *Cubitermes* sp. La théorie de la stigmergie: Essai d'interprétation du comportment des termites constructeurs. *Insectes Sociaux*, **6**, 41–83.

Grassé, P.-P. (1984). *Termitologia. Tome II.—Fondation des sociétés- construction*. Paris, Masson.

Griffin, D.R. (1998). *The question of animal awareness*. The Rockefeller University Press, New York.

Griffin, D.R. (1998). From cognition to consciousness. *Animal Cognition*, **1**, 3–16.

Grigg, G.C. (1973). Some consequences of the shape and orientation of 'magnetic' termite mounds. *Australian Journal of Zoology*, **21**, 231–237.

Grostal, P. and Walter, D.E. (1997). Kleptoparasitres or commensals? Effects of *Argyrodes antipodianus* (Araneae: Theridiidae) on *Nephila plumipes* (Araneae: Tetragnathidae). *Oecologia*, **111**, 570–574.

Grostal, P. and Walter, D.E. (1999). Host specificity and distribution of the kleptobiotic spider *Argyrodes antipodanus* (Araneae, Theridiidae) on orb webs in Queensland, Austrtalia. *Journal of Arachnology*, **27**, 522–530.

Guerette, P.A., Ginzinger, D.G., Weber, B.H.F., and Gosline, J.M. (1996). Silk properties determined by gland-specific expression of a spider fibroin gene family. *Science*, **272**, 112–115.

Guo, Q. (1996). Effects of bannertail kangaroo rat mounds on small-scale plant community structure. *Oecologia*, **106**, 247–256.

Gupta, J.L. and Deheri, G.M. (1990). A mathematical model of the effect of respiration on the gas concentration in an animal burrow. *National Academy of Science Letters*, **13**, 249–251.

Gurney, S.C. and Lawton, J.H. (1996). The population dynamics of ecosystem engineers. *Oikos*, 76, 273–283.

Gutiérrez, J.L. and Iribarne, O.O. (1998). The occurrence of juveniles of the grapsid crab *Chasmagnathus granulata* in siphon holes of the stout razor clam *Tagelus plebeius*. *Journal of Shellfish Research*, **17**, 925–929.

Guyer, C. and Hermann, S.M. (1997). Patterns of size and longevity of gopher tortoise (*Gopherus polyphemus*) burrows: implications for the longleaf pine ecosystem. *Chelonian Conservation and Biology*, **2**, 507–513.

Gwinner, H., Oltrogge, M., Trost, L., and Neinaber, U. (2000). Green plants in starling nests: effects on nestlings. *Animal Behaviour* **59**, 301–309.

Haddad, C.F.B., Pombal, J.P., and Gordo, M. (1990). Foam nesting in a hylid frog (Amphibia, Anura). *Journal of Herpetology*, **24**, 225–226.

Hall-Spencer, J.M. and Moore, P.G. (2000). *Limaria hians* (Mollusca: Limacea): a neglected reef-forming keystone species. *Aquatic Conservation: Marine and Freshwater Ecosystems*, **10**, 267–277.

Hamilton, W.D. (1964). The genetical theory of social behaviour, I, II. *Journal of Theoretical Biology*, **7**, 1–52.

Hansell, M.H. (1968*a*). The house building behaviour of the caddis fly larva *Silo pallipes* Fabricius. II: Description and analysis of the selection of small particles. *Animal Behaviour*, **16**, 562–577.

Hansell, M.H. (1968*b*). The house building behaviour of the caddis fly larva *Silo pallipes* Fabricius. I: the structure of the house and method of house extension. *Animal Behaviour* **16**, 558–561.

Hansell, M.H. (1972). Case building behaviour of the caddis fly larva *Lepidostoma hirtum*. *Journal of Zoology, London*, **167**, 179–192.

Hansell, M.H. (1973). Improvement and termination of house building in the caddis fly *Lepidostoma hirtum* Curtis. *Behaviour*, **46**, 141–153.

Hansell, M.H. (1974). Regulation of building unit size in the house building of the caddis larva *Lepidostoma hirtum*. *Animal Behaviour*, **22**, 133–143.

Hansell, M.H. (1981). Nest construction in the subsocial wasp *Parischnogaster mellyi* (Saussure) Stenogastrinae (Hymenoptera). *Insectes Sociaux*, **28**, 208–216.

Hansell, M.H. (1984). *Animal architecture and building behaviour*. Longman, London.

Hansell, M.H. (1986). The structure of the nest of the stenogastrine wasp *Holischnogaster gracilipes* (Van der Vecht). *Entomologists Monthly Magazine*, **122**, 185–188.

Hansell, M.H. (1987*a*). Nest building as a facilitating and limiting factor in the evolution of eusociality in the Hymenoptera. In *Oxford surveys of* *evolutionary biology* Vol. 4. (eds. P.H. Harvey and L. Partridge), pp. 155–181. Oxford University Press, oxford.

Hansell, M.H. (1987*b*). What's so special about using tools? *New Scientist*, 8 January, 54–56.

Hansell, M.H. (1993). The ecological impact of animal nests and burrows. *Functional Ecology*, **7**, 5–12.

Hansell, M.H. (1996*a*). The function of lichen flakes and white spider cocoons on the outer surface of birds' nests. *Journal of Natural History*, **30**, 303–311.

Hansell, M.H. (1996*b*). Wasps make nests: nests make conditions. In *Natural history and evolution of paper wasps* (eds. S. Turillazzi and M.J. West-Eberhard), pp. 272–289. Oxford University Press, Oxford.

Hansell, M.H. (2000). *Bird nests and construction behaviour*. Cambridge University Press, Cambridge, MA.

Hansell, M. and Ruxton, G.D. (2002). An experimental study of the availability of feathers for avian nest building. *Journal of Avian Biology*, **33**, 319–321.

Hansell, M.H. and Turillazzi, S. (1991). A new stenogastrine nest from Papua New Guinea probably belonging to the genus *Stenogaster* Guerin 1831 (Hymenoptera Vespidae). *Tropical Zoology*, **4**, 81–87.

Hansell, M.H. and Turillazzi, S. (1995). Nest structure and building material of three species of *Anischnogaster* (Vespidae Stenogastrinae) from Papua New Guinea. *Tropical Zoology*, **8**, 203–215.

Hashimoto, A. and Kitaoka, S. (1982). Composition of the wax secreted by a scale insect, *Drosicha corpulenta* Kuwana (Homoptera: Margarodidae). *Applied Entomology and Zoology*, **17**, 453–459.

Hasiotis, S.T. and Dubiel, R.F. (1995). Termite (Insecta: Isoptera) nest ichnofossils from the Upper Triassic Chinle Formation, Petrified Forest National Park, Arizona. *Ichnos*, **4**, 119–130.

Hauber, M.E. (2002). Is reduced clutch size a cost of parental care in eastern phoebes (*Sayornis phoebe*)? *Behavioural Ecology and Sociobiology*, **51**, 503–509.

Haussknecht, T. and Kuenzer, P. (1990). An experimental study of the building behaviour sequence of a shell-breeding cichlid-fish from Lake Tanganyika (*Lamprologus ocellatus*). *Behaviour*, **116**, 127–143.

Hayashi, C.Y. and Lewis, R.V. (2001). Spider flagelliform silk: lessons in protein design, gene structure, and molecular evolution. *BioEssays*, **23**, 750–756.

Hayashi, C.Y., Shipley, N.H. and Lewis, R.V. (1999). Hypotheses that correlate the sequence, structure and mechanical properties of spider silk proteins. *International Journal of Biological Macromolecules*, **24**, 271–275.

Hayward, P.J. and Ryland, J.S. (1990). *The marine fauna of the British Isles and North-west Europe. Vol. 1. Introduction and Protozooans through to arthropods*. Clarendon Press, Oxford.

Heath, J.D. (1985). Whorl overlap and the economical construction of the gastropod shell. *Biological Journal of the Linnean Society*, **24**, 165–174.

Heath, M. and Hansell, M.H. (2002). Weaving techniques in two species of Icterini, the yellow oriole and collared oropendola. In *Studies in Trinidad and Tobago: ornithology honouring Richard ffrench* (ed. F. Hayes), Occasional Papers of the Department of Life Sciences. St Augustine, University of the West Indies, Trinidad.

Heiling, A.M. and Herberstein, M.E. (1999). The role of experience in web-building spiders (Araneidae). *Animal Cognition*, **2**, 171–177.

Hellmund, W. and Hellmund, M. (2001). Ein Pflanzenmaterial besterhender Trichopterenköcher (Insecta, Phryganeidae) aus dem Oberoligozän von Rott im Siebengebirge (Rheinland, Deutschland). *Decheniana (Bonn)*, **154**, 101–107.

Henaut, Y. (2000). Host selection by a kleptoparasitic spider. *Journal of Natural History*, **34**, 747–753.

Henschel, J.R. (1995). Tool use by spiders: stone selection and placement by corolla spiders *Ariadna* (Segestriidae) of the Namib desert. *Ethology*, **101**, 187–199.

Henschel, J.R. and Jocqué, R. (1994). Bauble spiders: a new species of *Achaearanea* (Araneae: Theridiidae) with ingenious spiral retreats. *Journal of Natural History*, **28**, 1287–1295.

Hepburn, H.R. (1986). *Honeybees and wax*. Springer-Verlag, Berlin.

Hepburn, H.R. (1998). Reciprocal interactions between honeybees and combs in the integration of some colony functions in *Apis mellifera* L. *Apidologie*, **29**, 47–66.

Hepburn, H.R. and Kurstjens, S.P. (1984). On the strength of propolis (bee glue). *Naturwissenschaften*, **71**, 591.

Hepburn, H.R. and Kurstjens, S.P. (1988). The combs of honeybees as composite materials. *Apidologie*, **19**, 25–36.

Hepburn, H.R. and Muller, W.J. (1988). Wax secretion in honeybees. *Naturwissenschaften*, **75**, 628–629.

Herbers, J.M. (1986). Nest site limitation and facultative polygyny in the ant *Leptothorax longispinosus*. *Behavioural Ecology and Sociobiology*, **19**, 115–122.

Herbers, J.M. and Banschbach, V.S. (1999). Plasticity of social organisation in a forest ant species. *Behavioural Ecology and Sociobiology*, **45**, 451–465.

Herberstein, M.E. and Elgar, M.A. (1994). Foraging strategies of *Eriophora transmarina* and *Nephila plumipes* (Araneae: Araneoidea): nocturnal and diurnal orb-weaving spiders. *Australian Journal of Ecology*, **19**, 451–457.

Herberstein, M.E. and Heiling, A.M. (1998). Does mesh height influence prey length in orb-web spiders (Araneae)? *European Journal of Entomology*, **95**, 367–371.

Herberstein, M.E., Craig, C.L., Coddington, J.A., and Elgar, M.A. (2000). The functional significance of silk decorations of orb-web spiders: a critical review of the empirical evidence. *Biological Reviews*, **75**, 649–669.

Hershey, A.E., Hiltner, A.L., Hullar, M.A.J., Miller, M.C., Vestal, J.R., Lock, M.A., and Rundle, S. (1988). Nutrient influence on a stream grazer: *Orthocladius* microcommunities respond to nutrient input. *Ecology*, **69**, 1383–1392.

Heth, G., Golenberg, E.M. and Nevo, E. (1989). Foraging strategy in a subterranean rodent, *Spalax ehrenbergi*: a test case for optimal foraging theory. *Oecologia*, **79**, 496–505.

Heyer, W.R. and Rand, A.S. (1977). Foam nest construction in the leptodactylid frogs *Leptodactylus pentadactylus* and *Physalaemus pustulosus* (Amphibia, Anura, Leptodactylidae). *Journal of Herpetology*, **11**, 225–228.

Hickman, G.C. (1983). Burrow structure of the talpid mole *Parascalops breweri* from Oswego County, New York State. *Zeitschrift fur Saugetierkunde*, **48**, 265–269.

Hieber, C.S. (1992). Spider cocoons and their suspension systems as barriers to generalist and specialist predators. *Oecologia*, **91**, 530–535.

Higgins, L.E. and Buskirk, R.E. (1992). A trap-building predator exhibits different tactics for different aspects of foraging behaviour. *Animal Behaviour*, **44**, 485–499.

Hodkinson, I.D., Coulson, S.J., Harrison, J. and Webb, N.R. (2001). What a wonderful web they weave: spiders, nutrient capture and early ecosystem development in the high Arctic—some counter-intuitive ideas on community assembly. *Oikos*, **95**, 349–352.

Hoffmann, A.A. (1994). Behaviour genetics and evolution. In *Behaviour and evolution* (eds. P.J.B. Slater and T.R. Halliday), pp. 7–42. Cambridge University Press, Cambridge.

Hölldobler, B. and Wilson, E.O. (1983). The evolution of communal nest-weaving in ants. *American Scientist*, **71**, 490–499.

Hölldobler, B. and Wilson, E.O. (1990). *The ants*. Springer-Verlag, Berlin.

Hotta, M. (1994). Infanticide in little swifts taking over costly nests. *Animal Behaviour*, **47**, 491–493.

Howman, H.R.G. and Begg, G.W. (1995). Intra-seasonal and inter-seasonal nest renovation in the masked weaver, *Ploceus velatus*. *Ostrich*, **66**, 122–128.

Hozumi, S. and Yamane, S. (2001*a*). Incubation ability of the functional envelope in paper wasp nests (Hymenoptera, Vespidae, *Polistes*): 1. Field measurements of nest temperature using paper models. *Journal of Ethology*, **19**, 39–46.

Hozumi, S. and Yamane, S. (2001*b*). Effects of surface darkening on cell temperature in paper wasp nests: measurements using paper model nests. *Entomological Science*, **4**, 251–256.

Huber, S., Millesi, E. and Dittami, J.P. (2002). Paternal effort and its relation to mating success in the European ground squirrel. *Animal Behaviour*, **63**, 157–164.

Humphries, S. and Ruxton, G.D. (1999). Bower-building: coevolution of display traits in response to the costs of female choice? *Ecology Letters*, **2**, 404–413.

Hunt, G.R. (1996). Manufacture and use of hook tools by New Caledonian crows. *Nature*, **379**, 249–251.

Hunt, G.R. (2000*a*). Human-like, population-level specialization in the manufacture of pandanus tools by the New Caledonian crows *Corvus moneduloides*. *Proceedings of the Royal Society of London B*, **267**, 403–413.

Hunt, G.R. (2000*b*). Tool use by the New Caledonian crow *Corvus moneduloides* to obtain Cerambycidae from dead wood. *Emu*, **100**, 109–114.

Hunt, G.R. and Gray, R.D. (2003). Diversification and cumulative evolution in New Caledonian crow tool manufacture. *Proceedings of the Royal Society of London B*, **270**, 867–874.

Iglesia, H.O. de la, Rodriguez, E.M. and Dezi, R.E. (1994). Burrow plugging in the crab *Uca uruguayensis* and its synchronization with photoperiod and tides. *Physiology and Behaviour*, **5**, 913–919.

Ingold, T. (1986). Tools, and *Homo faber*: construction and authorship of design. In *The appropriation of nature: Essays on human ecology and social relations* (ed. T. Ingold), pp. 40–78. Manchester University Press, Manchester.

Ishay, J.S. and Barenholz-Paniry, V. (1995). Thermoelectric effect in hornet *Vespa orientalis* silk and thermoregulation in a hornet's nest. *Journal of Insect Physiology*, **41**, 753–759.

Ishay, J.S., Litinetsky, L., Pertsis, V., Linsky, D., Lusternik, V., and Voronel, A. (2002). Hornet silk; thermophysical properties. *Journal of Thermal Biology*, **27**, 7–15.

Ishimatsu, A., Hishida, Y., Takita, T., Kanda, T., Oikawa, S., Takeda, T., and Huat, K.K. (1998). *Nature*, **391**, 237–238.

Ito, Y. (1993). *Behaviour and social evolution of wasps: the communal aggregation hypothesis*. Oxford University Press, Oxford.

Jacklyn, P. (1991). Evidence of adaptive variation in the orientation of *Amitermes* (Isoptera: Termitinae) mounds from Northern Australia. *Australian Journal of Zoology*, **39**, 569–577.

Jacklyn, P.M. (1992). "Magnetic" termite mound surfaces are oriented to suit wind and shade conditions. *Oecologia*, **91**, 385–395.

Jackson, D.R. and Milstrey E.G. (1989). The fauna of gopher tortoise burrows. In *Gopher tortoise relocation symposium proceedings* (eds. J.E. Diemer, , D.R. Jackson, J.L. Landers, J.N. Layne, and D.A. Woods), pp. 86–98. *Florida Game and Freshwater Fish Commission, Nongame Wildlife Program Technical Report No. 5.*

Jackson, T.P., Roper, T.J., Conradt, L., Jackson, M.J., and Bennett, N.C. (2002). Alternative refuge strategies and their relation to thermophysiology in two sympatric rodents, *Parotomys brantsii* and *Otomys unisculatus*. *Journal of Arid Environments*, **51**, 21–34.

Jakobsson, S., Borg, B., Haux, C., and Hyllner, S.J. (1999). An 11-ketotestosterone induced kidney-secreted protein: the nest building glue from male three-spined stickleback, *Gasterosteus aculeatus*. *Fish Physiology and Biochemistry*, **20**, 79–85.

James, R., Atkinson, A., and Nash R.D.M. (1990). Some preliminary observations on the burrows of *Callianassa subterranea* (Montagu) (Decapoda: Thalassinidae) from the west coast of Scotland. *Journal of Natural History*, **24**, 403–413.

James, R., Atkinson, A., and Pullin, R.S. (1996). Observations on the burrows and burrowing behaviour of the red band fish, *Cepola rubescens* L. *Marine Ecology*, **17**, 23–40.

Janetos, A.C. (1982). Active foragers vs. sit-and-wait predators: a simple model. *Journal of Theoretical Biology*, **95**, 381–385.

Jarvis, J.U.M., O'Riain, M.J., Bennett, N.C., and Sherman, P.W. (1994). Mammalian eusociality: a family affair. *Trends in Ecology and Evolution*, **9**, 47–51.

Jeanne, R.L. (1970). Chemical defence of brood by a social wasp. *Science*, **168**, 1465–1466.

Jeanne, R.L. (1975). Adaptiveness of social wasp nest architecture. *Quarterly Review of Biology*, **50**, 267–287.

Jeanne, R.L. (1978). Intraspecific nesting associations in the neotropical social wasp *Polybia rejecta* (Hymenoptera: Vespidae). *Biotropica*, **10**, 234–235.

Jeanne, R.L. (1986). The organisation of work in *Polybia occidentalis*: costs and benefits of specialisation in a social wasp. *Behavioural Ecology and Sociobiology*, **19**, 333–341.

Jeanne, R.L. (1987). Do water foragers pace nest construction activity in *Polybia occidentalis*? *Experientia Supplementum, vol. 54, Behaviour in Social Insects*, 241–251.

Jeanne, R.L. (1991*a*). The swarm-founding polistinae. In *The social biology of wasps* (eds. K.G. Ross and R.W. Matthews), pp. 191–231. Comstock, Ithaca, NY.

Jeanne, R.L. (1991*b*). Polyethism. In *The social biology of wasps* (eds. K.G. Ross and R.W. Matthews), pp. 389–425. Comstock, Ithaca, NY.

Jeanne, R.L. (1996). Regulation of nest construction behaviour in *Polybia occidentalis.Animal Behaviour*, **52**, 473–488.

Jeanne, R.L., Downing, H.A., and Post, D.C. (1988). Age polyethism and individual variation in *Polybia occidentalis* an advanced eusocial wasp. In *Interindividual behavioural variability in social insects* (ed. R.L. Jeanne), pp. 323–357, Westview Press, Boulder, CO.

Jianping, S.U. (1992). Energy cost of foraging and optimal foraging in the fossorial rodent (*Myospalax baileyi*). *Acta Theriologica*, **12**, 117–125.

Johansson, A. (1991). Caddis larvae cases (Trichoptera, Limnephilidae) as anti-predatory devices against brown trout and sculpin. *Hydrobiologia*, **211**, 185–194.

Johansson, A. and Nilsson, A.N. (1992). *Dytiscus latissimus* and *D. circumcinctus* (Coleoptera, Dytiscidae) larvae as predators on three case-making caddis larvae. *Hydrobiologia*, **248**, 201–213.

Johnson, K.S., Eischen, F.A. and Giannasi, D.E. (1994). Chemical composition of North American bee propolis, and biological activity towards larvae of the greater wax moth (Lepidoptera: Pyralidae). *Journal of Chemical Ecology*, **20**, 1783–1791.

Johnston, C.A. and Naiman, R.J. (1990). Aquatic patch creation in relation to beaver population trends. *Ecology*, 71, 1617–1621.

Jones, C.G., Lawton, J.H., and Shachak, M. (1994). Organisms as ecosystem engineers. *Oikos*, **69**, 373–386.

Jones, C.G., Lawton, J.H., and Shachak, M. (1997). Positive and negative effects of organisms as physical ecosystem engineers. *Ecology*, **78**, 1946–1957.

Jonkman, J.C.M. (1980). The external and internal structure and growth f the nests of the leaf-cutting ant *Atta vollenweideri* Forel, 1893 (Hym. Formicidae), part I. *Zeitschrift für Angewandte Entolologie*, **89**, 158–173.

Jordăn, F., Takăcs-Sănta, A., and Molnăr, I. (1999). A reliability theoretical quest for keystones. *Oikos*, **86**, 453–462.

Kabisch, K., Herrmann, H.J., Klossek, P., Luppa, H. and Brauer, K. (1998). Foam gland and chemical analysis of the foam of *Polypedates leucomystax* (Gravenhorst 1829) (Anura: Rhacophoridae). *Russian Journal of Herpetology*, **5**, 10–14.

Kaib, M., Hussender, C., Epplen, C., Epplen, J.T., and Brandl, R. (1996). Kin-biased foraging in a termite. *Proceedings of the Royal Society of London B*, **263**, 1527–1532.

Kappner, I., Al-Moghrabi, S.M., and Richter, C. (2000). Mucus-net feeding by a vermetid gastropod *Dendropoma maxima* in coral reefs. *Marine Ecology Progress Series*, **204**, 309–313.

Karino, K. (1996). Tactic for bower acquisition by male cichlids, *Cyathopharynx furcifer*, in Lake Tanganyika. *Ichthyological Research*, **43**, 125–132.

Karplus, I. (1987). The association between gobiid fishes and burrowing alpheid shrimps. *Oceanography and Marine Biology Annual Review*, **25**, 507–562.

Karplus, I., Szlep, R., and Tsurnamal, M. (1972). Associative behaviour of the fish *Cryptocentrus cryptocentrus* (Gobiidae) and the pistol shrimp *Alpheus djiboutiensis* (Alpheidae) in artificial burrows. *Marine Biology*, **15**, 95–104.

Karsai, I. and Pénzes, Z. (1993). Comb building in social wasps: Self-organisation and stigmergic script. *Journal of Theoretical Biology*, **161**, 505–525.

Karsai, I. and Pénzes, Z. (2000). Optimality of cell arrangement and rules of thumb of cell initiation in *Polistes dominulus*: a modelling approach. *Behavioural Ecology*, **11**, 387–395.

Karsai, I. and Theraulaz, G. (1995). Nest building in a social wasp: Postures and constraints (Hymenoptera: Vespidae). *Sociobiology*, **26**, 83–114.

Karsai, I and Wenzel, J.W. (1998). Productivity, individual-level and colony-level flexibility, and organisation of work as a consequences of colony size. *Proceedings of the National Academy of Sciences*, **95**, 8665–8669.

Kaufmann, E., Weissflog, A., Hashim, R., and Maschwitz, U. (2001). Ant-gardens on the giant bamboo *Gigantochla scortechinii* (Poaceae) in West-Malaysia. *Insectes Sociaux*, **48**, 125–133.

Keltner, J. and McCafferty, P. (1986). Functional morphology in the mayflies *Hexagenia limbata* and *Pentagenia vittigera*. *Zoological Journal of the Linnean Society*, **87**, 139–162.

Kerley, G.I.H., Whitford, W.G., and Kay, F.R. (1997). Mechanisms for the key-stone status of kangaroo rats: graminivory rather than granivory. *Oecologia*, **111**, 422–428.

Kilgore, D.L. and Knudsen, K.L. (1977). Analysis of materials in cliff and barn swallow nests: Relationship between mud selection and architecture. *Wilson Bulletin*, **89**, 562–571.

Klaas, B.A., Moloney, K.A., and Danielson, B.J. (2000). The tempo and mode of gopher mound production in a tallgrass prairie remnant. *Ecography*, **23**, 246–256.

Kleeman, K. (1984). Lebensspurum von *Upogebia operculata* (Crustacea, Decapoda) in karibischen Steinkorallen (Madreporaria, Anthozoa). *Beiträge zur Paläontollogie von Österreich*, **11**, 35–49.

Kleineidam, C., Ernst, R., and Roces, F. (2001). Wind-induced ventilation of the giant nests of the leaf-cutting ant *Atta vollenweideri*. *Naturwissenschaften*, **88**, 301–305.

Kohler, T. and Vollrath, F. (1995). Thread biomechanics in the two orb-weaving spiders *Araneus diadematus* (Araneae, Araneidae) and *Uloborus walkenaerius* (Araneae, Uloboridae). *Journal of Experimental Zoology*, **27**, 1–17.

Kojima, J. (1983). Defense of the pre-emergence colony against ants by means of a chemical barrier in *Ropalidia fasciata* (Hymenoptera, Vesoidae). *Japanese Journal of Ecology*, **33**, 213–223.

Kojima, J. (1992). The ant repellent function of the rubbing substance in an Old World polistine *Parapolybia indica* (Hymenoptera, Vespidae). *Ethology, Ecology and Evolution*, **4**, 183–185.

König, B. (1985). Plant sources of propolis. *Bee World*, **66**, 136–139.

König, B. and Dustmann, J.H. (1989). Tree resins, bees and antiviral chemotherapy. *Animal Research and Development*, **29**, 21–42.

Korb, J. and Linsenmair, K.E. (1998). The effects of temperature on the architecture and distribution of *Macrotermes bellicosus* (Isoptera, Macrotermitinae) mounds in different habitats of a West African Guinea savannah. *Insectes Sociaux*, **45**, 51–65.

Korb, J. and Linsenmair, K.E. (1999). The architecture of termite mounds: a result of a trade-off between thermoregulation and gas exchange. *Behavioural Ecology*, **10**, 312–316.

Korb, J. and Linsenmair, K.E. (2000). Ventilation of termite mounds: new results require a new model. *Behavioural Ecology*, **11**, 486–494.

Kotliar, N.B. (2000). Application of the new keystone-species concept to prairie dogs: how well does it work? *Conservation Biology*, **14**, 1715–1721.

Kotliar, N.B., Baker, B.W., Whicker, A.D., and Plumb, G. (1999). A critical review of assumptions about the prairie dog as a keystone species. *Environmental Management*, **24**, 177–192.

Krantzberg, G. (1985). The influence of bioturbation on physical, chemical and biological parameters in aquatic environments: a review. *Environmental Pollution*, **39**, 99–122.

Krink, T. and Vollrath, F. (1997). Analysing spider web-building behaviour with rule-based simulations and genetic algorithms. *Journal of Theoretical Biology*, **185**, 321–331.

Krink, T. and Vollrath, F. (1998). Emergent properties in the behaviour of a virtual spider robot. *Proceedings of the Royal Society of London B*, **265**, 2051–2055.

Krink, T. and Vollrath, F. (1999). A virtual robot to model the use of regenerated legs in a web-building spider. *Animal Behaviour*, **57**, 223–232.

Krombein, K.V. (1978). Biosystematic studies of Celonese wasps. III. Life history, nest and associates of *Paraleptomenes mephitis* (Cameron) (Hymenoptera: Eumenidae). *Journal of the Kansas Entomological Society*, **51**, 721–734.

Krombein, K.V. (1991). Biosystematic studies of Celonese wasps, XIX: natural history notes in several families (Hymenoptera: Eumenidae, Vespidae, Pompilidae, and Crabronidae). *Smithsonian Contributions to Zoology*, **515**, 1–41.

Kudo, K. and Yamane, S. (1996). Occurrence of a binding matrix in a nest of a primitively eusocial wasp, *Eustenogaster calyptodoma* (Hymenoptera, Vespidae). *Japanese Journal of Entomology*, **64**, 891–895.

Kudo, K., Yamane, S. and Yamamoto, H. (1998). Physiological ecology of nest construction and protein flow in pre-emergence colonies of *Polistes chinensis* (Hymenoptera Vespidae): effects of rainfall and microclimates. *Ethology, Ecology and Evolution*, **10**, 171–183.

Kudo, K., Yamamoto, H., and Yamane, S. (2000). Amino acid composition of the protein in the pre-emergence nests of a paper wasp, *Polistes chinensis* (Hymenoptera, Vespidae). *Insectes Sociaux*, **47**, 371–375.

Kudo, S.I. (1994). Observations on lepidopteran leaf-shelters as moulting refuges for the stink bug *Elasmucha putoni* (Hymenoptera: Acanthosomatidae). *Psyche*, **101**, 183–186.

Kuhn, S.L. and Stiner, M.C. (1998). Middle Palaeolithic 'creativity'. In *Creativity in human evolution and prehistory* (ed. S. Mithen), pp. 143–164. Routledge, London.

Kunz, T.H., Fujita, M.S., Brooke, A.P., and McCracken, G.F. (1994). Convergence in tent architecture and tent making behaviour among Neotropical and Paleotropical bats. *Journal of Mammalian Evolution*, **2**, 57–58.

Kurczewski, F.E. and Spofford, M.G. (1986). Observations on the nesting behaviour of *Tachetes parvus* and *T. obductus* Fox (Hymenoptera: Sphecidae). *Proceedings of the Entonological Society of Washington*, **8**, 13–24.

Kurstjens, S.P. and McClain, E. (1990). The proteins of beeswax. *Naturwissenschaften*, **77**, 34–35.

Kurstjens, S.P., Hepburn, H.R., Schoening, F.R.L., and Davidson, B.C. (1985). The conversion of wax scales into comb wax by African honeybees. *Journal of Comparative Physiology B*, **156**, 95–102.

Kusmierski, R., Borgia, G., Uy, A., and Crozier, R.H. (1997). Labile evolution of display traits in bowerbirds indicates reduced effects of phylogenetic constraint *Proceedings of the Royal Society of London B*, **264**, 307–313.

Lafuma, L., Lambrechts, M.M., and Raymond, M. (2001). Aromatic plants in bird nests as a protection against blood-sucking flying insects? *Behavioural Processes*, **56**, 113–120.

Laland, K.N., Odling-Smee, F.J., and Feldman, M.W. (1996). The evolutionary consequence of niche construction: a theoretical investigation using two-locus theory. *Journal of Evolutionary Biology*, **9**, 293–316.

Laland, K.N., Odling-Smee, F.J., and Feldman, M.W. (1999). Evolutionary consequences of niche construction and their implications for ecology. *Proceedings of the National Academy of Science, USA*, **96**, 10242–10247.

Laland, K.N., Odling-Smee, J., and Feldman, M.W. (2000). Niche construction, biological evolution, and cultural change. *Behavioural and Brain Sciences*, **23**, 131–175.

Lavelle, P. (1997). Faunal activities and soil processes : adaptive strategies that determine ecosystem function. *Advances in Ecological Research*, **27**, 93–132.

Laville, E. (1989). Etude cinématique du fouissage chez *Arvicola terrestris* Scherman (Rodentia, Arvicolidae). *Mammalia*, **53**, 177–189.

Lawick-Goodall, J. Van (1968). The behaviour of free-living chimpanzees in the Gombe stream reserve. *Animal Behaviour Monographs*, **165**, 194–210, 298–300.

Lawton, J.H. and Jones, C.G. (1995). Linking species and ecosystems: organisms as ecosystem engineers. In *Linking species and ecosystems* (eds. C.G. Jones and J.H. Lawton), pp. 141–150. Chapman and Hall, New York.

Leader, N. and Yom-Tov, Y. (1998). Possible function of stone ramparts at the nest entrance of the blackstart. *Animal Behaviour*, **56**, 207–217.

Lee, K.E. and Wood, T.G. (1971). *Termites and soils*. Academic Press, London.

Lee, P.L.M., Clayton, D.H., Griffiths, R., and Page, R.D.M. (1996). Does behaviour reflect phylogeny in swiftlets (Aves: Apodidae)? A test using cytochrome *b* mitochondrial DNA sequences. *Proceedings of the National Academy of Science, USA*, **93**, 7091–7096.

Lefebvre, L., Nicolakakis, N., and Boire, D. (2002). Tools and brains in birds. *Behaviour*, **139**, 939–973.

Levington, J. (1995). Bioturbators as ecosystem engineers: control of the sediment fabric, inter-individual interactions, and material fluxes. In *Linking species and ecosystems* (eds. C.G. Jones and J.H. Lawton), pp. 29–36. Chapman and Hall, New York.

Li, D. and Lee, W.S. (2004). Predator-induced plasticity in web-building behaviour. *Animal Behaviour*, **67**, 309–318.

Lin, L.H., Edmonds, D.T., and Vollrath, F. (1995). Structural engineering of an orb-spider's web. *Nature*, **373**, 146–148.

Liversidge, R. (1962). The breeding biology of the little sparrowhawk *Accipiter minullus*. *Ibis*, **104**, 399–406.

Lombardo, M.P., Bosman, R.M., Faro, C.A., Houtteman, S.G., and Kulisza, T.S. (1995). Effect of feathers as nest insulation on incubation behaviour and reproductive performance of tree swallows (*Tachycineta bicolor*). *Auk*, **112**, 973–981.

London, K.B. and Jeanne, R.L. (2000). The interaction between mode of colony founding, nest architecture and ant defence in polistine wasps. *Ethology, Ecology and Evolution*, **12**, 13–25.

Loucks, O.L., Plumb-Mentjes, M.L., and Rogers, D. (1985). Gap processes and large-scale disturbances in sand prairies. In *The ecology of natural disturbance and patch dynamics* (eds. S.T.A. Pickett and P.S. White), pp. 72–83. Academic Press, Orlando, FL.

Loudon, C. and Alstad, D.N. (1990). Theoretical mechanisms of particle capture: Predictions for hydropsychid caddisfly distributional ecology. *American Naturalist*, **135**, 360–381.

Lounibos, L.P. (1975). The cocoon spinning behaviour of the Chinese oak silkworm *Antheraea pernyi*. *Animal Behaviour*, **23**, 843–853.

Lovegrove, B.G. and Siegfried, W.R. (1989). Spacing and origin(s) of Mima-like earth mounds in the Cape Province of South Africa. *South African Journal of Science*, **85** (7), 108–112.

Lubin, Y.D. (1995). Is there division of labour in the social spider *Achaearanea 1wau* (Theridiidae)? *Animal Behaviour*, **49**, 1315–1323.

Lucas, J.R. (1982). The biophysics of pit construction by antlion larvae (*Myrmeleon*, Neuroptera). *Animal Behaviour*, **30**, 651–664.

Luscher, M. (1955). Die Lufterneuerung im nest der Termite *Macrotermes natalensis* (Haviland). *Insectes Sociaux*, **3**, 273–276.

Lyon, B.E. and Cartar, R.V. (1996). Functional significance of the cocoon in two arctic *Gynaephora* moth species. *Proceedings of the Royal Society of London B*, **263**, 1159–1163.

MacArthur, R.H. (1955). Fluctuations of animal populations, and a measure of community stability. *Ecology*, **36**, 533–536.

MacKay, W.P. (1982). The effect of predation on western widow spiders (Araneae: Theridiidae) on harvester ants (Hymenoptera: Formicidae). *Oecologia*, **53**, 406–411.

MacKay, W.P. and MacKay, E.E. (1984). Why do harvester ants store seeds in their nests? *Sociobiology*, **9** (6), 31–47.

Madden, J. (2001). Sex, bowers and brains. *Proceedings of the Royal Society of London B*, **268**, 833–838.

Madden, J. (2002). Bower decorations attract females but provoke other male spotted bowerbirds: bower owners resolve this trade-off. *Proceedings of the Royal Society of London B*, **269**, 1347–1351.

Madden, J.R. (2003*a*). Male spotted bowerbirds preferentially choose, arrange and proffer objects that are good predictors of mating success. *Behavioural Ecology and Sociobiology*, **53**, 263–268.

Madden, J.R. (2003*b*). Bower decorations are good predictors of mating success in the spotted bowerbird. *Behavioural Ecology and Sociobiology*, **53**, 269–277.

Madge, S.G. (1970). Nest of the long-billed spiderhunter *Arachnothera robusta*. *Malay Naturalists Journal*, **23**, 125.

Magoshi, J., Magoshi, Y., and Nakamura S. (1994). Mechanism of fiber formation of silkworm. In *Silk polymers: materials science biotechnology, ACS Symposium Series 544* (eds. D. Kaplan, W.W. Adams, B. Farmer, and C. Viney), pp. 292–310. American Chemical Society, Washington.

Malas, D. and Wallace, J.B. (1977). Strategies for coexistence in three species of net-spinning caddisflies (Trichoptera) in second-order southern Appalachian streams. *Canadian Journal of Zoology*, **55**, 1829–1840.

Mallon, E.B. and Franks, N.R. (2000). Ants estimate area using Buffon's needle. *Proceedings of the Royal Society of London B*, **267**, 765–770.

Manson, D.C.M. and Gerson, U. (1996). Web spinning, wax secretion and liquid secretion by eriophyoid mites. In *Eriophyoid mites—their biology, natural enemies and control* (eds. E.E. Lindquist, M.W. Sabelis, and J. Bruin), pp. 251–258. Elsevier-Science Amsterdam.

Marcucci, M.C., Rodriguez, J., Ferres, F., and Bankova, V. (1998). Chemical composition of Brazilian propolis from Sao Paulo State. *Zeitschrift für Naturforschung C*, **53**, 117–119.

Martin, L.D. and Bennett, D.K. (1977). The burrows of the Miocene beaver *Palaeocastor*, Western Nebraska, U.S.A. *Palaeoecography, Palaeoclimatology, Palaeoecology*, **22**, 173–193.

Martin, T.E. (1993). Nest predation and nest sites. *Bioscience*, **43**, 523–532.

Martinsen, G.D., Floate, K.D., Waltz, A.M., Wimp, G.M., and Whitham, T.G. (2000). Positive interactions between leafrollers and other arthropods enhance biodiversity on hybrid cottonwoods. *Oecologia*, **123**, 82–89.

Martius, C. (1994). Termite nests as structural elements of the Amazon floodplain forest. *Andrias*, **13**, 137–150.
Martius, C. and Weller, M. (1998). Observations on dynamics of foraging hole construction of two leaf-feeding, soil-inhabiting *Syntermes* species (Insecta: Isoptera) in an Amazonian rainforest, Brazil. *Acta Amazonica*, **28**, 325–330.
Martius, C., Höfer, H., Verhaagh, M. Adis, J., and Mahnert, V. (1994). Terrestrial arthropods colonising an abandoned termite nest in a floodplain forest of the Amazon River during the flood. *Andrias*, **13**, 17–22.
Maschwitz, U., Dorow, W.H.O., and Botz, T. (1990). Chemical composition of the nest walls, and nesting behaviour, of *Ropalidia (Icarielia) opifex* van der Vecht, 1962 (Hymenoptera: Vespidae), a Southeast Asian social wasp with translucent nests. *Journal of Natural History*, **24**, 1311–1319.
Masters, W. and Moffat, A.J.M. (1983). A functional explanation of top-bottom asymmetry in vertical orbweds. *Animal Behaviour*, **31**, 1043–1046.
Matsuzawa, T. (1994). Field experiments on use of stone tools by chimpanzees in the wild. In *Chimpanzee cultures* (eds. R.W. Wrangham, W.C. McGrew, F.B.M. de Waal, and P.G. Heltne), pp. 351–370. Harvard University Press, Cambridge, MA.
Matthews, R.W. (1968). Nesting biology of the social wasp *Microstigmus comes. Psyche*, **75**, 23–45.
Matthews, R.W. (1991). Evolution of social behaviour in sphecid wasps. In *The social biology of wasps* (eds. K.G. Ross and R.W. Matthews), pp. 570–602. Comstock, Ithaca, NY.
Matthews, R.W. and Starr, C.K. (1984). *Microstigmus comes* wasps have a method of nest construction unique among social insects. *Biotropica*, **16**, 55–58.
Mattson, S. and Cedhagen, T. (1989). Aspects of the behaviour and ecology of *Dyopedos monacanthus* (Metzger) and *D. porrectus* Bate, with comparative notes on *Dulichia tuberculata* Boeck (Crustacea: Amphipoda: Podoceridae). *Journal of Experimental Marine Biology and Ecology*, **127**, 253–272.
Mayer, M.S., Schaffner, L., and Kemp, W.M. (1995). Nitrification potentials of benthic macrofaunal tubes and burrow walls: effects of sediment NH4+ and animal irrigation behaviour. *Marine Ecology Progress Series*, **121**, 157–169.
McDaniel, N. and Banse, K. (1979). A novel method of suspension feeding by the maldanid polychaete *Praxillura maculata.Marine Biology*, **55**, 129–132.
McGavin, G.C. (2001). *Essential entomology*. Oxford, Oxford University Press.
McGrew, W.C. (1974). Tool use by wild chimpanzees in feeding upon driver ants. *Journal of Human Evolution*, **3**, 501–508.
McGrew, W.C. (1992). *Chimpanzee material culture.* Cambridge University Press, Cambridge.

McGrew, W.C. (1993). The intelligent use of tools: Twenty propositions. In *Tools language and cognition in human evolution* (eds. K.R. Gibson and T. Ingold), pp. 151–170. Cambridge University Press, Cambridge.

McGrew, W.C. (1998). Culture in non-human primates? *Annual Review of Anthropology*, **27**, 301–328.

McGrew, W.C., Ham, R.M., White, L.J.T., Tutin, C.E.G., and Fernandez, M. (1997). Why don't chimpanzees in Gabon crack nuts? *International Journal of Primatology*, **18**, 353–374.

McNab, B.K. (1966). The metabolism of fossorial rodents: a study of convergence. *Ecology*, **47**, 712–733.

McNab, B.K. (1979). The influence of body size on the energetics and distribution of fossorial and burrowing animals. *Ecology*, **60**, 1010–1021.

Medway, Lord (1962). The swiftlets (*Collocalia*) of the Niah Cave, Sarawak. *Ibis*, **104**, 45–66.

Melo, G.A.R. and Matthews, R.W. (1997). Six new species of *Microstigmus* wasps (Hymenoptera: Sphecidae), with notes on their biology. *Journal of Natural History*, **31**, 421–437.

Mello, M.L.S., Pimentel, E.R., Yamada, A.T., and Storopoli-Neto, A. (1987). Composition and structure of the froth of the spittlebug, *Deois* sp. *Insect Biochemistry*, **17**, 493–502.

Merino, S. and Potti, J. (1995). Pied flycatchers prefer to nest in clean nest boxes in an area with detrimental nest parasites. *Condor*, **97**, 828–831.

Michener, C.D. (1974). *The social behaviour of the bees*. Belknap Press, Cambridge, MA.

Milborrow, B.V., Kennedy, J.M., and Dollin, A. (1987). Composition of wax made by the Australian stingless bee *Trigona australis.Australian Journal of Biological Science*, **40**, 15–25.

Miller, G. (2000). *The mating mind*. Heinemann, London.

Miller, J.S. and Zwickel, F.C. (1972). Characteristics and ecological significance of hay piles of pikas. *Mammalia*, **36**, 657–667.

Mills, L.S., Soulé, M.E., and Doak, D.F. (1993). The keystone-species concept in ecology and conservation. *Bioscience*, **43**, 219–224.

Miyashita, T. (1997). Factors affecting the difference in foraging success in three co-existing *Cyclosa* spiders. *Journal of Zoology, London*, **242**, 137–149.

Møller, A.P. (1990). Nest predation selects for small nest size in the blackbird. *Oikos*, **57**, 237–240.

Møller, A.P. (1991). The effect of feather nest lining on reproduction in the swallow *Hirundo rustica*. *Ornis Scandinavica*, **22**, 396–400.

Molles, M.C. and Nislow, K.H. (1990). Geographic variation in the structure of caddisfly cases: clues and influences of competition and predation. *Proceedings of the 6th International Symposium on Trichoptera*, 177–180.

Monson, G. (1943). Food habits of the banner-tailed Kangaroo rat in Arizona. *Journal of Wildlife Management*, **7**, 98–102.

Moore, J.M. and Picker, M.D. (1991). *Heuweltjies* (earth mounds) in the Clanwilliam district, Cape Province, South Africa 4000-year old termite nests. *Oecologia*, **86**, 424–432.

Moreno, J., Soler, M., Møller, A.P., and Linden, M. (1994). The function of stone carrying in the black wheatear, *Oenanthe leucura. Animal Behaviour*, **47**, 1297–1309.

Mugford, S.T., Mallon, E.B., and Franks, N.R. (2001). The accuracy of Buffon's needle: a rule of thumb used by ants to estimate area. *Behavioural Ecology*, **12**, 655–658.

Myles, T.G. (1988). Resource inheritance in social evolution from termites to man. In *The ecology of social behaviour* (ed. C.N. Slobodchikoff), pp. 379–423. Academic Press, New York.

Nakata, K. and Ushimaru, A. (1999). Feeding experience affects web relocation and investment in web threads in an orb-web spider, *Cyclosa argenteoalba. Animal Behaviour*, **57**, 1251–1255.

Nalesso, R.C., Duarte, L.F.L., Pierozzi, I., and Enumo, E.F. (1995). Tube epifauna of the polychaete *Phyllochaetopterus socialis* Claparède. *Esturine, Coastal and Shelf Science*, **41**, 91–100.

Nash, R.D.M., Chapman, C.J., Atkinson, R.J.A., and Morgan, P.J. (1984). Observations on the burrows and burrowing behaviour of *Calocaris macandreae* (Crustacea: Decapoda: Thalassinoidea). *Journal of Zoology*, **202**, 425–439.

Naug, D. and Gadagkar, R. (1998). The role of age in temporal polyethism in a primitively eusocial wasp. *Behavioural Ecology and Sociobiology*, **42**, 37–47.

Neal, E.G. (1986). *The natural history of badgers*. Croom Helm, London.

Nehring, S., Jensen, P., and Lorenzen, S. (1990). Tube-dwelling nematodes: tube construction and possible ecological effects on sediment-water interfaces. *Marine Ecology Progress Series* **6**, 123–128.

Nentwig, W. and Heimer, S. (1987). Ecological aspects of spider webs. In *Ecophysiology of spiders* (ed. W. Nentwig), pp. 211–225. Springer-Verlag, Berlin.

Nepomnyashchikh, V.A. (1993). Storage of information on the quality of particles for the case building by caddis fly larvae *Chaetopteryx villosa* (Trichoptera, Insecta). *Zhurnal Obnsshcei Biologii*, **53**, 317–323.

Nevo, E. (1979). Adaptive convergence and divergence of subterranean mammals. *Annual Review of Ecology and Systematics*, **10**, 269–308.

Nickell, L.A. and Atkinson, R.J.A. (1995). Functional morphology of burrows and trophic modes of three thalassinidean shrimp species, a new approach to the classification of thalassinidean burrow morphology. *Marine Ecology Progress Series*, **128**, 181–197.

Nickell, L.A., Hughes, D.J., and Atkinson, J.A. (1997). Megafaunal bioturbation in organically enriched Scottish sea lochs. *Biology and Ecology of Shallow Coastal Waters, 28 EMBS Symposium*, 315–322.

Nickell, L.A., Atkinson, R.J.A., and Pinn, E.H. (1998). Morphology of thalassinidean (Crustacea: Decapoda) mouthparts and periopods in relation to feeding, ecology and grooming. *Journal of Natural History*, **32**, 733–761.

Nicol, E.A.T. (1931). The feeding mechanism, formation of the tube, and physiology of digestion in *Sabella pavonina*. *Transactions of the Royal Society of Edinburgh*, **56**, 537–597.

Nislow, K.H. and Molles, M.C. (1993). The influence of larval case design on vulnerability of *Limnephilus frijole* (Trichoptera) to predation. *Freshwater Biology*, **29**, 411–417.

Noble, J.C. (1993). Relict surface-soil features in semi-arid mulga (*Acacia aneura*) woodlands. *Rangelands Journal*, **15**, 48–70.

Noji, T.T., Bathmann, U.V., Bodungen, B. von, Voss, M., Anita, A., Krumbholz, M., Klein, B., Peeken, I., Noji, C.I.M., and Rey, F. (1997). Clearance of pikoplankton-sized particles and formation of rapidly sinking aggregates by a pteropod, *Limacina retroversa*. *Journal of Plankton Research*, **19**, 863–875.

Nores, A.I. and Nores, M. (1994). Old nest accumulation as anti-predator strategy in brown cacholotes. *Journal für Ornithologie*, **135**, 198.

Nowbahari, B. and Thibout, E. (1990). The cocoon and humidity in the development of *Acrolepiopsis assectella* (Lep.) pupae: consequences in adults. *Physiological Entomology*, **15**, 363–368.

Nummi, P. and Pöysä, H. (1997). Population and community level responses in *Anas* species to patch disturbance caused by an ecosystem engineer, the beaver. *Ecography*, **20**, 580–584.

Odling-Smee, F.J., Laland, K.N., and Feldman, M.W. (1996). Niche construction. *American Naturalist*, **147**, 641–648.

Odling-Smee, F.J., Laland, K.N., and Feldman, M.W. (2003). *Niche construction: the neglected process in evolution*. Princeton University Press, Princeton, NJ.

O'Donnell, S.O. and Jeanne, R.L. (1990). Forager specialisation and the control of nest repair in *Polybia occidentalis* Olivier (Hymenoptera: Vespidae). *Behavioural Ecology and Sociobiology*, **27**, 359–364.

Okamoto, H. and Hirowatari, T. (2000). Biology of *Vespina nielseni* Kozlov (Lepidoptera: Incurvariidae), with description of immature stages and redescription of adults. *Entomological Science*, **3**, 511–518.

Olberg, G. (1959). *Das Verhalten der solitaren Wespen Mitteleuropas (Vespidae, Pompilidae, Sphecidae)*. VEB Deutcher Verlag der Wissenschaften, Berlin.

Oldroyd, H. (1954). *The horse-flies (Diptera: Tabanidae) ofthe Ethiopian region*. British Museum, London.

Olivera, R.F., McGregor, P.K., Burford, F.R., Custodio, M.R., and Latruffe, C. (1998). Functions of mudballing in the European fiddler crab *Uca tangeri*. *Animal Behaviour*, **55**, 1299–1309.

Olsson, K. and Allander, K. (1995). Do fleas, and/or old nest material, influence nest site preference in hole-nesting passerines? *Ethology*, **101**, 160–170.

Opell, D. (1994). Factors governing the stickiness of cribellar prey capture threads in the spider family Uloboridae. *Journal of Morphology*, **221**, 111–119.

Opell, B.D. (1995). Ontogenetic changes in cribellum spigot number and cribellar prey capture thread stickiness in the spider family Uloboridae. *Journal of Morphology*, **224**, 47–56.

Opell, B.D. (1997). A comparison of capture thread and architectural features ofdeinopoid and araneoid orb-webs. *Journal of Arachnology*, **25**, 295–306.

Opell, B.D. (1998). Economics of spider orb-webs: the benefits of producing adhesive capture thread and recycling silk. *Functional Ecology*, **12**, 613–624.

Opell, B.D. and Bond, J.E. (2001). Changes in the mechanical properties of capture threads and the evolution of modern orb-weaving spiders. *Evolutionary Ecology Research*, **3**, 567–581.

Oppliger, A., Richner, H., and Christe, P. (1994). Effect of an ectoparasite on lay date, nest-site choice, desertion, and hatching success in the great tit (*Parus major*). *Behavioural Ecology*, **5**, 130–134.

Osaki, S. (1989). Seasonal change in colour of spiders' silk. *Acta Arachnologica*, **38**, 21–28.

Oster, G.F. and Wilson, E.O. (1978). *Caste and ecology in the social insects*. Princeton University Press, Princeton, NJ.

Ott, J.A., Fuchs, B., and Malasek, A. (1976). Observations on the biology of *Callianassa stebbeini* Borrodaille and *Upogebia litoralis* Risso and their effect upon the sediment. *Senckbergiana Maritima*, **8**, 61–79.

Otto, C. (1987). Behavioural adaptations by *Agrypnia pagetana* (Trichoptera) larvae to cases of different value. *Oikos*, **50**, 191–196.

Otto, C. and Johansson, A. (1995). Why do caddis larvae in running waters construct heavy, bulky cases? *Animal Behaviour*, **49**, 473–478.

Otto, C. and Svensson, B.S. (1980). The significance of case material selection for the survival of caddis larvae. *Journal of Animal Ecology*, **49**, 855–865.

Page, R.E., Robinson, G.E., Britton, D.S., and Fondrk, M.K. (1992). Genotypic variability for rates of behavioural development in worker honeybees (*Apis mellifera* L.). *Behavioural Ecology*, **3**, 173–180.

Paine, R.T. (1969). A note on trophic complexity and community stability. *American Naturalist*, **103**, 91–93.

Palomino, J.J., Martín-Vivaldi, M., Soler, M., and Soler, J.J. (1998). Functional significance of nest size variation in the rufous bush robin *Cercotrichas galactotes*. *Ardea*, **86**, 177–185.

Pasquet, A., Ridwan, A., and Leborgne, R. (1994). Presence of potential prey affects web-building in an orb-weaving spider *Zygiella x-notata. Animal Behaviour*, **47**, 477–480.

Patricelli, G.L., Uy, J.A.C., and Borgia, G. (2003). Multiple male traits interact: attractive bower decorations facilitate attractive behavioural displays in satin bowerbirds. *Proceedings of the Royal Society of London B*, **270**, 2389–2395.

Peachey, R.L. and Bell, S.S. (1997). The effects of mucous tubes on the distribution, behaviour and recruitment of seagrass meiofauna. *Journal of Experimental Marine Biology and Ecology*, **209**, 279–291.

Pearson, T.G. (1936). *Birds of America*. Garden City, New York.

Peckham, D.J. (1985). Ethological observations on *Oxybelus* (Hymenoptera: Sphecidae) in Southwestern New Mexico. *Annals of the Entomological Society of America*, **78**, 865–872.

Penna, M. and Solis, R. (1999). Extent and variation of sound enhancement inside burrows of the frog *Eupsophus emiliopugini* (Leptodactylidae). *Behavioural Ecology and Sociobiology*, **47**, 94–103.

Pestrong, R. (1991). Nature's angle. *Pacific Discovery*, **44**, 28–37.

Petal, J. (1978). The role of ants in ecosystems. In *Production ecology of ants and termites* (ed. M.V. Brian), pp. 293–325. Cambridge University Press, Cambridge.

Peters, H.M. (1987). Fine structure and function of capture threads. In *Ecophysiology of spiders* (ed. W. Nentwig), pp. 188–202. Springer-Verlag, Berlin.

Peters, H.M. (1995). Ultrastructure of orb spiders' gluey capture threads. *Naturwissenschaften*, **82**, 380–382.

Petersen, R.C., Lena, B.M., and Wallace, J.B. (1984). Influence of velocity and food availability on catchnet dimensions of *Neureclipsis bimaculata* (Trichoptera: Polycentropidae). *Holarctic Ecology*, **7**, 380–389.

Petit, C.P., Hossaert-McKey, M., Perret, P., Blondel, J., and Lamberchts, M.M. (2002). Blue tits use selected plants and olfaction to maintain an aromatic environment for nestlings. *Ecology Letters*, **5**, 585–589.

Plague, G.R. (1999). Evolution of net-spinning caddisflies: a hypothetical mechanism for the reproductive isolation of conspecific competitors. *Oikos*, **87**, 204–208.

Powell, E.N. (1977). The relationship of the trace fossil *Gyrolithes* (=*Xenohelix*) to the family Capitellidae (Polychaeta). *Journal of Palaeontology*, **51**, 552–556.

Power, M.E., Tilman, D., Estes, J.A., Menge, B.A., Bond, W.J., Mills, S., Daily, G., Castilla, J.C. Lubchenko, J., and Paine, R.T. (1996). Challenges in the quest for keystones. *Bioscience*, **46**, 609–620.

Prance, G. (1992). Ant association with *Parinari excelsa* (Chrysobalanaceae) in Marajo, Brazil. *Biotropica*, **24**, 102–104.

Pruett-Jones, M.A. and Pruett-Jones, S.G. (1982). Spacing and distribution of bowers in Macgregor's bowerbird (*Amblyornis macgregoriae*). *Behavioural Ecology and Sociobiology*, **11**, 25–32.

Pyke, G.H. (1984). Optimal foraging theory: a critical review. *Annual Review of Ecology and Systematics*, **15**, 523–575.

Raup, D.M. (1966). Geometric analysis of shell coiling: general problems. *Journal of Paleontology*, **40**, 1178–1190.

Read, A.T., McTeague, J.A., and Govind, C.K. (1991). Morphology and behaviour of an unusually flexible thoracic limb in the snapping shrimp, *Alpheus heterochelis*. *Biological Bulletin*, **181**, 158–168.

Reddy, M.V. (1983). Some physico-chemical properties of carton material of the structural-wood destroying *Coptotermes kishori* (Roonwall and Chhotani) in relation to the underlying soils. *Comparative Physiology and Ecology*, **8**, 345–348.

Redman, P., Selman, C., and Speakman, J.R. (1999). Male short-tailed voles (*Microtus agrestis*) build better insulated nests than females. *Journal of Comparative Physiology B*, **169**, 581–587.

Reichman, O.J. and Rebar, C. (1985). Seed preferences by desert rodents based on levels of mouldiness. *Animal Behaviour*, **33**, 726–729.

Reichman, O.J. and Seabloom, E.W. (2002). The role of pocket gophers as subterranean ecosystem engineers. *Trends in Ecology and Evolution*, **17**, 44–49.

Reichman, O.J., Benedix, J.H., and Seastedt, T.R. (1993). Distinct animal-generated edge effects in a tallgrass prairie community. *Ecology*, **74**, 1281–1285.

Reichman, O.J., Fattaey, A., and Fattaey, K. (1986). Management of sterile and mouldy seeds by a desert rodent. *Animal Behaviour*, **34**, 221–225.

Reid, J.M., Cresswell, W., Holts, S., Mellanby, R.J., Whitfield, D.P., and Ruxton, G.D. (2002). Nest scrape design and clutch heat loss in pectoral sandpipers (*Calidris melanotus*). *Functional Ecology*, **16**, 305–312.

Reid, J.M., Monaghan, P. and Ruxton, G.D. (1999). The effect of clutch cooling rate on starling, *Sturnus vulgaris*, incubation strategy. *Animal Behaviour*, **58**, 1161–1167.

Reynolds, P.C. (1993). The complementation theory of language and tool use. In *Tools, language and cognition in human evolution* (eds. K.R. Gibson and T. Ingold), pp. 407–428. Cambridge University Press, Cambridge.

Rhisiart, A. and Vollrath, F. (1994). Design features of the orb web of the spider, *Araneus diadematus*. *Behavioural Ecology*, **5**, 280–287.

Riakow, R.J. (1986). Why are there so many kinds of passerine birds? *Systematic Zoology*, **35**, 255–259.

Richard, B. (1961). Le déblaiement chez le castor: Rapport entre le déblaiement et la réalisation des canaux et des barrages. *Vie et Milieu*, **12**, 507–515.

Richard, P.B. (1955). Bièvres, constructeurs de barrages. *Mammalia*, **19**, 293–301.

Richard, P.B. (1958). Les pattes du bièvre (*Castor fiber*). *Mammalia*, **22**, 566–574.

Richner, H., Oppliger, A. and Christe, P. (1993). Effects of an ectoparasite on reproduction in great tits. *Journal of Animal Ecology*, **62**, 703–710.

Riechert, S.E. (1985). Why do some spiders co-operate? *Aelana consociata*, a case study. *Florida Entomologist*, **68**, 105–116.

Ripley, S.D. (1957). Notes on the horned coot, *Fulica cornuta* Bonaparte. *Postilla, Yale Peabody Museum of Natural History*, **30**, 1–8.

Risch, S.J. and Carroll, C.R. (1986). Effects of seed predation by a tropical ant on competition among weeds. *Ecology*, **67**, 1319–1327.

Roberts, R., Walsh, G., Murray, A., Olley, J., Jones, R., Morwood, M., Tuniz, C., Lawson, E., Macphall, M., Bowrery, D., and Naumann, I. (1997). Luminescence dating of rock art and past environments using mud-wasp nests in Northern Australia. *Nature*, **387**, 696–699.

Robinson, G.E., Page, R.E. and Huang, Z.-Y. (1994). Temporal polyethism in social insects in a developmental process. *Animal Behaviour*, **48**, 467–469.

Robinson, M.H. and Robinson, B. (1970). The stabilimentum of the orb web spider, *Argiope argentata*: an improbable defence against predators. *Canadian Entomologist*, **102**, 641–655.

Robinson, M.H. and Robinson, B. (1972). The structure, possible function and origin of the remarkable ladder-web built by a New Guinea orb-web spider (Araneae-Araneidae). *Journal of Natural History*, **6**, 687–694.

Roper, T.J., Bennett, N.C., and Molteno, A.J. (2001). Environmental conditions in burrows of two species of African mole-rat, *Georhychus capensis* and *Cryptomys damarensis*. *Journal of Zoology, London*, **254**, 101–107.

Roper, T.J., Jackson, T.P., Conradt, L., and Bennett, N.C. (2002). Burrow use and the influence of ectoparasites in Brants' Whistling rat *Parotomys brantsii*. *Ethology*, **108**, 557–564.

Rose, K.D. and Emry, R.J. (1983). Extraordinary adaptations in the Oligocene palaeodonts *Epoicotherium* and *Xenocranium* (Mammalia). *Journal of Morphology*, **175**, 33–56.

Rosell, F. and Parker, H. (1996). Beverens innvirkning på økosystemet en nøkkelart vendertilbake. *Fauna (Oslo)*, **49** 192–211.

Rosenheim, J.A. (1987). Nesting biology and bionomics of a solitary ground-nesting wasp, *Ammophila dysmica* (Hymenoptera: Sphecidae): Influence of parasite pressure. *Annals of the Entomological Society of America*, **80**, 739–749.

Roubik, D.W. (1989). *Ecology and natural history of tropical bees*. Cambridge University Press, Cambridge, MA.

Rowden, A.A. and Jones, M.B. (1997). Recent mud shrimp burrows and bio-turbation. *Porcupine Newsletter*, **6**, 154–158.

Rowell, D.M. and Avilés, L. (1995). Sociality in a bark-dwelling huntsman spider from Australia, *Delena cancerides* Walckenaer (Araneae: Sparassidae). *Insectes Sociaux*, **42**, 287–302.

Rowlands, M.L.J. (1985). Factors controlling case building behaviour in *Glyphotaelius pellucidus* (Limnephilidae). PhD. thesis, University of Glasgow.

Rowley, I. (1970). The use of mud in nest building—a review of incidence and taxonomic importance. *Ostrich*, Suppl. 8, 139–148.

Ruf, C. and Fiedler, K. (2000). Thermal gains through collective metabolic heat production in social caterpillars of *Eriogaster lanestris*. *Naturwissenschaften*, **87**, 193–196.

Ryan, M.J. (1997). Sexual selection and mate choice. In *Behavioural ecology: an evolutionary approach, 4th Edition*: (eds. J.R. Krebs and N.B. Davies), pp. 179–202. Blackwell Scientific, Oxford.

Rypstra, A.L. (1982). Building a better insect trap; an experimental investigation of prey capture in a variety of spider webs. *Oecologia*, **52**, 31–36.

Sakagami, S.F., Tanno, K., Tsutsui, H., and Honma, K. (1985). The role of cocoons in overwintering of the soybean pod borer *Leguminivora glycinivorella* (Lepidoptera: Tortricidae). *Journal of the Kansas Entomological Society*, **58**, 240–247.

Sakagami, S.F., Inoue, T., Yamane, S., and Salmah, S. (1989). Nests of myrmecophilous stingless bee, *Trigona moorei*: how do bees initiate their nests within an arboreal ant nest? *Biotropica*, **21**, 265–274.

Sakurai, K. (1988*a*). Leaf size recognition and evaluation of some attelabid weevils. (2) *Apoderus balteatus*. *Behaviour*, **106**, 301–316.

Sakurai, K. (1988*b*). Leaf size recognition and evaluation of some attelabid weevils. (1) *Chonostropheus chujoi*. *Behaviour*, **106**, 279–300.

Salvadori, M. (1980). *Why buildings stand up*. Norton & Co. New York.

Sandoval, C.P. (1994). Plasticity in web design in the spider *Parawixia bistriata*: a response to variable prey type. *Functional Ecology*, **8**, 701–707.

Sato, R., Tanaka, Y. and Ishimaru, T. (2001). House production by *Oikopleura dioica* (Tunicata, Appendicularia) under laboratory conditions. *Journal of Plankton Research*, **23**, 415–423.

Sattaur, O. (1991). Termites change the face of Africa. *New Scientist*, 26 January, 27.

Schaller, G.B. (1998). *Wildlife of the tibetan steppe*. University of Chicago Press, Chicago, IL.

Schauensee, R.M. de (1984). *The birds of China*. Smithsonian Institution Press, Washington.

Schmidt, G.H. and Gürsch, E. (1971). Anerlyse der Spinnbewegungen der Larve von *Formica pratensis* Retz. (Form. Hym. Ins.). *Zeitschrift fur Tierpsychologie*, **28**, 19–32.

Schmitt, W.L. (1975). *Crustaceans*. University of Michigan Press, Ann Arbor, MI.

Schneider, J.M. and Vollrath, F. (1998). The effect of prey type on the geometry of the capture web of *Araneus diadematus*. *Naturwissenschaften*, **85**, 391–394.

Schremmer, F. (1977). Das Baumrinden-Nest der neotropischen Faltenwespe *Nectarinella championi*, umgeben von einem Leimring als Ameisen-Abwehr (Hymenoptera: Vespidae). *Entomologia Germanica*, **3**, 344–355.

Schremmer, F. (1978). Zum Einfluss verschiedener Nestunterlagen-Neigungen aus nestform und Wabengrosse bei zwei neotropischen Arten sozialer Faltenvespen der Gattung *Parachartergus* (Hymenoptera: Vespidae). *Entomologia Germanica*, **4**, 356–367.

Schremmer, F. (1983). Das nest der neotropischen faltenvespe *Leipomeles dorsata*. Ein beitrag zur kenntnis der nestarchitektur der sozialen faltenvespen (Vespidae, Polistini, Polybiini). *Zoologische Anziger*, **211**, 95–107.

Schremmer, F., Marz, L., and Simonsberger, P. (1985). Chitin im Speichel der Papierwespen (soziale Faltenwespen, Vespidae): *Biologie, Chemismus, Feinstruktur und Mikroskopie*, **42**, 52–56.

Schütt, K. (1996). Wie Spinnen ihre Netz befestigen. *Mikrokosmos*, **85**, 274–278.

Scott Turner, J. (1994). Ventilation and thermal constancy of a colony of a southern African termite (*Odontotermes transvaalensis*: Macrotermitinae). *Journal of Arid Environments*, **28**, 231–248.

Scott Turner, J. (2000). *The extended organism: the physiology of animal-built structures*. Harvard, Cambridge, MA.

Seah, W.K. and Li, D. (2001). Stabilimenta attract unwelcome predators to orbwebs. *Proceedings of the Royal Society of London B*, **268**, 1553–1558.

Seeley, T. (1977). Measurement of nest cavity volume by the honey bee (*Apis mellifera*). *Behavioural Ecology and Sociobiology*, **2**, 201–227.

Seeley, T.D. (1982). Adaptive significance of age polyethism schedule in honeybee colonies. *Behavioural Ecology and Sociobiology*, **11**, 287–293.

Seeley, T.D. and Kolmes, S.A. (1991). Age polyethism for hive duties in honey bees—Illusion or reality? *Ethology*, **87**, 284–297.

Sehnal, F. and Akai, H. (1990). Insect silk glands: their types, development and function, and effects of environmental factors and morphogenetic hormones on them. *International Journal of Insect Morphology and Embryology*, **19**, 79–132.

Sendall, K.A., Fontaine, A.R., and O'Foighil, D. (1995). Tube morphology and activity patterns related to feeding and tube building in the polychaete *Mesochaetopterus taylori* Potts. *Canadian Journal of Zoology*, **73**, 509–517.

Sendova-Franks, A.B. and Franks, N.R. (1999). Self-assembly, self-organisation and division of labour. *Philosophical Transactions of the Royal Society B*, **354**, 1395–1405.

Shaw, W.T. (1934). The ability of the giant kangaroo rat as a harvester of seeds. *Journal of Mammalogy*, **15**, 275–286.

Sherman, P.M. (1994). The orb-web: an energetic and behavioural estimator of the spider's dynamic foraging and reproductive strategies. *Animal Behaviour*, **48**, 19–34.

Shivashankar, T. (1994). Advanced sub-social behaviour in the scorpion *Heterometrus fulvipes* Brunner (Arachnida). *Journal of Bioscience*, **19**, 81–90.

Sick, H. (1993). *Birds in Brazil.* Princeton University Press, Princeton, NJ.

Silver, M.W., Coale, S.L., Pilskaln, C.H., and Steinberg, D.K. (1998). Giant aggregates: Importance as microbial centres and agents of material flux in the mesopelagic zone. *Limnology and Oceanography*, **43**, 498–507.

Simmons, A.H., Michal, C.A., and Jelinski, L.W. (1996). Molecular orientation and two-component nature of the crystalline fraction of spider dragline silk. *Science*, **271**, 84–87.

Skarka, V., Deneubourg, J.L., and Belic, M.R. (1990). Mathematical model of building behaviour of *Apis mellifera. Journal of Theoretical Biology*, **147**, 1–16.

Skutch, A.F. (1954). *Life histories of Central American birds: families Fringillidae, Thraupidae, Icteridae, Parulidae and Coerebidae*, No. 31, Cooper Ornithological Society, Berkeley, CA.

Skutch, A.F. (1960). *Life histories of Central American birds Vol. 2: families Fringillidae, Vireonidae, Sylviidae, Turdidae, Trogloditidae, Paridae, Corvidae, irundinidae and Tyrannidae*, No. 34, Cooper Ornithological Society Berkeley, CA.

Skutch, A.F. (1969). *Life histories of Central American birds Vol. 3: Families Cotingidae, Pipridae, Formicariidae, Furnariidae, Dendrocolaptidae and Picidae*, No. 35, Cooper Ornithological Society, Berkeley, CA.

Slagsvold, T. (1989*a*). Experiments on clutch size and nest size in passerine birds. *Oecologia*, **79**, 297–302.

Slagsvold, T. (1989*b*). On the evolution of clutch size and nest size in passerine birds. *Oecologia*, **79**, 300–3005.

Smallwood, P.D. (1993). Web-site tenure in the long-jawed spider: is it risk-sensitive foraging, or conspecific interactions. *Ecology*, **74**, 1826–1835.

Smith, A.T. and Foggin, J.M. (1999). The plateau pika (*Ochotona curzoniae*) is a keystone species for biodiversity on the Tibetan plateau. *Animal Conservation*, **2**, 235–240.

Smith, F.A. (1995). Den characteristics and survivorship of woodrats (*Neotoma lepida*) in the Mojave desert. *The Southwestern Naturalist*, **40**, 366–372.

Soler, J.J., Møller, A.P., and Soler, M. (1998). Nest building, sexual selection and parental investment. *Evolutionary Ecology*, **12**, 427–441.

Soler, J.J., De Neve, L., Martinez, J.G., and Soler, M. (2001). Nest size affects clutch size and the start of incubation in magpies: an experimental study. *Behavioural Ecology*, **12**, 301–307.

Solís, J.C. and Lope, F. de (1995). Nest and egg crypsis in the ground-nesting stone curlew *Burhinus oedicnemus. Journal of Avian Biology*, **26**, 135–138.

Sousa, W.P. (1985). Disturbance and patch dynamics on rocky intertidal shores. In *The ecology of natural disturbance and patch dynamics* (eds. S.T.A. Pickett and P.S. White), pp. 101–124. Academic Press, Orlando, FL.

Spada, F., Steen, H., Troedsson, C., Kallesøe, T., Spriet, E., Mann, M., and Thompson, E.M. (2001). Molecular patterning of the oikoplastic epithelium of the larvacean tunicate *Oikopleura dioica*. *Journal of Biological Chemistry*, **276**, 20624–20632.

Spradbery, J.P. (1973). *Wasps. An account of the biology and natural history of solitary and social wasps*. Packard Publishing Ltd, Chichester.

Stamhuis, E.J., Schreurs, C.E., and Videler, J.J. (1997). Burrow architecture and turbative activity of the thalassinid shrimp *Callianassa subterranea* from the central North Sea. *Marine Ecology Progress Series*, **151**, 155–163.

Stanley, T.R. (2002). How many kilojoules does a black-billed magpie nest cost? *Journal of Field Ornithology*, **73**, 292–297.

Statzner, B., Arens, M.-F., Champagne, J.-Y., Morel, R., and Herouin, E. (1999). Silk-producing stream insects and gravel erosion: significant biological effects on critical shear stress. *Water Resources Research*, **35**, 3495–3506.

Steinberg, D.K., Silver, M.W., Pilskaln, C.H., Coale, S.L., and Paduan, J.B. (1994). Midwater zooplankton communities on pelagic detritus (giant larvacean houses) in Monterey Bay, California. *Limnology and Oceanography*, **39**, 1606–1620.

Steinberg, D.K., Silver, M.W., and Pilskaln, C.H. (1997). Role of mesopelagic zooplankton in the community metabolism of giant larvacean house detritus in Monterey Bay, California, USA. *Marine Ecology Progress Series*, **147**, 167–179.

Stevens, D.J., Hansell, M.H., Freel, J.A., and Monaghan, P. (1999). Developmental trade-offs in caddis flies: increased investment in larval defence alters adult resource allocation. *Proceedings of the Royal Society, London B*, **266**, 1049–1054.

Stevens, D.J., Hansell, M.H., and Monaghan, P. (2000). Developmental trade-offs and life histories: strategic allocation of resources in caddis flies. *Proceedings of the Royal Society, London B*, **267**, 1511–1515.

Stowe, M.K. (1986). Prey specialisation in the Araneidae. In *Spiders: webs, behaviour, and evolution* (ed. W.A. Shear), pp. 101–131. Stanford University Press, Stanford, CA.

Strassmann, J.E., Hughes, C.R., Turillazzi, S., Solís, C.R. and Queller, D.C. (1994). Genetic relatedness and incipient eusociality in stenogastrine wasps. *Animal Behaviour* **48**, 813–821.

Streng, R. (1974). Theoretische Betrachtung der Achtertour, ein bei kokonspinnenden Insektenlarven häufiges Bewegungsmuster. *Zeitschrift fur Tierpsychologie*, **35**, 157–172.

Stuart, A.E. and Currie, D.C. (2002). Behaviour is not reliably inferred from end-product structure in caddisflies. *Ethology*, **108**, 837–856.

Stuart, A.E. and Hunter, F.F. (1995). A re-description of the cocoon-spinning behaviour of *Simulium vittatum* (Diptera Simuliidae). *Ethology, Ecology and Evolution*, **7**, 363–377.

Stuart, A.E. and Hunter, F.F. (1998). End products of behaviour versus behavioural characters: a phylogenetic investigation of pupal cocoon construction and form in some North American black flies (Diptera: Simuliidae). *Systematic Entomology*, **23**, 387–398.

Sugiyama, Y. (1985). The brush-stick of chimpanzees found in South-west Cameroon and their cultural characteristics. *Primates*, **26**, 361–374.

Suzuki, S., Kuroda, S. and Nishihara, T. (1995). Tool-set for termite fishing by chimpanzees in the Ndoki forest, Congo. *Behaviour*, **132**, 219–235.

Taffe, C.A. (1983). The biology of the mud-wasp *Zeta abdominale* (Drury) (Hymenoptera: Eumenidae). *Zoological Journal of the Linnean Society*, **77**, 385–393.

Tagawa, J. (1996). The function of the cocoon of the parasitoid wasp, *Cotesia glomerata*L. (Hymenoptera: Braconidae): protection against dessication. *Applied Entomological Zoology*, **31**, 99–103.

Takagi, S. (1992). A contribution to conchaspidid systematics (Homoptera: Coccoidea). *Insecta Matsumurana*, **46**, 1–71.

Takagi, S. and Miyatake, Y. (1993). SEM observations on two lerp-forming psylloids (Homoptera). *Insecta Matsumurana*, **49**, 69–104.

Takegaki, T. and Nakazono, A. (2000). The role of mounds in promoting water-exchange in the egg-tending burrows of monogamous goby, *Valenciennea longipinnis* (Lay et Bennett). *Journal of Marine Biology and Ecology*, **253**, 149–163.

Tanaka, K. (1989). Energetic cost of web construction and its effect on web relocation in the web-building spider *Agelana limbata*. *Oecologia*, **81**, 459–464.

Taylor, M.I., Turner, G.F., Robinson, R.L., and Stauffer, J.R. (1998). Sexual selection, parasites and bower height skew in a bower-building cichlid fish. *Animal Behaviour*, **56**, 379–384.

Tebbich, S. and Bshary, R. (2004). Cognitive abilities related to tool use in the woodpecker finch, *Castospiza pallida*. *Animal Behaviour*, **67**, 289–297.

Tebbich, S., Taborsky. M., Ferssl, B., and Blomqvist, D. (2001). Do woodpecker finches acquire tool-use by social learning? *Proceedings of the Royal Society of London*, **268**, 2189–2193.

Tevis, L. (1958). Interrelations between the harvester ant *Veromessor pergandei* (Mayr) and some desert ephemerals. *Ecology*, **39**, 695–704.

Thaler, E. (1976). Nest und Nestbau von Winter- und Sommergoldhühnchen (*Regulus regulus und R. ignicappillus*). *Journal für Ornithologie*, **117**, 121–144.

Theraulaz, G. and Bonabeau, E. (1995). Modelling the collective building of complex architectures in social insects with lattice swarms. *Journal of Theoretical Biology*, **177**, 381–400.

Theraulaz, G., Bonabeau, E., and Deneubourg, J.-L. (1998). The origin of nest complexity in social insects. *Complexity*, **3** (6), 15–25.

Thiel, M., Sampson, S., and Watling, L. (1997). Extended parental care in two endobenthic amphipods. *Journal of Natural History*, **31**, 713–725.

Thies, K.M., Thies, M.L., and Caire, W. (1996). House construction by the southern plains woodrat (*Neotoma micropus*) in Southwestern Oklahoma. *The Southwestern Naturalist*, **41**, 116–122.

Thirunavukarasu, P., Nicholson, M., and Elgar, M.A. (1996). Leaf selection by the leaf-curling spider *Phonognatha graeffei* (Keyserling) (Araneoidea: Araneae). *Bulletin of the British Arachnological Society*, **10**: 187–189.

Tho, Y.P. (1981). Unique defence strategy in the termite *Prohabitermes mirabilis* (Haviland) of Peninsular Malaysia. *Biotropica*, **13**, 236–238.

Thompson, A.W. D. (1942). *On growth and form*. Cambridge University Press, Cambridge.

Tillinghast, E.K. and Townley, M. (1987). Chemistry, physical properties, and synthesis of araneidae orb webs. In *Ecophysiology of spiders* (ed. W. Nentwig), pp. 203–210. Berlin, Springer-Verlag.

Timm, R.M. (1984). Tent construction by *Vampyressa* in Costa Rica. *Journal of Mammalogy*, **65**, 166–167.

Timm, R.M. (1987). Tent construction by bats of the genera *Artibeus* and *Uroderma*. In *Studies in neotropical mammology: Essays in honour of Philip Herskgovitz* (eds. B.D. Patterson and R.M. Timm), *Fieldiana Zoology, New Series* **39**, 187–212.

Timm, R.M. and Mortimer, (1976). Selection of roost sites by Honduran white bats, *Ectophylla alba* (Chiroptera: Phyllostomatidae). *Ecology*, **57**, 385–389.

Tinbergen, N. (1953). Specialists in nest building. *Country Life*, 30 January, 270–271.

Tofts, C. and Franks, N.R. (1992). Doing the right thing: ants, honeybees and naked mole-rats. *Trends in Ecology and Evolution*, **7** (c4), 346–349.

Tokeshi, M. and Townsend, C.R. (1987). Random patch formation and weak competition: coexistence in epiphitic chironomid community. *Journal of Animal Ecology*, **56**, 833–845.

Townley, M.A., Bernstein, D.T., Gallagher, K.S., and Tillinghast, E.K. (1991). Comparative study of orb web hygroscopicity and adhesive spiral composition in three araneid spiders. *Journal of Experimental Zoology*, **259**, 154–165.

Tso, I.M. (1998). Stabilimentum-decorated webs spun by *Cyclosa conica* (Araneae, Araneidae) trapped more insects than undecorated webs. *Journal of Arachnology* **26**, 101–105.

Tulloch, A.P. (1980). Beeswax-Composition and analysis. *Bee World*, **61**, 47–62.

Turillazzi, S. (1989). The origin and evolution of social life in the Stenogastrinae (Hymenoptera, Vespidae). *Journal of Insect Behaviour*, **2**, 649–661.

Turillazzi, S. (1991). The Stenogastrinae. In *The social biology of wasps* (eds. K.G. Ross and R.W. Matthews), pp. 74–98. Comstock Publishing Associates, Ithaca, NY.

Turillazzi, S. (1999). New species of *Liostenogaster* van der Vecht 1969, with keys to adults and nests (Hymenoptera Vespidae Stenogastrinae). *Tropical Zoology*, **12**, 335–358.

Turillazzi, S. and Pardi, L. (1981). Ant guards on nests of *Parischnogaster nigricans serrei* (Buysson) (Stenogastrinae). *Monitore Zoologico Italiano*, **15**, 1–7.

Tusculescu, R., Topoff, H.,and Wolfe, S. (1975). Mechanisms of pit construction by antlion larvae. *Annals of the Entomological Society of America*, **68**, 719–720.

Tweddle, D., Eccles, D.H., Frith, C.B., Fryer, G., Jackson, P.B.N., Lewis, D.S.C., and Lowe-McConnell, R.H. (1998). Cichlid spawning structures—bowers or nests? *Environmental Biology of Fishes*, **51**, 107–109.

Ubilla, M. and Altuna, C.A. (1990). Analyse de la morphologie de la main chez des espèces de *Ctenomys* de l'Uruguay (Rodentia: Octodontidae). Adaptations au fouissage et implications évolutives. *Mammalia*, **54**, 107–117.

Uttal, L. and Buck, K.R. (1996). Dietry study of the midwater polychaete *Poeobius meseres* in Monterey Bay, California. *Marine Biology* **125**, 333–343.

Uy, J.A.C. and Borgia, G. (2000). Sexual selection drives rapid divergence in bowerbird display traits. *Evolution*, **54**, 273–278.

Van der Kloot, W.G. and Williams, C.M. (1953). Cocoon construction by the *Cecropia* silkworm. I: the role of the external environment. *Behaviour*, **5**, 141–156.

Vander Wall, S.B. (1990). *Food hoarding in animals*. Chicago University Press, Chicago, IL.

Vellenga, R.E. (1970). Behaviour of the male satin bowerbird at the bower. *The Australian Birdbander*, March, 3–11.

Venner, S., Pasquet, A. and Leborgne, R. (2000). Web-building behaviour in the orb-weaving spider *Zygiella x-notata*: influence of experience. *Animal Behaviour*, **59**, 603–611.

Vitousek, P.M., Mooney, H.A., Lubchenko, J., and Melillo, J.M. (1997). Human domination of Earth's ecosystems. *Science*, **277**, 494–499.

Vleck, D. (1979). The energy cost of burrowing by the pocket gopher *Thomomys bottae*. *Physiological Zoology*, **52**, 122–136.

Vleck, D. (1981). Burrow structure and foraging costs in the fossorial rodent, *Thomomys bottae*. *Oecologia*, **49**, 391–396.

Vogel, S. (1978). Organisms that capture currents. *Scientific American*, **239** (2), 108–117.

Vogel, S., Ellington, C.P., and Kilgore, D.L. (1973). Wind-induced ventilation of the burrow of the prairie-dog, *Cynomys ludovicianus*. *Journal of Comparative Physiology*, **85**, 1–14.

Vollrath, F. (1985). Web spider's dilemma: a risky move or site dependent growth. *Oecologia*, **68**, 69–72.

Vollrath, F. (1987). Altered geometry of webs in spiders with regenerated legs. *Nature*, **328**, 247–248.

Vollrath, F. (1999). Biology of spider silk. *International Journal of Biological Macromolecules*, **24**, 81–88.

Vollrath, F. and Knight, D.P. (2001). Liquid crystalline spinning of spider silk. *Nature*, **110**, 541–548.

Vollrath, F. and Tillingast, E.K. (1991). Glycoprotein glue beneath a spider web's aqueous coat. *Naturwissenschaften*, **78**, 557–559.

Vollrath, F., Fairbrothwer, W.J., Williams, R.J.P., Tillinghast, E.K., Bernstein, D.T., Gallagher, K.S., and Townley, M.A. (1990). Compounds in the droplets of the orb spider's viscid spiral. *Nature*, **345**, 526–528.

Vovelle, J. (1973). Evolution de la taille des grains du tube arénacé en function de la croissance chez *Pectinaria (Lagis) koreni* Malmgren (Polychète sédentaire). *Ophelia*, **10**, 169–184.

Vovelle, J. (1997). Organes constructeurs et matériaux sécretés chez les polychètes tubicoles: homologies et convergences. *Bulletin de la Société Zoologique de France: Evolution et Zoologie*, **122**, 59–66.

Wakano, J.Y., Nakata, K., and Yammura, N. (1998). Dynamic model of optimal age polyethism in social insects under stable and fluctuating environments. *Journal of Theoretical Biology*, **193**, 153–165.

Wallace, E.D. and Bennett, N.C. (1998). The colony structure and social organisation of the giant Zambian mole-rat, *Cryptomys mechowi*. *Journal of Zoology, London*, **244**, 51–61.

Wallace, J.B. and Malas, D. (1976). The fine structure of capture nets of larval Philopotamidae (Trichoptera), with special emphasis on *Dolophilodes distinctus*. *Canadian Journal of Zoology*, **54**, 1788–1802.

Wallace, J.B. and Sherberger, F.F. (1975). The larval dwelling and feeding structure of *Macronema transversum* (Walker) (Trichoptera): Hydropsychidae). *Animal Behaviour*, **23**, 592–596.

Wallis, D.I. (1960). Spinning movements in the larvae of the ant *Formica fusca Insectes Sociaux*, **7**, 187–199.

Watanabe, T. (1999*a*). Prey attraction as a possible function of silk decoration of the uloborid spider *Octonoba sybotides*. *Behavioural Ecology*, **10**, 607–611.

Watanabe, T. (1999*b*). The influence of energetic state on the form of stabilimentum built by *Octonoba sybotides* (Araneae: Uloboridae). *Ethology*, **105**, 719–725.

Watanabe, T. (2000). Web tuning of an orb-web spider, *Octonoba sybotides*, regulates prey-catching behaviour. *Proceedings of the Royal Society of London B*, **267**, 565–569.

Weeks, H.P. (1977). Nest reciprocity in eastern phoebes ans barn swallows. *Wilson Bulletin*, **89**, 632–635.

Weeks, H.P. (1978). Clutch size variation in the eastern phoebe in Southern Indiana. *Auk*, **95**, 656–666.

Weir, A.A.S., Chappell, J. and Kacelnik, A. (2002). Shaping of hooks in New Caledonian crows. *Science*, **297**, 981.

Wenzel, J.W. (1989). Endogenous factors, external cues, and eccentric construction in *Polistes annularis* (Hymenoptera: Vespidae). *Journal of Insect Behaviour*, **2**, 679–699.

Wenzel, J.W. (1991). Evolution of Nest Architecture. In *The social biology of wasps* (eds. K.G. Ross and R.W. Matthews), pp. 480–519. Cornell University Press, Ithaca, NY.

Wenzel, J.W. (1993). Application of the biogenetic law to behavioural ontogeny: a test using nest architecture in paper wasps. *Journal of Evolutionary Biology*, **6**, 229–247.

Wenzel, J.W. (1998). A generic key to the nests of hornets, yellowjackets, and paper wasps worldwide (Vespidae: Vespinae, Polistinae). *American Museum Novitates*, **3224**, 1–39.

West Eberhard, M.J. (1969). The social biology of polistine wasps. *Miscellaneous Publications of the Museum of Zoology of the University of Michigan* **140**, 1–101.

Westoby, M., Cousins, J.M., and Grice, A.C. (1982). Rate of decline of some soil seed populations during drought in western New South Wales. In *Ant-plant interactions in Australia* (ed. R.C. Buckley), pp. 7–10. Junk, The Hague.

Wheeler, A.G. and Hoebeke, E.R. (1988). *Apterona helix* (Lepidoptera: Psychidae), a palearctic bagworm moth in North America: new distribution records, seasonal history, and host plants. *Proceedings of the Entomological Society of Washington*, **90**, 20–27.

Wheeler, W.M. (1910). *Ants, their structure, development and behaviour*. New York, Colombia University Press.

White, P.S. and Pickett, S.T.A. (1985). Natural disturbance and patch dynamics: An introduction. In *The ecology of natural disturbance and patch dynamics* (eds. S.T.A. Pickett and P.S. White), pp. 3–13. Academic Press, Orlando, FL.

Whitehouse, M.E.A. and Jackson, R.R. (1998). Predatory behaviour and parental care in *Argyrodes flavipes*, a social spider from Queensland. *Journal of Zoology, London*, **244**, 95–105.

Whiten, A., Goodall, J., McGrew, W.C., Nishida, T., Reynolds, V., Sugiyama, Y., Tutin, C.E.G., Wrangham, R.W., and Boesch, C. (1999). Cultures in chimpanzees. *Nature*, **399**, 682–685.

Whitford, W.G. and Kay, F.R. (1999). Biopedturbation by mammals in deserts: a review. *Journal of Arid Environments*, **41** (7), 203–230.

Whitney, B.M.,Pacheco, J.F., Fonseca, P.S.M. da, and Barth, R.H. (1996). The nest and nesting ecology of *Acrobatornis fonsecai* (Furnariidae), with implications for intrafamilial relationships. *Wilson Bulletin*, **108**, 434–448.

Wiggins, G.B. (1977). *Larvae of the North American caddisfly genera (Trichoptera).* University of Toronto Press, Toronto.

Wildish, D. and Kristmanson, D. (1997). *Benthic suspension feeders and flow.* Cambridge University Press, Cambridge.

Williams, D.D., Read, A.T., and Moore, K.A. (1983). The biology and zoogeography of *Helicopsyche borealis* (Trichoptera: Helicopsychidae): a Nearctic representative of a tropical genus. *Canadian Journal of Zoology*, **61**, 2288–2299.

Wilson, E.O. (1971). *The Insect Societies.* Belknap Press, Cambridge, MA.

Wilson, E.O. (1980). Caste and division of labour in leaf-cutter ants (Hymenoptera: Formicidae: *Atta*), I: the overall pattern in *Atta sexdens. Behavioural Ecology and Sociobiology*, **7**, 143–156.

Wilson, E.O. (1983*a*). Caste and division of labour in leaf-cutter ants (Hymenoptera: Formicidae: *Atta*), III: ergonomic resiliency in foraging by *A.cephalotes*. *Behavioural Ecology and Sociobiology*, **14**, 47–54.

Wilson, E.O. (1983*b*). Caste and division of labour in leaf-cutter ants (Hymenoptera: Formicidae: *Atta*), IV: colony ontogeny of *A.cephalotes. Behavioural Ecology and Sociobiology*, **14**, 55–60.

Wilson, J.B. and Agnew, A.D.Q. (1992). Positive-feedback switches in plant communities. *Advances in Ecological Research*, **23**, 263–325.

Wilson, K.J. and Kilgore, D.L. (1978). The effects of location and design on the diffusion of respiratory gasses in mammal burrows. *Journal of Theoretical Biology*, **71**, 73–101.

Wilson, W.H. (1979). Community structure and species diversity of the sedimentry reefs constructed by *Petaloproctus socialis* (Polychaeta Malonidae). *Journal of Marine Research*, **37**, 623–641.

Wilsson, L. (1976). Observations and experiments on the ethology of the European beaver (*Castor fiber*). In *External construction by animals* (eds. N.E. Collias and E.C. Collias), pp. 374–403. Dowden Hutchinson and Ross, Stroudsburg.

Winkler D.W. and Sheldon, F.H. (1993). Evolution of nest construction in swallows (Hirundinidae): a molecular phylogenetic perspective. *Proceedings of the National Academy of Sciences USA*, **90**, 5705–5707.

Winkler, S. and Kaplan, D.L. (2000). Molecular biology of spider silk. *Reviews in Molecular Biotechnology*, **74**, 85–93.

Wirth, E. and Barth, F.G. (1992). Forces in the spider orb web. *Journal of Comparative Physiology A*, **171**, 359–371.

Withers, P.C. (1977). Energetic aspects of reproduction by the cliff swallow. *Auk*, **94**, 718–725.

Withers, P.C. (1978). Models of diffusion-mediated gas exchange in animal burrows. *American Naturalist*, **112**, 1101–1113.

Wojtusiak, J. and Raczka, P. (1983). Silking behaviour of the swallowtail, *Papilio machaon* L. and peacock, *Vanessa io* L. (Lepidoptera, Papilionidae, Nymphalidae) buttetflies before pupation. *Folia Biologica (Kraków)*, **31**, 41–49.

Wolters, V. and Schaefer, M. (1993). Effects of burrowing by the earthworm *Aporrectodea caliginosa* (Savigny) on beech litter decomposition in an agricultural and in a forest soil. *Geoderma*, **56**, 627–632.

Wood, T.G. (1988). Termites and the soil Environment. *Biology and Fertility of Soils*, **6**, 228–236.

Wood, T.G. and Sands, W.A. (1978). The role of termites in ecosystems. In *Production ecology of ants and termites* (ed. M.V. Brian), pp. 245–292. Cambridge University Press, Cambridge.

Wyatt, T.D. (1986). How a subsocial intertidal beetle, *Bledius spectabilis*, prevents flooding and anoxia in its burrow. *Behavioural Ecology and Sociobiology*, **19**, 323–331.

Yamane, S., Kudo, K., Tajima, T., Nihon'Yanagi, K., Shinoda, M., Saito, K., and Yamamoto, H. (1999). Comparison of investment in nest construction by foundresses of consubgeneric Polistes wasps, *P. (Polistes) riparius* and *P. (P.) chinensis* (Hymenoptera, Vespidae). *Journal of Ethology*, **16**, 97–104.

Yeargan, K.V. and Quate, L.W. (1997). Adult male bolas spiders retain juvenile hunting tactics. *Oecologia*, **112**, 572–576.

Zeibis, W., Foster, S., Huettel, M., and Jorgensen, B.B. (1996*a*). Complex burrows of the mud shrimp *Callianassa truncata* and their geochemical impact on the seabed. *Nature*, **382**, 619–622.

Zeibis, W., Huettel, M. and Foster, S. (1996*b*). Impact of biogenic sediment topography on oxygen fluxes in permeable seabeds. *Marine Ecology Progress Series*, **140**, 227–237.

Zschokke, S. (1996). Early stages of orb web construction in *Araneus diadematus* Clerck. *Revue Suisse de Zoologie*, **2**, 709–720.

Zschokke, S. (2000). Radius construction and structure in the orb-web of *Zila diodia* (Araneidae). *Journal of Comparative Physiology A*, **186**, 999–1005.

Zschokke, S. and Vollrath, F. (1995). Unfreezing the behaviour of two orb spiders. *Physiology and Behaviour*, **58**, 1167–1173.

Zubay, G.L., Parson, W.W., and Vance, D.E. (1995). *Principles of biochemistry*. Wm. C. Brown Publishers, Dubuque.

Zuppa, A., Osella, G. and Biondi, S. (1994). Parental care in Attelabidae (Coleoptera Curculionoidea). *Ethology, Ecology and Evolution*, **3** (Special Issue), 113–118.

Zyskowski, K. and Prum, R.O. (1999). Phylogenetic analysis of the nest architecture of neotropical ovenbirds (Furnariidae). *Auk*, **116**, 891–911.

Author Index

Species Index: Scientific Names

Subject Index